Advances in Food Traceability Techniques and Technologies

Related titles

New Analytical Approaches for Verifying the Origin of Food
(ISBN 978-0-85709-274-8)

Food Chain Integrity: A Holistic Approach to Food Traceability, Safety, Quality and Authenticity
(ISBN 978-0-85709-068-3)

Ensuring Global Food Safety
(ISBN 978-0-12374-845-4)

Woodhead Publishing Series in Food Science, Technology and Nutrition: Number 301

Advances in Food Traceability Techniques and Technologies

Improving Quality Throughout the Food Chain

Edited by

Montserrat Espiñeira

Francisco J. Santaclara

ELSEVIER

AMSTERDAM • BOSTON • CAMBRIDGE • HEIDELBERG
LONDON • NEW YORK • OXFORD • PARIS • SAN DIEGO
SAN FRANCISCO • SINGAPORE • SYDNEY • TOKYO
Woodhead Publishing is an imprint of Elsevier

WP
WOODHEAD PUBLISHING

Woodhead Publishing is an imprint of Elsevier
The Officers' Mess Business Centre, Royston Road, Duxford, CB22 4QH, UK
50 Hampshire Street, 5th Floor, Cambridge, MA 02139, USA
The Boulevard, Langford Lane, Kidlington, OX5 1GB, UK

British Library Cataloguing-in-Publication Data
A catalogue record for this book is available from the British Library

Library of Congress Cataloging-in-Publication Data
A catalog record for this book is available from the Library of Congress

ISBN: 978-0-08-100310-7 (print)
ISBN: 978-0-08-100321-3 (online)

For information on all Woodhead Publishing publications visit our website at https://www.elsevier.com/

Publisher: Nikki Levy
Acquisition Editor: Rob Sykes
Editorial Project Manager: Karen Miller
Production Project Manager: Susan Li
Designer: Victoria Pearson

Contents

List of Contributors

R. Accorsi Alma Mater Studiorum - University of Bologna, Bologna, Italy

M. Aceto Università degli Studi del Piemonte Orientale, Viale Teresa Michel, Alessandria, Italy

S. Armenta University of Valencia, Burjassot, Valencia, Spain

R. Badia-Melis Polytechnic University of Madrid, Madrid, Spain

A. Beulens LEI Wageningen UR, Wageningen, The Netherlands

M. Borit University of Tromsø (UiT) – The Arctic University of Norway, Tromsø, Norway

D. Cozzolino Central Queensland University, Rockhampton, QLD, Australia

F. Dabbene IEIIT–CNR, Politecnico di Torino, Turin, Italy

M. de la Guardia University of Valencia, Burjassot, Valencia, Spain

T.M.L. Diallo University of Lyon, University Lyon 1, Villeurbanne, France

H. Dong Guangzhou Quality Supervision and Testing Institute, Guangzhou, Guangdong, China

M. Espiñeira ANFACO-CECOPESCA, Vigo, Spain

A.K. Evizal Islamic University of Riau, Pekanbaru, Riau, Indonesia; Universiti Teknologi Malaysia, Johor Bahru, Malaysia

E. Ferrari Alma Mater Studiorum - University of Bologna, Bologna, Italy

M. Gamberi Alma Mater Studiorum - University of Bologna, Bologna, Italy

P. Gay IEIIT–CNR, Politecnico di Torino, Turin, Italy; Università degli Studi di Torino, Grugliasco, Italy

R. Ghizzoni Council for Agricultural Research and Economics, Genomics Research Center, CREA-GPG, Fiorenzuola d'Arda (PC), Italy

S. Henry University of Lyon, University Lyon 1, Villeurbanne, France

J.E. Hobbs University of Saskatchewan, Saskatoon, SK, Canada

E. Ilie-Zudor Institute for Computer Science and Control, Hungarian Academy of Sciences, Budapest, Hungary

K.M. Karlsen Norwegian Institute of Food, Fisheries and Aquaculture Research (Nofima), Tromsø, Norway

Zs. Kemény Institute for Computer Science and Control, Hungarian Academy of Sciences, Budapest, Hungary

D. Luo Guangzhou Quality Supervision and Testing Institute, Guangzhou, Guangdong, China

H. Luo Guangzhou Quality Supervision and Testing Institute, Guangzhou, Guangdong, China

R. Manzini Alma Mater Studiorum - University of Bologna, Bologna, Italy

G. Mirabelli University of Calabria, Rende, Italy

C. Morcia Council for Agricultural Research and Economics, Genomics Research Center, CREA-GPG, Fiorenzuola d'Arda (PC), Italy

P. Olsen Norwegian Institute of Food, Fisheries and Aquaculture Research (Nofima), Tromsø, Norway

Y. Ouzrout University of Lyon, University Lyon 2, Lyon, France

T. Pizzuti University of Calabria, Rende, Italy

S.K.A. Rahim Universiti Teknologi Malaysia, Johor Bahru, Malaysia

T.A. Rahman Universiti Teknologi Malaysia, Johor Bahru, Malaysia

A. Regattieri Alma Mater Studiorum - University of Bologna, Bologna, Italy

S.L. Rosa Islamic University of Riau, Pekanbaru, Riau, Indonesia

L. Ruiz-Garcia Polytechnic University of Madrid, Madrid, Spain

F.J. Santaclara IDIS CHUS, Santiago de Compostela, Spain

H. Scholten Wageningen University, Wageningen, The Netherlands

V. Terzi Council for Agricultural Research and Economics, Genomics Research Center, CREA-GPG, Fiorenzuola d'Arda (PC), Italy

C. Tortia Università degli Studi di Torino, Grugliasco, Italy

G. Tumino Council for Agricultural Research and Economics, Genomics Research Center, CREA-GPG, Fiorenzuola d'Arda (PC), Italy

J.G.A.J. van der Vorst Wageningen University, Wageningen, The Netherlands

C.N. Verdouw LEI Wageningen UR, The Netherlands; Wageningen University, Wageningen, The Netherlands

L. Zach Food System Solutions, Mt Horeb, Wisconsin, WI, United States; Food System Solutions, Christchurch, New Zealand

Woodhead Publishing Series in Food Science, Technology and Nutrition

1 **Chilled foods: A comprehensive guide**
Edited by C. Dennis and M. Stringer
2 **Yoghurt: Science and technology**
A. Y. Tamime and R. K. Robinson
3 **Food processing technology: Principles and practice**
P. J. Fellows
4 **Bender's dictionary of nutrition and food technology Sixth edition**
D. A. Bender
5 **Determination of veterinary residues in food**
Edited by N. T. Crosby
6 **Food contaminants: Sources and surveillance**
Edited by C. Creaser and R. Purchase
7 **Nitrates and nitrites in food and water**
Edited by M. J. Hill
8 **Pesticide chemistry and bioscience: The food-environment challenge**
Edited by G. T. Brooks and T. Roberts
9 **Pesticides: Developments, impacts and controls**
Edited by G. A. Best and A. D. Ruthven
10 **Dietary fibre: Chemical and biological aspects**
Edited by D. A. T. Southgate, K. W. Waldron, I. T. Johnson and G. R. Fenwick
11 **Vitamins and minerals in health and nutrition**
M. Tolonen
12 **Technology of biscuits, crackers and cookies Second edition**
D. Manley
13 **Instrumentation and sensors for the food industry**
Edited by E. Kress-Rogers
14 **Food and cancer prevention: Chemical and biological aspects**
Edited by K. W. Waldron, I. T. Johnson and G. R. Fenwick
15 **Food colloids: Proteins, lipids and polysaccharides**
Edited by E. Dickinson and B. Bergenstahl
16 **Food emulsions and foams**
Edited by E. Dickinson
17 **Maillard reactions in chemistry, food and health**
Edited by T. P. Labuza, V. Monnier, J. Baynes and J. O'Brien
18 **The Maillard reaction in foods and medicine**
Edited by J. O'Brien, H. E. Nursten, M. J. Crabbe and J. M. Ames
19 **Encapsulation and controlled release**
Edited by D. R. Karsa and R. A. Stephenson

20 **Flavours and fragrances**
Edited by A. D. Swift
21 **Feta and related cheeses**
Edited by A. Y. Tamime and R. K. Robinson
22 **Biochemistry of milk products**
Edited by A. T. Andrews and J. R. Varley
23 **Physical properties of foods and food processing systems**
M. J. Lewis
24 **Food irradiation: A reference guide**
V. M. Wilkinson and G. Gould
25 **Kent's technology of cereals: An introduction for students of food science and agriculture Fourth edition**
N. L. Kent and A. D. Evers
26 **Biosensors for food analysis**
Edited by A. O. Scott
27 **Separation processes in the food and biotechnology industries: Principles and applications**
Edited by A. S. Grandison and M. J. Lewis
28 **Handbook of indices of food quality and authenticity**
R. S. Singhal, P. K. Kulkarni and D. V. Rege
29 **Principles and practices for the safe processing of foods**
D. A. Shapton and N. F. Shapton
30 **Biscuit, cookie and cracker manufacturing manuals Volume 1: Ingredients**
D. Manley
31 **Biscuit, cookie and cracker manufacturing manuals Volume 2: Biscuit doughs**
D. Manley
32 **Biscuit, cookie and cracker manufacturing manuals Volume 3: Biscuit dough piece forming**
D. Manley
33 **Biscuit, cookie and cracker manufacturing manuals Volume 4: Baking and cooling of biscuits**
D. Manley
34 **Biscuit, cookie and cracker manufacturing manuals Volume 5: Secondary processing in biscuit manufacturing**
D. Manley
35 **Biscuit, cookie and cracker manufacturing manuals Volume 6: Biscuit packaging and storage**
D. Manley
36 **Practical dehydration Second edition**
M. Greensmith
37 **Lawrie's meat science Sixth edition**
R. A. Lawrie
38 **Yoghurt: Science and technology Second edition**
A. Y. Tamime and R. K. Robinson
39 **New ingredients in food processing: Biochemistry and agriculture**
G. Linden and D. Lorient
40 **Benders' dictionary of nutrition and food technology Seventh edition**
D. A. Bender and A. E. Bender
41 **Technology of biscuits, crackers and cookies Third edition**
D. Manley

42 **Food processing technology: Principles and practice Second edition**
P. J. Fellows
43 **Managing frozen foods**
Edited by C. J. Kennedy
44 **Handbook of hydrocolloids**
Edited by G. O. Phillips and P. A. Williams
45 **Food labelling**
Edited by J. R. Blanchfield
46 **Cereal biotechnology**
Edited by P. C. Morris and J. H. Bryce
47 **Food intolerance and the food industry**
Edited by T. Dean
48 **The stability and shelf-life of food**
Edited by D. Kilcast and P. Subramaniam
49 **Functional foods: Concept to product**
Edited by G. R. Gibson and C. M. Williams
50 **Chilled foods: A comprehensive guide Second edition**
Edited by M. Stringer and C. Dennis
51 **HACCP in the meat industry**
Edited by M. Brown
52 **Biscuit, cracker and cookie recipes for the food industry**
D. Manley
53 **Cereals processing technology**
Edited by G. Owens
54 **Baking problems solved**
S. P. Cauvain and L. S. Young
55 **Thermal technologies in food processing**
Edited by P. Richardson
56 **Frying: Improving quality**
Edited by J. B. Rossell
57 **Food chemical safety Volume 1: Contaminants**
Edited by D. Watson
58 **Making the most of HACCP: Learning from others' experience**
Edited by T. Mayes and S. Mortimore
59 **Food process modelling**
Edited by L. M. M. Tijskens, M. L. A. T. M. Hertog and B. M. Nicolaï
60 **EU food law: A practical guide**
Edited by K. Goodburn
61 **Extrusion cooking: Technologies and applications**
Edited by R. Guy
62 **Auditing in the food industry: From safety and quality to environmental and other audits**
Edited by M. Dillon and C. Griffith
63 **Handbook of herbs and spices Volume 1**
Edited by K. V. Peter
64 **Food product development: Maximising success**
M. Earle, R. Earle and A. Anderson
65 **Instrumentation and sensors for the food industry Second edition**
Edited by E. Kress-Rogers and C. J. B. Brimelow

66 **Food chemical safety Volume 2: Additives**
Edited by D. Watson
67 **Fruit and vegetable biotechnology**
Edited by V. Valpuesta
68 **Foodborne pathogens: Hazards, risk analysis and control**
Edited by C. de W. Blackburn and P. J. McClure
69 **Meat refrigeration**
S. J. James and C. James
70 **Lockhart and Wiseman's crop husbandry Eighth edition**
H. J. S. Finch, A. M. Samuel and G. P. F. Lane
71 **Safety and quality issues in fish processing**
Edited by H. A. Bremner
72 **Minimal processing technologies in the food industries**
Edited by T. Ohlsson and N. Bengtsson
73 **Fruit and vegetable processing: Improving quality**
Edited by W. Jongen
74 **The nutrition handbook for food processors**
Edited by C. J. K. Henry and C. Chapman
75 **Colour in food: Improving quality**
Edited by D. MacDougall
76 **Meat processing: Improving quality**
Edited by J. P. Kerry, J. F. Kerry and D. A. Ledward
77 **Microbiological risk assessment in food processing**
Edited by M. Brown and M. Stringer
78 **Performance functional foods**
Edited by D. Watson
79 **Functional dairy products Volume 1**
Edited by T. Mattila-Sandholm and M. Saarela
80 **Taints and off-flavours in foods**
Edited by B. Baigrie
81 **Yeasts in food**
Edited by T. Boekhout and V. Robert
82 **Phytochemical functional foods**
Edited by I. T. Johnson and G. Williamson
83 **Novel food packaging techniques**
Edited by R. Ahvenainen
84 **Detecting pathogens in food**
Edited by T. A. McMeekin
85 **Natural antimicrobials for the minimal processing of foods**
Edited by S. Roller
86 **Texture in food Volume 1: Semi-solid foods**
Edited by B. M. McKenna
87 **Dairy processing: Improving quality**
Edited by G. Smit
88 **Hygiene in food processing: Principles and practice**
Edited by H. L. M. Lelieveld, M. A. Mostert, B. White and J. Holah
89 **Rapid and on-line instrumentation for food quality assurance**
Edited by I. Tothill
90 **Sausage manufacture: Principles and practice**
E. Essien

91 **Environmentally-friendly food processing**
Edited by B. Mattsson and U. Sonesson
92 **Bread making: Improving quality**
Edited by S. P. Cauvain
93 **Food preservation techniques**
Edited by P. Zeuthen and L. Bøgh-Sørensen
94 **Food authenticity and traceability**
Edited by M. Lees
95 **Analytical methods for food additives**
R. Wood, L. Foster, A. Damant and P. Key
96 **Handbook of herbs and spices Volume 2**
Edited by K. V. Peter
97 **Texture in food Volume 2: Solid foods**
Edited by D. Kilcast
98 **Proteins in food processing**
Edited by R. Yada
99 **Detecting foreign bodies in food**
Edited by M. Edwards
100 **Understanding and measuring the shelf-life of food**
Edited by R. Steele
101 **Poultry meat processing and quality**
Edited by G. Mead
102 **Functional foods, ageing and degenerative disease**
Edited by C. Remacle and B. Reusens
103 **Mycotoxins in food: Detection and control**
Edited by N. Magan and M. Olsen
104 **Improving the thermal processing of foods**
Edited by P. Richardson
105 **Pesticide, veterinary and other residues in food**
Edited by D. Watson
106 **Starch in food: Structure, functions and applications**
Edited by A.-C. Eliasson
107 **Functional foods, cardiovascular disease and diabetes**
Edited by A. Arnoldi
108 **Brewing: Science and practice**
D. E. Briggs, P. A. Brookes, R. Stevens and C. A. Boulton
109 **Using cereal science and technology for the benefit of consumers: Proceedings of the 12PthP International ICC Cereal and Bread Congress, 24 – 26PthP May, 2004, Harrogate, UK**
Edited by S. P. Cauvain, L. S. Young and S. Salmon
110 **Improving the safety of fresh meat**
Edited by J. Sofos
111 **Understanding pathogen behaviour: Virulence, stress response and resistance**
Edited by M. Griffiths
112 **The microwave processing of foods**
Edited by H. Schubert and M. Regier
113 **Food safety control in the poultry industry**
Edited by G. Mead
114 **Improving the safety of fresh fruit and vegetables**
Edited by W. Jongen

115 **Food, diet and obesity**
Edited by D. Mela
116 **Handbook of hygiene control in the food industry**
Edited by H. L. M. Lelieveld, M. A. Mostert and J. Holah
117 **Detecting allergens in food**
Edited by S. Koppelman and S. Hefle
118 **Improving the fat content of foods**
Edited by C. Williams and J. Buttriss
119 **Improving traceability in food processing and distribution**
Edited by I. Smith and A. Furness
120 **Flavour in food**
Edited by A. Voilley and P. Etievant
121 **The Chorleywood bread process**
S. P. Cauvain and L. S. Young
122 **Food spoilage microorganisms**
Edited by C. de W. Blackburn
123 **Emerging foodborne pathogens**
Edited by Y. Motarjemi and M. Adams
124 **Benders' dictionary of nutrition and food technology Eighth edition**
D. A. Bender
125 **Optimising sweet taste in foods**
Edited by W. J. Spillane
126 **Brewing: New technologies**
Edited by C. Bamforth
127 **Handbook of herbs and spices Volume 3**
Edited by K. V. Peter
128 **Lawrie's meat science Seventh edition**
R. A. Lawrie in collaboration with D. A. Ledward
129 **Modifying lipids for use in food**
Edited by F. Gunstone
130 **Meat products handbook: Practical science and technology**
G. Feiner
131 **Food consumption and disease risk: Consumer–pathogen interactions**
Edited by M. Potter
132 **Acrylamide and other hazardous compounds in heat-treated foods**
Edited by K. Skog and J. Alexander
133 **Managing allergens in food**
Edited by C. Mills, H. Wichers and K. Hoffman-Sommergruber
134 **Microbiological analysis of red meat, poultry and eggs**
Edited by G. Mead
135 **Maximising the value of marine by-products**
Edited by F. Shahidi
136 **Chemical migration and food contact materials**
Edited by K. Barnes, R. Sinclair and D. Watson
137 **Understanding consumers of food products**
Edited by L. Frewer and H. van Trijp
138 **Reducing salt in foods: Practical strategies**
Edited by D. Kilcast and F. Angus
139 **Modelling microorganisms in food**
Edited by S. Brul, S. Van Gerwen and M. Zwietering

140 **Tamime and Robinson's Yoghurt: Science and technology Third edition**
A. Y. Tamime and R. K. Robinson
141 **Handbook of waste management and co-product recovery in food processing Volume 1**
Edited by K. W. Waldron
142 **Improving the flavour of cheese**
Edited by B. Weimer
143 **Novel food ingredients for weight control**
Edited by C. J. K. Henry
144 **Consumer-led food product development**
Edited by H. MacFie
145 **Functional dairy products Volume 2**
Edited by M. Saarela
146 **Modifying flavour in food**
Edited by A. J. Taylor and J. Hort
147 **Cheese problems solved**
Edited by P. L. H. McSweeney
148 **Handbook of organic food safety and quality**
Edited by J. Cooper, C. Leifert and U. Niggli
149 **Understanding and controlling the microstructure of complex foods**
Edited by D. J. McClements
150 **Novel enzyme technology for food applications**
Edited by R. Rastall
151 **Food preservation by pulsed electric fields: From research to application**
Edited by H. L. M. Lelieveld and S. W. H. de Haan
152 **Technology of functional cereal products**
Edited by B. R. Hamaker
153 **Case studies in food product development**
Edited by M. Earle and R. Earle
154 **Delivery and controlled release of bioactives in foods and nutraceuticals**
Edited by N. Garti
155 **Fruit and vegetable flavour: Recent advances and future prospects**
Edited by B. Brückner and S. G. Wyllie
156 **Food fortification and supplementation: Technological, safety and regulatory aspects**
Edited by P. Berry Ottaway
157 **Improving the health-promoting properties of fruit and vegetable products**
Edited by F. A. Tomás-Barberán and M. I. Gil
158 **Improving seafood products for the consumer**
Edited by T. Børresen
159 **In-pack processed foods: Improving quality**
Edited by P. Richardson
160 **Handbook of water and energy management in food processing**
Edited by J. Klemeš, R. Smith and J.-K. Kim
161 **Environmentally compatible food packaging**
Edited by E. Chiellini
162 **Improving farmed fish quality and safety**
Edited by Ø. Lie
163 **Carbohydrate-active enzymes**
Edited by K.-H. Park

164 **Chilled foods: A comprehensive guide Third edition**
Edited by M. Brown
165 **Food for the ageing population**
Edited by M. M. Raats, C. P. G. M. de Groot and W. A. Van Staveren
166 **Improving the sensory and nutritional quality of fresh meat**
Edited by J. P. Kerry and D. A. Ledward
167 **Shellfish safety and quality**
Edited by S. E. Shumway and G. E. Rodrick
168 **Functional and speciality beverage technology**
Edited by P. Paquin
169 **Functional foods: Principles and technology**
M. Guo
170 **Endocrine-disrupting chemicals in food**
Edited by I. Shaw
171 **Meals in science and practice: Interdisciplinary research and business applications**
Edited by H. L. Meiselman
172 **Food constituents and oral health: Current status and future prospects**
Edited by M. Wilson
173 **Handbook of hydrocolloids Second edition**
Edited by G. O. Phillips and P. A. Williams
174 **Food processing technology: Principles and practice Third edition**
P. J. Fellows
175 **Science and technology of enrobed and filled chocolate, confectionery and bakery products**
Edited by G. Talbot
176 **Foodborne pathogens: Hazards, risk analysis and control Second edition**
Edited by C. de W. Blackburn and P. J. McClure
177 **Designing functional foods: Measuring and controlling food structure breakdown and absorption**
Edited by D. J. McClements and E. A. Decker
178 **New technologies in aquaculture: Improving production efficiency, quality and environmental management**
Edited by G. Burnell and G. Allan
179 **More baking problems solved**
S. P. Cauvain and L. S. Young
180 **Soft drink and fruit juice problems solved**
P. Ashurst and R. Hargitt
181 **Biofilms in the food and beverage industries**
Edited by P. M. Fratamico, B. A. Annous and N. W. Gunther
182 **Dairy-derived ingredients: Food and neutraceutical uses**
Edited by M. Corredig
183 **Handbook of waste management and co-product recovery in food processing Volume 2**
Edited by K. W. Waldron
184 **Innovations in food labelling**
Edited by J. Albert
185 **Delivering performance in food supply chains**
Edited by C. Mena and G. Stevens

186 **Chemical deterioration and physical instability of food and beverages**
Edited by L. H. Skibsted, J. Risbo and M. L. Andersen
187 **Managing wine quality Volume 1: Viticulture and wine quality**
Edited by A. G. Reynolds
188 **Improving the safety and quality of milk Volume 1: Milk production and processing**
Edited by M. Griffiths
189 **Improving the safety and quality of milk Volume 2: Improving quality in milk products**
Edited by M. Griffiths
190 **Cereal grains: Assessing and managing quality**
Edited by C. Wrigley and I. Batey
191 **Sensory analysis for food and beverage quality control: A practical guide**
Edited by D. Kilcast
192 **Managing wine quality Volume 2: Oenology and wine quality**
Edited by A. G. Reynolds
193 **Winemaking problems solved**
Edited by C. E. Butzke
194 **Environmental assessment and management in the food industry**
Edited by U. Sonesson, J. Berlin and F. Ziegler
195 **Consumer-driven innovation in food and personal care products**
Edited by S. R. Jaeger and H. MacFie
196 **Tracing pathogens in the food chain**
Edited by S. Brul, P. M. Fratamico and T. A. McMeekin
197 **Case studies in novel food processing technologies: Innovations in processing, packaging, and predictive modelling**
Edited by C. J. Doona, K. Kustin and F. E. Feeherry
198 **Freeze-drying of pharmaceutical and food products**
T.-C. Hua, B.-L. Liu and H. Zhang
199 **Oxidation in foods and beverages and antioxidant applications Volume 1: Understanding mechanisms of oxidation and antioxidant activity**
Edited by E. A. Decker, R. J. Elias and D. J. McClements
200 **Oxidation in foods and beverages and antioxidant applications Volume 2: Management in different industry sectors**
Edited by E. A. Decker, R. J. Elias and D. J. McClements
201 **Protective cultures, antimicrobial metabolites and bacteriophages for food and beverage biopreservation**
Edited by C. Lacroix
202 **Separation, extraction and concentration processes in the food, beverage and nutraceutical industries**
Edited by S. S. H. Rizvi
203 **Determining mycotoxins and mycotoxigenic fungi in food and feed**
Edited by S. De Saeger
204 **Developing children's food products**
Edited by D. Kilcast and F. Angus
205 **Functional foods: Concept to product Second edition**
Edited by M. Saarela
206 **Postharvest biology and technology of tropical and subtropical fruits Volume 1: Fundamental issues**
Edited by E. M. Yahia

207 **Postharvest biology and technology of tropical and subtropical fruits Volume 2: Açai to citrus**
Edited by E. M. Yahia
208 **Postharvest biology and technology of tropical and subtropical fruits Volume 3: Cocona to mango**
Edited by E. M. Yahia
209 **Postharvest biology and technology of tropical and subtropical fruits Volume 4: Mangosteen to white sapote**
Edited by E. M. Yahia
210 **Food and beverage stability and shelf life**
Edited by D. Kilcast and P. Subramaniam
211 **Processed Meats: Improving safety, nutrition and quality**
Edited by J. P. Kerry and J. F. Kerry
212 **Food chain integrity: A holistic approach to food traceability, safety, quality and authenticity**
Edited by J. Hoorfar, K. Jordan, F. Butler and R. Prugger
213 **Improving the safety and quality of eggs and egg products Volume 1**
Edited by Y. Nys, M. Bain and F. Van Immerseel
214 **Improving the safety and quality of eggs and egg products Volume 2**
Edited by F. Van Immerseel, Y. Nys and M. Bain
215 **Animal feed contamination: Effects on livestock and food safety**
Edited by J. Fink-Gremmels
216 **Hygienic design of food factories**
Edited by J. Holah and H. L. M. Lelieveld
217 **Manley's technology of biscuits, crackers and cookies Fourth edition**
Edited by D. Manley
218 **Nanotechnology in the food, beverage and nutraceutical industries**
Edited by Q. Huang
219 **Rice quality: A guide to rice properties and analysis**
K. R. Bhattacharya
220 **Advances in meat, poultry and seafood packaging**
Edited by J. P. Kerry
221 **Reducing saturated fats in foods**
Edited by G. Talbot
222 **Handbook of food proteins**
Edited by G. O. Phillips and P. A. Williams
223 **Lifetime nutritional influences on cognition, behaviour and psychiatric illness**
Edited by D. Benton
224 **Food machinery for the production of cereal foods, snack foods and confectionery**
L.-M. Cheng
225 **Alcoholic beverages: Sensory evaluation and consumer research**
Edited by J. Piggott
226 **Extrusion problems solved: Food, pet food and feed**
M. N. Riaz and G. J. Rokey
227 **Handbook of herbs and spices Second edition Volume 1**
Edited by K. V. Peter
228 **Handbook of herbs and spices Second edition Volume 2**
Edited by K. V. Peter
229 **Breadmaking: Improving quality Second edition**
Edited by S. P. Cauvain

230 **Emerging food packaging technologies: Principles and practice**
Edited by K. L. Yam and D. S. Lee
231 **Infectious disease in aquaculture: Prevention and control**
Edited by B. Austin
232 **Diet, immunity and inflammation**
Edited by P. C. Calder and P. Yaqoob
233 **Natural food additives, ingredients and flavourings**
Edited by D. Baines and R. Seal
234 **Microbial decontamination in the food industry: Novel methods and applications**
Edited by A. Demirci and M.O. Ngadi
235 **Chemical contaminants and residues in foods**
Edited by D. Schrenk
236 **Robotics and automation in the food industry: Current and future technologies**
Edited by D. G. Caldwell
237 **Fibre-rich and wholegrain foods: Improving quality**
Edited by J. A. Delcour and K. Poutanen
238 **Computer vision technology in the food and beverage industries**
Edited by D.-W. Sun
239 **Encapsulation technologies and delivery systems for food ingredients and nutraceuticals**
Edited by N. Garti and D. J. McClements
240 **Case studies in food safety and authenticity**
Edited by J. Hoorfar
241 **Heat treatment for insect control: Developments and applications**
D. Hammond
242 **Advances in aquaculture hatchery technology**
Edited by G. Allan and G. Burnell
243 **Open innovation in the food and beverage industry**
Edited by M. Garcia Martinez
244 **Trends in packaging of food, beverages and other fast-moving consumer goods (FMCG)**
Edited by N. Farmer
245 **New analytical approaches for verifying the origin of food**
Edited by P. Brereton
246 **Microbial production of food ingredients, enzymes and nutraceuticals**
Edited by B. McNeil, D. Archer, I. Giavasis and L. Harvey
247 **Persistent organic pollutants and toxic metals in foods**
Edited by M. Rose and A. Fernandes
248 **Cereal grains for the food and beverage industries**
E. Arendt and E. Zannini
249 **Viruses in food and water: Risks, surveillance and control**
Edited by N. Cook
250 **Improving the safety and quality of nuts**
Edited by L. J. Harris
251 **Metabolomics in food and nutrition**
Edited by B. C. Weimer and C. Slupsky
252 **Food enrichment with omega-3 fatty acids**
Edited by C. Jacobsen, N. S. Nielsen, A. F. Horn and A.-D. M. Sørensen
253 **Instrumental assessment of food sensory quality: A practical guide**
Edited by D. Kilcast

254 **Food microstructures: Microscopy, measurement and modelling**
Edited by V. J. Morris and K. Groves
255 **Handbook of food powders: Processes and properties**
Edited by B. R. Bhandari, N. Bansal, M. Zhang and P. Schuck
256 **Functional ingredients from algae for foods and nutraceuticals**
Edited by H. Domínguez
257 **Satiation, satiety and the control of food intake: Theory and practice**
Edited by J. E. Blundell and F. Bellisle
258 **Hygiene in food processing: Principles and practice Second edition**
Edited by H. L. M. Lelieveld, J. Holah and D. Napper
259 **Advances in microbial food safety Volume 1**
Edited by J. Sofos
260 **Global safety of fresh produce: A handbook of best practice, innovative commercial solutions and case studies**
Edited by J. Hoorfar
261 **Human milk biochemistry and infant formula manufacturing technology**
Edited by M. Guo
262 **High throughput screening for food safety assessment: Biosensor technologies, hyperspectral imaging and practical applications**
Edited by A. K. Bhunia, M. S. Kim and C. R. Taitt
263 **Foods, nutrients and food ingredients with authorised EU health claims: Volume 1**
Edited by M. J. Sadler
264 **Handbook of food allergen detection and control**
Edited by S. Flanagan
265 **Advances in fermented foods and beverages: Improving quality, technologies and health benefits**
Edited by W. Holzapfel
266 **Metabolomics as a tool in nutrition research**
Edited by J.-L. Sébédio and L. Brennan
267 **Dietary supplements: Safety, efficacy and quality**
Edited by K. Berginc and S. Kreft
268 **Grapevine breeding programs for the wine industry**
Edited by A. G. Reynolds
269 **Handbook of antimicrobials for food safety and quality**
Edited by T. M. Taylor
270 **Managing and preventing obesity: Behavioural factors and dietary interventions**
Edited by T. P. Gill
271 **Electron beam pasteurization and complementary food processing technologies**
Edited by S. D. Pillai and S. Shayanfar
272 **Advances in food and beverage labelling: Information and regulations**
Edited by P. Berryman
273 **Flavour development, analysis and perception in food and beverages**
Edited by J. K. Parker, S. Elmore and L. Methven
274 **Rapid sensory profiling techniques and related methods: Applications in new product development and consumer research**
Edited by J. Delarue, J. B. Lawlor and M. Rogeaux
275 **Advances in microbial food safety: Volume 2**
Edited by J. Sofos
276 **Handbook of antioxidants for food preservation**
Edited by F. Shahidi

277 **Lockhart and Wiseman's crop husbandry including grassland: Ninth edition**
H. J. S. Finch, A. M. Samuel and G. P. F. Lane
278 **Global legislation for food contact materials**
Edited by J. S. Baughan
279 **Colour additives for food and beverages**
Edited by M. Scotter
280 **A complete course in canning and related processes 14th Edition: Volume 1**
Revised by S. Featherstone
281 **A complete course in canning and related processes 14th Edition: Volume 2**
Revised by S. Featherstone
282 **A complete course in canning and related processes 14th Edition: Volume 3**
Revised by S. Featherstone
283 **Modifying food texture: Volume 1: Novel ingredients and processing techniques**
Edited by J. Chen and A. Rosenthal
284 **Modifying food texture: Volume 2: Sensory analysis, consumer requirements and preferences**
Edited by J. Chen and A. Rosenthal
285 **Modeling food processing operations**
Edited by S. Bakalis, K. Knoerzer and P. J. Fryer
286 **Foods, nutrients and food ingredients with authorised EU health claims Volume 2**
Edited by M. J. Sadler
287 **Feed and feeding practices in aquaculture**
Edited by D. Allen Davis
288 **Foodborne parasites in the food supply web: Occurrence and control**
Edited by A. Gajadhar
289 **Brewing microbiology: design and technology applications for spoilage management, sensory quality and waste valorisation**
Edited by A. E. Hill
290 **Specialty oils and fats in food and nutrition: Properties, processing and applications**
Edited by G. Talbot
291 **Improving and tailoring enzymes for food quality and functionality**
Edited by R. Yada
292 **Emerging Technologies for Promoting Food Security: Overcoming the World Food Crisis**
Edited by C. Madramootoo
293 **Innovation and Future Trends in Food Manufacturing and Supply Chain Technologies**
Edited by C. E. Leadley
294 **Functional Dietary Lipids: Food Formulation, Consumer Issues and Innovation for Health**
Edited by T. Sanders
295 **Handbook on Natural Pigments in Food and Beverages: Industrial Applications for Improving Color**
Edited by R. Carle and R. M. Schweiggert
296 **Integrating the Packaging and Product Experience in Food and Beverages: A Road-Map to Consumer Satisfaction**
Edited by P. Burgess
297 **The Stability and Shelf Life of Food 2nd edition**
Edited by Persis Subramaniam and Peter Wareing

298 **Multisensory Flavor Perception: From Fundamental Neuroscience through to the Marketplace**
Edited by Betina Piqueras-Fiszman and Charles Spence

299 **Flavor: From food to behaviors, wellbeing and health**
Edited by Andrée Voilley, Christian Salles, Elisabeth Guichard and Patrick Etiévant

300 **Developing Food Products for Consumers with Specific Dietary Needs**
Edited by Wayne Morley and Steve Osborn

301 **Advances in Food Traceability Techniques and Technologies: Improving Quality Throughout the Food Chain**
Edited by Montserrat Espiñeira and J. Francisco Santaclara

302 **Innovative Food Processing Technologies: Extraction, Separation, Component Modification and Process Intensification**
Edited by Kai Knoerzer, Pablo Juliano, and Geoffrey Smithers

Part One

Introduction to Food Traceability

What Is Food Traceability?

1

M. Espiñeira
ANFACO-CECOPESCA, Vigo, Spain

F.J. Santaclara
IDIS CHUS, Santiago de Compostela, Spain

1. Introduction

The traceability of the food sector has become a priority. It is essential to provide transparency and security to consumers who are demanding healthier products with a higher quality and the best nutritional characteristics. But it is also important for producers, because it ensures the quality of the raw material which is introduced into the food chain, allowing certification and accreditation of their products, quickly locating problematic items, and implementing control systems, preventing fraud and unfair competition between producers.

The legislation developed and the implementation of specific regulations on traceability have established control systems, both quality and safety, along the food chain, ensuring the quality of the products which are available in the market. This, coupled with the implementation of traceability systems in the food industry, has increased the transparency of the market, facilitating the opening of markets, the internationalization of many products, and the overall growth of the food industry.

2. What Is Meant by Traceability in the Food Industry

Traceability involves the ability to identify at any specified stage of the food chain (from production to distribution) from where the food came (one step back) and to where the food went (one step forward). Traceability allows the early warning for quality problems and hazards and efficient product recall when needed. Each trade unit of each food material is given a unique identifier which accompanies it and is recorded at all stages of its progress through its food chain. It allows the detailed monitoring of properties of food products as they are made and as they move through supply chains, from harvest through transport, storage, processing, distribution, and sales (IUFoST, 2012; Van der Vorst, 2006). Chapter 2 defines and analyzes the food supply chains, focusing in systematically mapping food supply chains and designing the information architecture of food traceability systems.

Despite the importance of traceability in food supply chains, there are several constraints and problems that make its implementation and application to long food supply chains difficult. Some of them are analyzed in Chapter 3, including the prevention of information loss, importance of standardization of information between different links

Advances in Food Traceability Techniques and Technologies. http://dx.doi.org/10.1016/B978-0-08-100310-7.00001-6

in the supply chain, or the effect of granularity levels and heterogeneous mixture of lots. In order to overcome these difficulties, new technologies have been incorporated to traceability systems, and specific applications which facilitate and expedite the collection and exchange of information between the different operators of the food chain continue their development.

3. The Need to Incorporate New Technologies in Traceability Systems

Currently, regional, national, and global networks and services are being rolled out to enable end-to-end traceability in food supply chains. To implement traceable food supply chains, technological innovations are needed for product identification, process and characterization, information capture, analysis, storage and transmission, as well as overall system integration (Chapter 4). This book describes and discusses the current technologies applied to food industry traceability, including future trends and examples of the effective use of this technology in the food sector.

A major feature of any traceable supply chain is the ability to trace back the history and consistently track the physical location of the products in the overall supply chain. Innovations in geospatial science and technology, such as radio frequency technology and mobile tracking devices, have the potential for collecting and transmitting data to distant locations for storage and analysis. Advances in identifiers and sensor and indicator technology, as well as changes in consumer communities, are setting the stage for further development within the food industry, as well as in the interaction of industry and end consumers. A summary of these recent developments is given in Chapter 4, highlighting potentials, critical points, expected trends, and interdependencies of this changing domain.

Among the technology for automated monitoring of processing, displacement, and storage of food items is the RFID (Radio Frequency Identification) technology. It has a great potential for traceability, logistics, supply chain management, and quick response systems. A review of challenges and opportunities of RFID technology applied to food traceability is showed in Chapter 5.

An effective traceability chain is an integrated system of distinct components involving data collection, the analysis, storage, and transmission of the recorded information, and a full backward and forward control system that permits the tracking of the product history. Such an integration of technologies relies on the application of appropriate information and computer systems and links the traceability chain to a central database (Opara, 2003). The software integration in food traceability is described in Chapter 10, where the benefits and specific needs for the food industry, both consumers and suppliers, are discussed.

In order to improve and optimize food traceability, it is recommended to implement in the traceability systems real-time tracking and remote monitoring at some stages. Chapter 11 shows how real-time tracking and remote monitoring can reveal, in many cases, the unseen issues, reducing the undesired situations, the handling of the goods, and the food waste. It also shows the implementations of the new information technologies will provide intelligence to the traceability system which, combined with the most recent advancements, can create solid and reliable food traceability.

In the case of animals, two main components are required in the process of tracing: the technologies for the identification of animals and technologies for the storage and management of information. The working principles of these technologies are described in Chapter 9, with a focus on the main issues that deal with the implementation of effective systems for the traceability of animals.

Another fundamental aspect in the traceability systems applied to the food industry is the control of the authenticity of the raw material or a processed product. Several analytical methods have been or are currently used to authenticate or trace foods (Chapter 6). These analytical methods are based on different technologies such as molecular methods, Near-Infrared (NIR) spectroscopy, and Inductively Coupled Plasma–Mass Spectrometry (ICP-MS), among others. Recent advances in genetic methods currently applied to traceability are described in the Chapter 6, and its application in several food products is discussed. Also, Chapters 7 and 8 are focused in the use of NIR and ICP-MS, respectively, including examples on the applications and uses related with authenticity and traceability in food products.

Recent advances in the mentioned technologies are included in each chapter, including future trends for the application of these to food traceability.

4. Regulations and Legislation on Food Traceability

In this growing global market, food supply chains are becoming longer and more complex. As a result, the tools of traceability become necessary in order to offer transparency to both consumers and food manufacturers (Chapter 13). For the same purpose, the regulation of traceability requirements and restrictions have been developed. Chapters 12 and 13 summarize the current legislation in food traceability in the European Union and the United States, respectively.

Chapter 12 describes the EU food traceability legislation, how it came into being, what the legal text entails, what the consequences of this text are, and if the approach of the legislature is consistent throughout the legal norms. It also evaluates if the legal text has weaknesses, because traceability systems implemented as minimum requirements of this text can be ineffective, and supply chains may thus become vulnerable to fraud.

On the other hand, Chapter 13 summarizes the current legislation in force in the United States, as well as the ongoing development of legal requirements, guidance, and rule-making at the US Food and Drug Administration (FDA) to meet the legislative requirements in the Food Safety Modernization Act of 2011 (FSMA). The chapter also shows the perspective on future developments in this area, working to develop further standards in a global context.

One of the complications in food traceability affects products internationally, since legal requirements and regulations are different in each world area. This examination led the authorities to note the importance of harmonization of traceability requirements and regulations to minimize the potential for misunderstanding and delays due to the difficulties in understanding the practices of each country. At present, efforts are being made to strengthen interoperability in order to overcome unintended trade restrictions and to improve traceability of food products globally (Charlebois et al., 2014).

5. Effective Use of Food Traceability

One of the most important applications of traceability is the return of a dangerous product that has already been supplied or made available to consumers by the producer or distributor (Chapter 14). Traceability is essential to guarantee the location of products and facilitate their recall when safety and quality standards have been breached (Opara, 2003). Chapter 14 provides an overview of the product recall process and explains how food traceability can be effectively used in product recall. It also explains how the recall process can be optimized by reducing the batch dispersion and root cause analysis through the traceability system.

Other effective use of traceability in the assurance of food quality is the authentication of the origin and composition of foods. It guarantees the origin and good production practices in the elaboration of foods for human consumption. Chapter 15 is focused in the use of traceability in evaluating origin claims. It describes and discusses the target analytes and analytical procedures used to discriminate foods from different areas or production methods. It includes examples of some of the most commonly evaluated foods for origin claim.

In addition to the development of international sourcing and markets, additional factors such as increased out of season demand and ethnic diversity of resident populations, have resulted in the increased requirements for a wider and deeper (specialist product) range on offer in more countries across the world, leading to the need to access a wider source origin of supply. This book describes the application of the traceability through different food chains, showing the greatest difficulties associated with each sector, the methodologies used, the most recent advances, and future prospects focused on that sector or particular application.

Another staple are cereals, and high demand generates the substitution of the cheaper cereal grain with the more expensive one or adding inferior cereal grain to the premium one, resulting in extremely widespread adulteration or mislabeling of cereal grain. Chapter 16 provides an overview concerning the current and the most effective uses of different analytical techniques that have been developed for determining the geographical origin in order to discriminate different origins of cereal grain, confirming their authenticity. Its advantages and limitations are discussed, and future trends are described.

The meat (and livestock) sector has been at the forefront of policy and industry developments concerning traceability. In a large part, this is due to the outbreak of several food safety crises that have affected this sector in recent years. It had served to focus public attention on food safety but also stimulated demand for more information on the source of food products and the production and processing methods in use within food supply chains. Chapter 17 examines the roles, outcomes, and implications of traceability in meat and livestock supply chains. It examines public and private sector approaches to meat and livestock traceability and discusses the implications for its effective use in meat supply chains.

The authenticity assessment in dairy products is an important issue not only to the economic point of view, but also to medical requirements, food allergies, or religious practices (Mafra et al., 2008). Traceability monitoring systems allow the safety and quality of food products across processing, packaging, storage, and distribution to be controlled properly. Chapter 18 focuses on the preservation of the quality of dairy products along the food chain. It introduces a monitoring protocol based on temperature sensors to track the environmental stresses experienced by dairy products during milk-run delivery tours. The aim of this traceability protocol is to assess the environmental and physical conditions experienced by dairy products during the transport and storage activities, identify criticalities, and measure the effectiveness of logistics and material-handling operations in addressing the quality and the shelf life of the dairy products.

In the case of animal feed industries, traceability can mean several different actions, such as the possibility to track a feed during production phases and to follow the distribution routes until the final destination. However, a key point for the safety and quality of a feed is represented by the type and characteristics of ingredients. For this purpose, one of the most innovative and popular technology is the DNA-based traceability, now widely accepted as playing a key role. Chapter 19 discusses the most commonly used methods to obtain an effective use of traceability in the feed chain, focusing on the identification of animal and plant species and/or genetically modified plants and on the presence of toxin-producing microorganisms. It also describes the future perspectives for feed traceability based on new technology advances.

6. Future Trends

The advances in food traceability and the effective use of traceability in different supply chains which are described in this book are a reflection of the ability of traceability systems as a tool to ensure safety and quality in the food industry. The constant improvement of the methodologies currently being used, the incorporation of new technologies, and the adaptation to the needs of the different food sectors will overcome the current limitations, providing greater efficiency, speed, and reliability in the traceability systems. All this will increase transparency and ensure that products reach the international market, promoting consumer confidence and the growth of the food industry.

References

Charlebois, S., Sterling, B., Haratifar, S., Naing, S.K., 2014. Comparison of global food traceability regulations and requirements. Comprehensive Reviews in Food Science and Food Safety 13, 1104–1123.

IUFoST, March 2012. The International Union of Food Science and Technology. IUFoST Scientific Information Bulletin (SIB).

Mafra, I., Ferreira, I.M., Oliveira, M.B., 2008. Food authentication by PCR-based methods. European Food Research and Technology 227, 649–665.

Opara, L.U., 2003. Traceability in agriculture and food supply chain: a review of basic concepts, technological implications, and future prospects. Journal of Food Agriculture and Environment 1, 101–106.

Van der Vorst, J.G., 2006. Product traceability in food-supply chains. Accreditation and Quality Assurance 11, 33–37.

Defining and Analyzing Traceability Systems in Food Supply Chains

2

H. Scholten
Wageningen University, Wageningen, The Netherlands

C.N. Verdouw
LEI Wageningen UR, The Netherlands; Wageningen University, Wageningen, The Netherlands

A. Beulens
LEI Wageningen UR, Wageningen, The Netherlands

J.G.A.J. van der Vorst
Wageningen University, Wageningen, The Netherlands

1. Introduction

Traceability is considered to be a vital issue for all stakeholders in food supply chains. The most important driver is the increasing societal need to guarantee food quality and provenance (Beulens et al., 2005; Trienekens et al., 2014). Because consumers cannot know in detail what processing steps are executed in the production of food and what ingredients or resources are used in these steps, they want to be assured that food products are safe, healthy, sustainable, and of high and consistent quality. The need for guarantees is strengthened by the continuing sequence of food calamities, which have required massive product recalls, sometimes even on a European scale. Recent examples include the horse meat scandal and the *Escherichia coli* outbreak. The societal concern about food safety has resulted in a lot of legislation, including, for example, the obligation for traceability in the European General Food Law, article 18 (EC, 2002). Food companies have to comply with consumers' demands and legislation. Moreover, they want to minimize costs and image damage when incidents occur. On the other hand, traceability optimizes business processes along the supply chain in order to improve efficiency and reduce lead times and food waste, among other things (van der Vorst, 2006).

The required ability to trace products back to their origin and production methods and to track the ongoing location of products has forced companies worldwide to implement coordinated traceability systems along the food supply chain (Beulens et al., 2005; Folinas et al., 2006; Fritz and Schiefer, 2009). However, the realization of such systems is a challenging task due to the specific characteristic of the food domain (van der Vorst et al., 2005; Trienekens et al., 2014; Verdouw et al., 2014b). In particular, there is high uncertainty regarding fresh product quality as well as available volumes due to variations in the natural production process. Furthermore, the sector is characterized by high supply network and process complexities, including

Advances in Food Traceability Techniques and Technologies. http://dx.doi.org/10.1016/B978-0-08-100310-7.00002-8

divergent and convergent processes, by-products, high risks for contamination, and many small companies. As a consequence of the demanding requirements and the specific characteristics of the food domain, there are many issues to be addressed in the development and realization of traceability systems. Traceability is still often achieved by conventional systems, and it focuses on a company or a specific part of the supply chain. Information systems of many supply chain participants are poorly integrated in the supply chain, which hampers end-to-end visibility and real-time tracking and tracing (Trienekens et al., 2012a,b). Furthermore, the usage is mainly limited to well-established functions, such as container tracking, while affordable solutions for more advanced functionalities are missing (Verdouw et al., 2015). A main challenge is to assure real-time traceability: (1) for all stakeholders, ie, including producers (farms), processors, transporters, retailers (supermarkets), consumers, and authorities/regulators; (2) for different types of fresh food; and (3) that can be extended to all kind of fresh food-related properties beyond whereabouts, such as information about allergens, safety, nutritional value, place of origin or provenance, animal welfare, use of antibiotics, sustainability, and taste.

This chapter proposes a method to define and analyze traceability systems in food supply chains. More specifically, it aims to support food traceability professionals by systematically mapping food supply chains and designing the information architecture of food traceability systems. The remainder of this chapter will first introduce the basic concepts and terminologies of traceability in food supply chains in Section 2. Subsequently, the two parts of the method are presented. Section 3 will describe how to systematically map food supply chains as a starting point of the definition and analysis of food traceability systems. Section 4 will introduce two distinctive architectures of food traceability systems that are based on fundamentally different approaches, ie, the one-step-forward and one-step-back principle versus end-to-end traceability. The next section will show how these architectures can be applied by introducing two case studies in meat supply chains. The chapter concludes with addressing challenges for future development and summarizing the main findings.

2. Traceability in Food Supply Chains

2.1 Food Supply Chains

The term supply chain is relatively new. It originated in logistics in the beginning of the 1980s as an inventory management approach that optimizes the physical flow from raw material to end products as one integrated system with a shared objective (Cooper and Ellram, 1993; Christopher, 1998; van der Vorst, 2000). In the 1990s Supply Chain Management (SCM) evolved toward an integrated process approach in which the concepts of logistics management were extended to incorporate the integration of firms in its supply chain. This development was enhanced by the introduction of the term "value chain" (Porter and Millar, 1985). A firm's value chain is a system of interlinked processes, in which every activity adds value to the product or service. Parallel to the evolvement of SCM from logistics, network approaches had come up from organizational and management studies. Different types of network relations were proposed between hierarchy and markets (Powell, 1990; Webster, 1992). The emphasis in SCM research on the network dimension finally resulted in increasing cooperation between these two streams (Lazzarini et al., 2001). As a result, a supply chain can be defined as "an integrated system of processes performed by a set of networks, comprised of horizontal and vertical ties between connected and interdependent organizations mutually and co-operatively working together to control, manage and improve the flow of materials and information from suppliers to end consumers" (Wiersinga et al., 2010).

Food supply chains comprise organizations that are responsible for the production and distribution of vegetable or animal-based products. In general, two main types can be distinguished (van der Vorst et al., 2005):

1. Supply chains for fresh agricultural products (such as fresh vegetables, flowers, and fruit). In general, these chains may comprise growers, auctions, wholesalers, importers and exporters, retailers, and specialty shops and their input and service suppliers. Basically, all of these stages leave the intrinsic characteristics of the product grown or produced in the countryside untouched. The main processes are the handling, conditioned storing, packing, transportation, and especially the trading of these goods.
2. Supply chains for processed food products (such as portioned meats, snacks, desserts, and canned food products). In these chains, agricultural products are used as raw materials for producing consumer products with higher added value. In most cases, conservation and conditioning processes extend the shelf life of the agricultural and consumer products.

2.2 What Is Traceability?

Traceability is the ability to determine the ongoing location of products and to trace products back to their origin and used production methods (Trienekens et al., 2014). Traceability information is information that may be used to inform stakeholders about the whereabouts of particular (physical) items or products, about their history, dynamic properties, content, and relationships with other products (van der Vorst et al., 2005). Traceability is closely related to the wider concept of supply chain transparency, which can be defined as "...the extent to which all its stakeholders have a shared understanding of, and access to, the product-related information that they request, without loss, noise, delay and distortion" (Trienekens et al., 2012a,b).

The basic idea of traceability is the possibility to track where a certain item is located and to trace the history of that item (Fig. 2.1):

- *Tracking*: the determination of the ongoing location of items during their way through the supply chain (Trienekens et al., 2014).
- *Tracing*: the ability to know the historical locations, the time spent at each location, record of ownership, packaging configurations, and environmental storage conditions for a particular item (Schuster et al., 2007). Tracing aims at defining the composition of an item and the treatments that item has received during the various stages of the product life cycle (Trienekens et al., 2014). Chain upstream (backward) tracing aims at determining the history of items and is used to determine the source of a problem of a defective item. Chain downstream (forward) tracing aims at the determination of the location of items that were produced using, for example, a contaminated batch of raw materials.

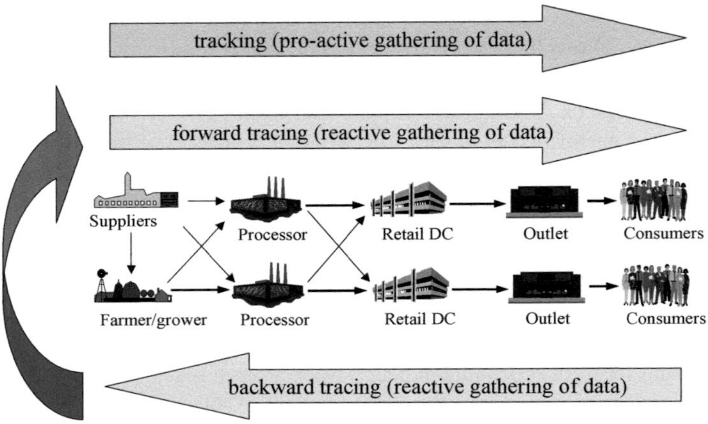

Figure 2.1 Tracking and tracing.
Reproduced from van der Vorst, J., Beulens, A., Beek, P.V., 2005. Innovations in logistics and ICT in food Supply chain networks. In: Jongen, W.M.F., Meulenberg, M.T.G., (Eds.), Innovations in Agri-food Systems: Product Quality and Consumer Acceptance. 2005, Wageningen Academic Publishers.

2.3 Traceability in Food Supply Chains

Traceability in food supply chains is a challenging issue. There are many sector-specific characteristics that complicate tracking and tracing, such as the following (Trienekens et al., 2014):

- High flow complexities characterize the food sector, as food supply chains consist of continuous product flows (bulk products traced in volumes or masses) and discrete product flows (packages traced in units). Traceability systems have to combine bulk products downstream with packaged products upstream.
- Batches and lots of food products have to be processed separately to prevent cross-contamination, which is not easy for continuous flows in food supply chains.
- Diverging and converging processes and (by-)product flows characterize many food supply chains. Often food products are composed of many ingredients (including packaging) that all have to be tracked and traced, including many waste and by-products.
- Food supply chains are often highly connected to each other, as waste/by-products of one supply chain are sometime input in other supply chains.
- Food products are often highly perishable, especially fresh foods. Consumers and society set high demands, including food safety regulations and quality standards. Traceability data must therefore also refer to best-before-dates, country of origin of all ingredients, and certification of quality, veterinary, phytosanitary, and ecological checks.
- Food supply chains are complex networks consisting of small- and medium-sized enterprises (ie, farms, departments of processing companies) that interact with multinational companies providing input to the supply chain and doing its retail. To ensure year-round supply of seasonable products, the supply chains are international, which makes tracking and tracing even more complex.

To enable tracking and tracing, the detailed registration of process, resource, and product characteristics, such as origin and history of products, quality variation, etc., is essential for the food industry (Trienekens et al., 2014). *Product information* concerns the composition and sensory aspects of products, as well as residue information (eg, pesticides and hormones). *Process information* to be exchanged includes the origin of the product (and its components in the case of a composite product), the history of the product (where the product has been on its trajectory through the food supply chain, who has dealt with it, and what resources have been used), storage time and quality decay, quality variation within and between different product lots, and resources used and waste.

A large actuality gap between the occurrence of events and the recording of data is often a problem when systems have to be used more or less in real time, like in the case of product recalls. In many cases, it has been a (sometimes too) big challenge to timely and adequately link a food incident to data recorded earlier in time.

2.4 Why Traceability in Food Supply Chains

Food supply chain information systems to enable traceability and transparency have various types of users, each requiring different traceability analyses. These users can be categorized in three main groups: (1) supply chain partners, ranging from food producers such as farmers, to retailers, (2) consumers, the *raison d'être* of food production and supply chains, and (3) authorities and regulators, aiming at guaranteeing safe food, controlling, empowering food laws and regulations, preventing contaminated food and other food related hazards, and aspiring fast and surgical responses to any food crisis.

As a result of the series of food crises, summarized in Section 1, a number of transparency measures are incorporated in food regulations, such as the European regulation Reg. No. 178/2002 (also referred to as the General Food Law or GFL), a series of additional regulations (including topics such as genetically modified organisms, hygiene, official controls, food contact materials, allergen labeling requirements, nutrition and health claims, issues related to nutrition, overweight, and obesity, etc.), and the more recent regulation Reg. No. 1169/2011, which regulates informing consumers on food (van der Meulen, 2013). In the United States and other countries, we also see fresh food regulations, directives, and laws becoming stricter, triggered by fresh food incidents on the one hand and by increasing food-related awareness of consumers on the other hand.

According to European legislation the minimum each partner should know is which supply chain partner delivered the incoming (half) products and to which supply chain partner the resulting products were delivered. This principle is often referred to as the *one-step-forward and one-step-back* principle. In this way, data passes from one partner to the next, and each partner should record where all products come from and go to. This simple and basic requirement tracks and traces all (intermediate) products along the supply chain, but information systems based on this principle make tracking and tracing a cumbersome process, as is shown for instance in the EHEC crisis of 2011[1] or in the horse meat scandal in 2013.[2] There is a long list of such food-related

[1]https://en.wikipedia.org/wiki/2011_Germany_E._coli_O104:H4_outbreak.

[2]https://en.wikipedia.org/wiki/2013_meat_adulteration_scandal.

problems and all show that tracking and tracing with information systems built on this simple requirement are very slow and inaccurate. One of the reasons is that the requirement does not prescribe how the information passes from partner to the next, so it is often based on outdated technologies (Fig. 2.2).

If one aims to enable fast and accurate traceability in both directions along the supply chain, ie, from producer to retailer and from retailer to producer, other designs of information systems are able to capture and store all data, in a way that serves a wider range of questions, eg, "where are all products that are related to a specific (contaminated) products," "how can I improve my inventory management," or "what do I eat."

The remainder of this section will discuss *enabling technologies* and *standards* that can be used for traceability systems and what kind of *analysis* can be performed with traceability systems.

2.5 Enabling Technologies

There are three technologies that enable the development of traceability systems: Automatic Identification and Data Capture (AIDC) technology, connectivity technology, and business intelligence on traceability (eg, online analytical processing, analytics, data mining, process mining, complex event processing, business performance management, benchmarking, text mining, predictive analytics, and prescriptive analytics).

AIDC refers to the methods of automatically identifying objects, collecting data about them, and entering that data directly into computer systems (ie, without human involvement). Technologies typically considered as part of AIDC include *bar codes*, *radio frequency identification* (RFID), biometrics, magnetic stripes, optical character

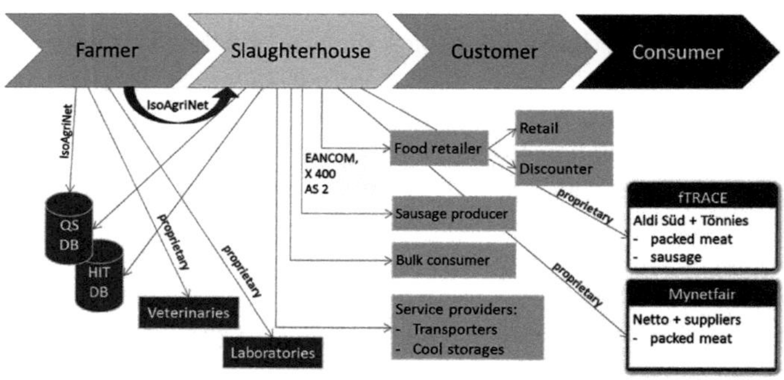

Figure 2.2 Data flow in a German meat supply chain (courtesy of the representative of the slaughterhouse; translated from German, Kassahun et al., 2014). QS (*Qualität und Sicherheit*) is a German quality assurance scheme with an associated company by the same name that does the bulk of meat quality assurance audits in Germany (Albersmeier et al., 2009). HIT (*Herkunftssicherungs- und Informationssystem Tiere*) is a German national database for registration of movement of bovine animals established in accordance with the EC Regulation 1760/2000. Mynetfair is a trade fair web portal. fTRACE is a third-party fresh food transparency system offered by GS1 Germany.[3]

[3]https://www.ftrace.com/en/gb/.

recognition (OCR), smart cards, and voice recognition. AIDC is also commonly referred to as *automatic identification*, *auto-ID*, and *automatic data capture*.

The most well-known Auto-ID technologies are bar code scanning and RFID. A bar code is a standardized code formed by a pattern of parallel lines of varying widths that can be read automatically by bar code scanners. The bar code standards of GS1 (comprising, among others, the former EAN International and Uniform Code Council) are internationally common practice. Bar codes are traditionally used for identification. However, new bar code standards, including the GS1 Databar and multidimensional bar codes (especially quick response (QR) codes and GS1 Datamatrix), can contain additional information like batch numbers, weight, and best-before-dates.

RFID technology is based on the wireless use of electromagnetic fields to transfer data and automatically identify data on tags attached to objects. RFID uses electronic labels (tags, transponders) that comprise a microchip, that can store and process data, and an antenna, that can wirelessly communicate these data with a reader. RFID tags can be passive or active. A passive RFID tag uses the electromagnetic energy to activate the tag when passing a reader. An active RFID tag includes a small battery. They can have a larger range and can include more advanced processing capabilities.

2.6 Standards

A crucial prerequisite of effective traceability systems is a rapid, error-free, efficient, and secure exchange of information between organizations in supply chains. This is only possible if information systems of the different chain actors "speak the same language." Information standards ensure this interoperability. Information standards are composed of common agreements on the format, content, and meaning of the information to be shared (ie, messages, codes), as well as on the method by which they are to be communicated technically. The most important supply chain information standards are the following (Verdouw et al., 2014a):

- *Identification standards (codes)*: digital number plates of, eg, products, load carriers, crates, and locations. These codes are used as information keys in applications through the entire chain, like scanning bar codes, reading RFID tags, and tracking and tracing.
- *Auto-ID standards*: agreements about the size of bar codes, RFID chips, etc., so that they can be read consistently.
- *Data communication standards*: technical agreements about how information (via the internet) should be exchanged between parties in the chain.
- *Standard messages*: agreements on the functionality and structure of electronic messages, for example: which information should be contained in an order?

The availability of commonly accepted identification standards is an essential basis of traceability systems. In food supply chains the identification standards of GS1 are internationally broadly adopted. These standards are on different levels of aggregation. The Global Trade Item Number (GTIN) is an identifier for trade items developed by GS1. The GTIN, also called the GS1 article code, is use for identification on a trade item level, eg, single products or cartons. GTINs are used to look up product information in a database by entering the number through a bar code scanner, QR scanner, or RFID. GTINs are unique and universal. An example is a GTIN for beef steak. Next to

GTINs, there are two other GTINs that include more detailed IDs. First, LGTIN (Lot GTIN), referring to specific batches and lots (eg, beef steak from 100 animals slaughtered in the morning of August 17, 2015; these are originating from a single farm or a group of farms). Second, SGTIN (serial GTIN) refers to a single, specific beef steak. The more detailed the identity of a food item, the more specifications can be connected to that piece of food. One can trace back a beef steak with an SGTIN to a specific farm and a specific animal.

Other GS1 identification standards include the Serial Shipping Container Code (SSCC), the Global Returnable Asset Identifier (GRAI), and Global Location Numbers (GLN).

As discussed in the previous section, GS1 is also leading in the standards for Auto-ID. Important GS1 bar code standards include EAN13, GS1-128, and GS1-Databar. However, in two-dimensional barcodes, the QR code is most widely used. The Electronic Product Code (EPC) standards of GS1 are most widely used for RFID applications. EPC goes beyond the Auto-ID standards, because it includes a broad set of specific standards for the communication of event data. The EPCglobal network consist of an Object Naming Service (ONS, enabling the discovery of object information on the basis of an EPC), an EPC Discovery Service (an instrument to find EPC Information Services within the network), EPC Information Services (EPCIS, explained below) and EPC Security Services (tools allowing a secured access to information of the EPCglobal network in accordance to the access rights of the partners).

EPCIS is a standard that codifies the syntax and semantics for supply chain events and the secure method for selectively sharing supply chain events with trading partners. The recent version of the standard (EPCIS 1.1) defines four event types:

1. *Object event* occurs when an object is observed (or is not observed while it should be).
2. *Aggregation event* occurs when an object is added to or removed from a containment (mainly used to track palletized objects).
3. *Transaction event* occurs when an object is associated or disassociated with a business transaction.
4. *Transformation event* occurs when one or more (input) objects are consumed and transformed into (output) objects.

Events contain data about the identity of the product, the date and time of event occurrence, the location where it occurred, and the reason why the event occurred. These are conveniently abbreviated as the *what*, *when*, *where*, and *why* of the event (Fig. 2.3). The *what* (the identity) is represented by EPCs with GTIN, LGTIN, or SGTIN (explained above). The *where* (location) is represented by globally unique GLNs (also explained above). The *when* (date and time) is a local time and time zone or UTC (Universal Time Coordinated) timestamp. The *why* (reason) of the event is described using a predefined but extensible vocabulary of business process steps, the Core Business Vocabulary, or CBV (GS1 AISBL, 2014). As the CBV terms are rather

What (SGTIN)	Where (GLN)	When	Why
urn:epc:id:sgtin:4000001.011629.2	Dockdoor1, Vendor1	12-12-2013 09:00:00	Goods receiving

Figure 2.3 Basic scheme of data to be captured for each EPCIS event.

generic, details can be added by using a User Vocabulary, according to the format described in the CBV standard (GS1 EPCglobal, 2014).

The data to describe events of, for instance, meat products in supply chains (birth, treatment, slaughtering, etc.) can be divided into two types: *dynamic data*, describing the events itself, and more static data that are called *master data*. The latter describe static issues, eg, GLNs of companies (which farm, slaughterhouse, etc., and its characteristics, eg, organic type of farm or a halal type of slaughtering).

Quality-related issues in an EPCIS-based transparency system, eg, *test on BSE of cadavers*, *use of antibiotics*, and *maximum temperature of a food item during an event* can be handled as vendor extensions to events.

In this section, some basic concepts and terminologies of traceability in food supply chains are introduced, including enabling technologies and standards. The next section will describe how to systematically map food supply chains as a starting point of the definition and analysis of food traceability systems.

3. Food Supply Chain Mapping

As argued previously in this chapter, food traceability systems have to be able to deal with a high supply chain complexity. Food supply chains have diverse network structures where many small- and medium-sized enterprises trade with huge multinationals. They also include a large variety of business processes that handle many interrelated objects at different granularity levels. In order to get a grip on this complexity, it is essential to start the definition and analysis of food traceability systems with a systematic mapping of food supply chains.

Lambert et al. (2005) addressed two frameworks with a sufficient level of detail for supply chains mapping, ie, the Supply Chain Operations Reference-model (SCOR) of the Supply-Chain Council (SCC, 2008) and the framework of the Global Supply Chain Forum (GSCF) (Lambert and Cooper, 2000). The GSCF framework has a strategic perspective and focuses on integration via relation management. On the other hand, the SCOR model is based on a more operational perspective and consequently focuses on information sharing and connecting with other members of the supply chain through transactional processes. It provides generic definitions of supply chain processes for production and logistics at three aggregation levels (SCC, 2008). The first level defines the process plan, source, make, deliver, and return. The second level addresses different process categories per level one process. The third level defines detailed activities per process category. However, although SCOR is a valuable framework to map business processes, it does not explicitly map the objects that are handled by these business processes. As a result, it is not sufficient for the purpose of traceability system implementation. Verdouw et al. (2010a,b) have developed a framework based on SCOR that has addressed this shortcoming. This framework identifies three types of process models at different levels of abstraction (Verdouw, 2010):

1. *Product Flow Models*: allocation of the basic transformations in the supply chain from input material to fresh or processed end products for the consumer.

2. *Thread Diagrams*: control and coordination of the basic transformations in specific supply chain configurations.
3. *Business Process Diagrams*: sequence and information flows among detailed control and coordination activities. These diagrams may consist of multiple levels of abstraction. One can zoom in from main processes to various levels of subprocesses.

Product Flow Models visualize the allocation of basic transformations to supply chain actors and the related product flows from input material into end products. Transformations are business processes that contribute directly to the creation and movement of products by a company such as engineering, production, and distribution (Verdouw et al., 2011). In a supply chain, transformations are performed by multiple firms, especially if there is a high degree of specialization. Product Flow Models depict the allocation of transformations to the actors in a specific supply chain configuration. Fig. 2.4 shows an example of a Product Flow Diagram of a supply chain for processed fruit.

The product flows among transformations comprise several levels of aggregation. Based on the GS1 Global Traceability Standard (Ryu and Taillard, 2007), four different units are distinguished:

1. *Shipping Unit (SU)*: an item or group of items delivered to one party's location at one moment in time, which undergoes the same dispatch and receipt processes. SUs can be identified with standard Shipment Identification Numbers (SINs).
2. *Logistics Unit (LU)*: an item of any composition established for transport and/or storage that needs to be managed through the supply chain. LUs can be identified with standard Serial Shipping Container Codes (SSCCs).
3. *Trade Unit (TU)*: product unit as it is traded before the point-of-sales in the supply chain. TUs can be identified with standard Global Trade Item Numbers (GTINs), in combination with a serial number (SGTIN) or with a batch/lot number.
4. *Consumer Unit (CU)*: product as it is sold to the end customer. CUs can be identified in the same way as TUs.

Thread Diagrams provide a process overview of the complete supply chain configuration in scope. Fig. 2.5 shows an example of a Thread Diagram of a supply chain for processed fruit. They depict how order-driven and forecast-driven processes are decoupled in specific supply chain configurations (Positions Customer Order Decoupling Points) and how interdependencies between processes are coordinated. Basic reference components of such diagrams are business control cases and coordination mechanisms. A business control case represents a sequenced group of business processes that follow the same control strategy. Business control cases can be either responsive (to order) or anticipatory (to forecast). Customer Order Decoupling Points (CODPs) decouple a series of responsive and a series of forecast-driven control cases. A coordination mechanism manages the interdependencies among business control cases. Coordination of Product and Order precedences (P/O) occur at every interface of two basic supply chain roles, ie, when products are passed on from one actor to another according to an agreement about the requirements (order). Coordination of capacity usage, capacity precedence, and material consumption (C/M) manages the dependencies among multiple control cases per actor. Last, the exchange of Demand and Supply information (D/S) connects anticipatory cases of a supplier with responsive cases of a customer.

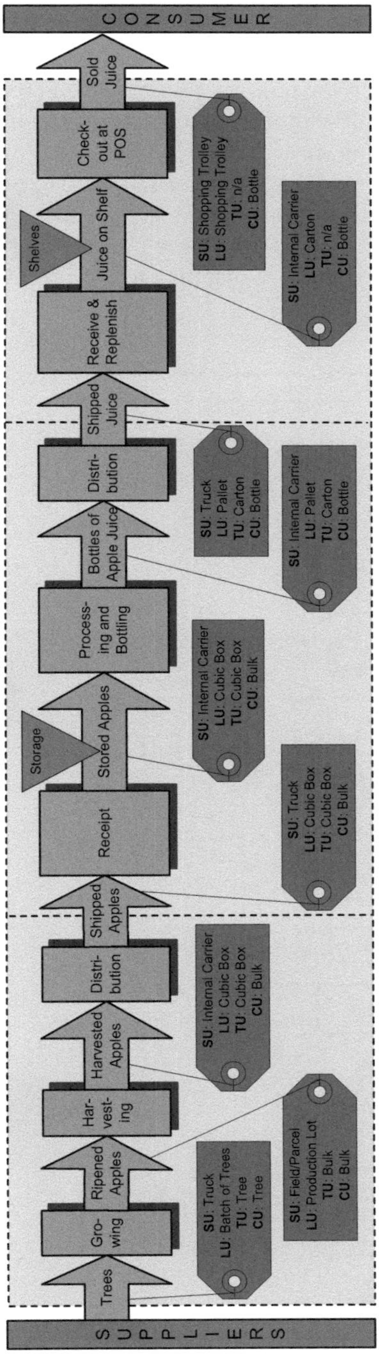

Figure 2.4 Example of a Product Flow Model for processed fruit: apple juice supplied via direct delivery to a retailer.

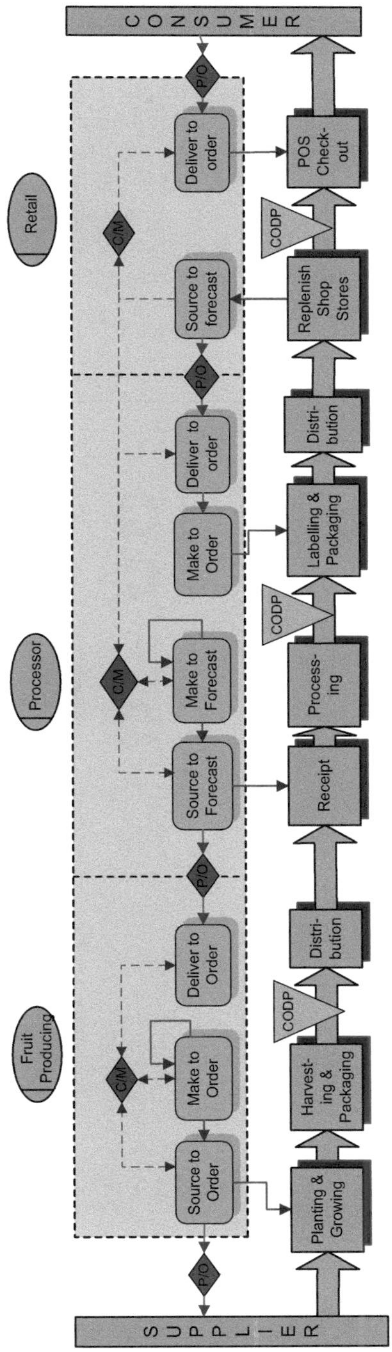

Figure 2.5 Example of a Thread Diagram for processed fruit: apple juice supplied via direct delivery to a retailer.

Business Process Diagrams depict the sequence and interaction of control and coordination activities (as identified in Thread Diagrams). Business Process Diagrams can be composed by zooming in on specific business control cases or coordination mechanisms. For example, Fig. 2.6 zooms in on the processor's "Make to Forecast" control case in Fig. 2.5 in order to illustrate how the different models are interlinked.

The basic building blocks for process modeling are activities. The Business Process Diagrams are modeled in the Business Process Modeling Notation (BPMN) to ensure smooth translation of supply chain designs to information systems architecture (OMG, 2010). BPMN has developed into the de facto standard for business process modeling in Service-Oriented Architectures (SOAs) and includes a mapping to the underlying

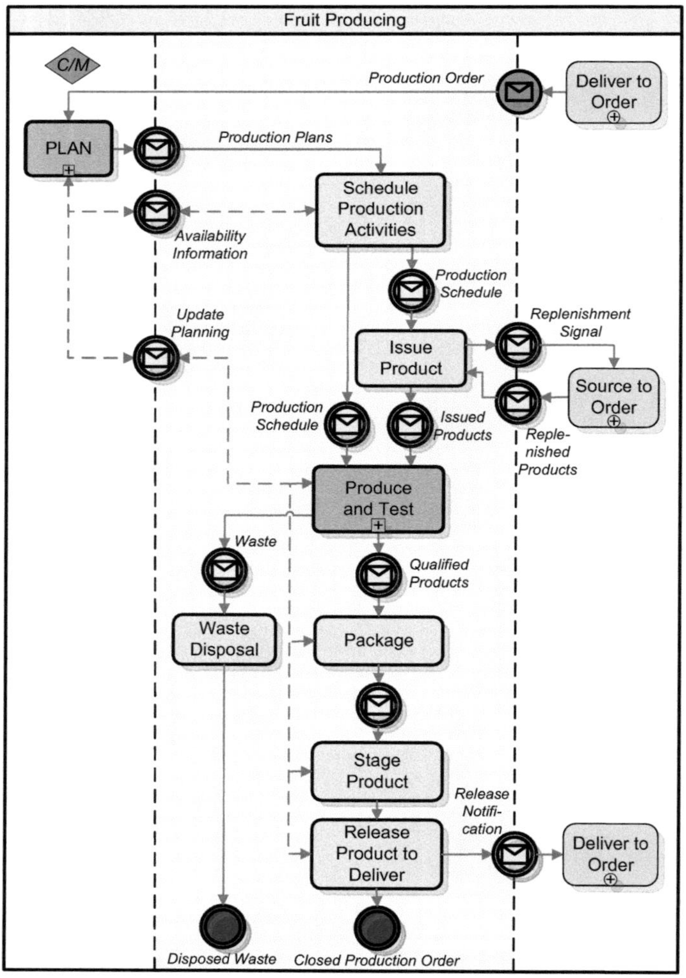

Figure 2.6 Example of a process model.

web service execution languages, ie, Business Process Execution Language and Business Process Modeling Language (www.bpmn.org).

This section has introduced the mapping of food supply chains, which is the first part of the method to define and analyze traceability systems in food supply chains. The next section will introduce the second part that can be used to design the information architecture of food traceability systems.

4. Food Traceability System Architectures

More detailed architectures of the distinct food traceability systems, introduced in Section 2.3, will be presented in this section. We distinguish two architectures that differ structurally. The first fits with the basic EU requirement that each food supply chain partner should know where the food products it uses comes from and where they go to (one-step-back and one-step-forward). Data on food items and associated processes are following the same paths as the food products. This enables tracking and tracing, even when parts of the data trail are paper-based. The second approach is based on sharing data with partners in the whole supply chain or (as is more often the case) in the whole supply network. The first approach will be discussed in Section 4.1, the second one in Section 4.2, and a more detailed discussion can be found in Kassahun et al. (2014).

4.1 One-Step-Forward and One-Step-Back Approach

At present, traceability systems often rely on basic technologies, eg, electronic data interchange (EDI), labeling, and "paper trails" left by email or fax (Fig. 2.7). Some

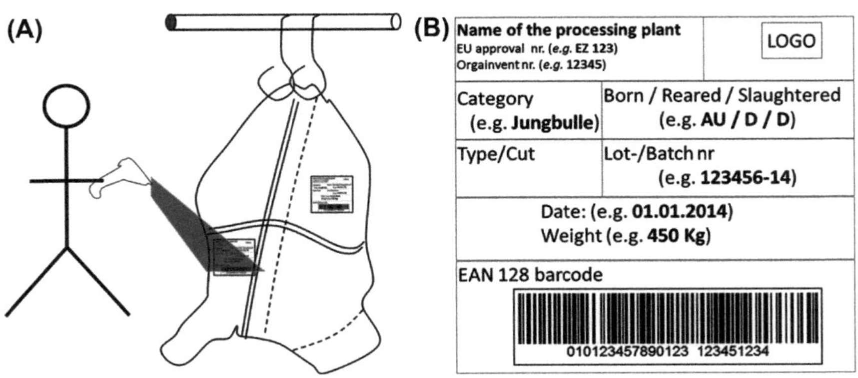

Figure 2.7 An example of an intermediate meat product, ie, a carcass quarter (A) and paper (printed) label placed on the intermediate product (B), based on real-life images obtained from the slaughterhouse.
From Kassahun, A., Hartog, R.J.M., Sadowski, T., Scholten, H., Bartram, T., Wolfert, J., Beulens, A.J.M., 2014. Enabling chain-wide transparency in meat supply chains based on the EPCIS global standard and cloud-based services. Computers and Electronics in Agriculture 109, 179–190. http://dx.doi.org/10.1016/j.compag.2014.10.002.

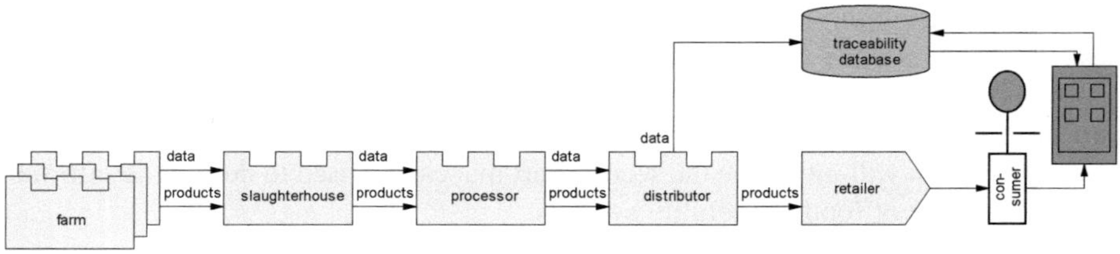

Figure 2.8 Outline of a traceability system for meat, based on the *one-step-forward and one-step-back* principle, in which data follow the same route as the food items (in this case meat). There are more farms depicted than just one, and the same is possible of the other supply chain partners, resulting in a food supply network. The distributor collects, arranges, and makes the information available for consumers.

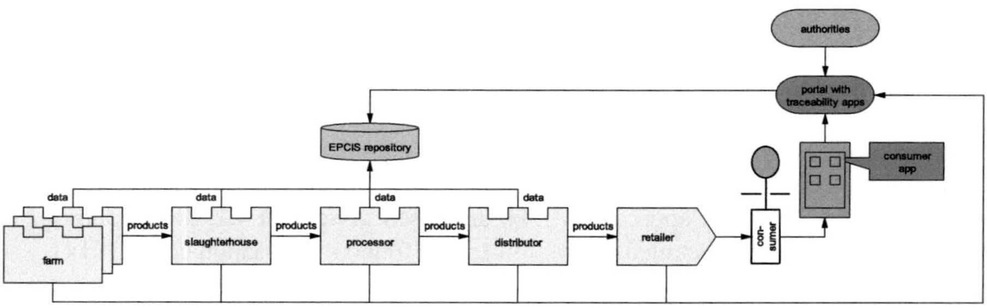

Figure 2.9 Outline of a traceability system for meat, based on sharing event data of meat items. Data are captured and stored in a single and shared repository. Some third-party solution provider provides tools that enable authorities/regulators, supply chain partners, and consumers to get relevant information from the system.

large food processing companies do have transparency systems in place as part of their enterprise system; however, the use of state of the art enterprise traceability systems rarely covers entire food supply chains (Trienekens et al., 2012a,b).

Fig. 2.8 depicts an example of a food supply chain, here for meat. Products start at the farm and pass to the next supply chain partner. Data follow the same path and every supply chain partner is responsible for the correct passing of these and his own data to the next right partner. The distributor collects these data, arranges them in the right format, and properly handles them for further use. For instance, if consumers should get access to (parts of) these data, they can be made available by storing these in a traceability database or repository. Consumers can get access through a PC or smartphone, at the retailer or at home.

4.2 Event-Oriented Approach

Figs. 2.9 and 2.10 outline the architecture of beyond state of the art transparency systems. This kind of system should best be based on capturing all events of food items passing through a supply chain/network, store these on one or more EPCIS repositories, make these available on a portal with the help of a series of apps and interfaces for capturing, querying, discovering, and aggregating for further use by

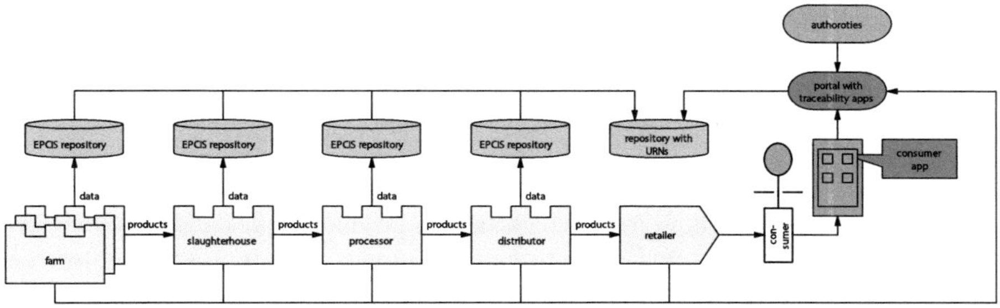

Figure 2.10 Outline of a traceability system for meat, based on sharing event data of all events meat items. Farm data are captured and stored in a single and shared repository. The other supply chain partners have their own EPCIS repository. Some third-party solution provider provides tools that enable authorities/regulators, *supply chain partners*, and consumers to get relevant information from the system.

authorities/regulators, supply chain partners, and consumers, as will be shown in Section 5.2.

A fundamental problem in this kind of innovative supply chain/network system is caused by the hesitation of supply chain partners to share data with others in the supply chain, as they see the risks of sharing business data to competitors, than the opportunity to benefit all of sharing data. They are not easily convinced that all supply chain partners can optimize their own business processes with the sharing approach. Guaranteeing that each supply chain party has only access to food item event data they are involved in may have a positive effect on this phenomenon.

5. Case Studies

This section describes two cases that are implementations of the architectures introduced in Section 4. Both cases denote fresh food supply chains and our examples will refer to meat, as this is a complex fresh food supply chain. The first one is implemented in a series of fresh food supply chains, aiming at providing information to consumers of supermarkets (see Section 5.1). The second one is about a prototype of a beef transparency system developed within a research project (see Section 5.2).

A beef supply chain in the perspective of a slaughterhouse in Germany is depicted in Fig. 2.2. It shows that not only information that is relevant for consumers has to be collected and sent forward along the supply chain, but also other data, enforced by European and German laws and regulations. Two third-party transparency providers are involved, ie, fTRACE and Mynetfair;[4] the former will be discussed in Section 5.1.

5.1 fTRACE

fTRACE is a service that enables consumers to trace fresh food products (meat, poultry, fish, fruit, and vegetables) from a growing number of supermarkets (Aldi, Lidl, Metro, etc.)

[4]http://www.mynetfair.com/.

in Northwest Europe, focusing on the origins and processing of fresh food. All information is provided by the manufacturers themselves about their own (intermediate) food products. Scanning an fTRACE code (Fig. 2.11) sends information about where the product comes from, when and how it was processed, and information about quality and recipe tips.

For many consumers, labeling of the country of origin for fresh food is not enough, as consumers want more and trustworthy information. A growing number of consumers want to know from which farm a product comes and how this farm operates, eg, is it organic, how does it treats its animals, how it uses antibiotics, and how the quality of meat is tested along the supply chain. Fresh food products that have an fTRACE code can be traced back to the individual batch, ie, the amount of a product that was processed in the same place at the same time. Farmers, where the fresh food items come from, show their homes and farms with pictures and videos, so consumers can see the living conditions of animals on a farm or how fruit and vegetables are harvested, how animals are kept, and which lakes or seas fish are caught in. Many supply chain partners voluntarily provide more information in fTRACE than is legally required, eg, quality control, allergy and nutrition information, and recipe tips can also be shared.

To get an overview of what kind of consumer information is provided by fTRACE, one can download the smartphone app of fTRACE and scan the QR-code of Fig. 2.11. Alternatively, Fig. 2.12 shows a summary of the information supermarket customers got in a test workshop in 2013.

Figure 2.11 fTRACE label of meat bought on August 1, 2015 in Aldi's supermarket, Zetten (NL) with the fTRACE code VRG 017545779. Because it is minced beef, it is not originating from a single farm, but from a series of farms in the Netherlands. Scanning the QR-code with the fTRACE app[5] shows information similar to the one of Fig. 2.12, but in Dutch, as the meat item was retailed in the Netherlands.

[5]Download from https://www.ftrace.com/en/gb.

5.2 FIspace's MIP Trial EPCIS-Based Transparency System

The Future Internet Public Private Partnership (FI-PPP) aimed at advancing Europe's competitiveness in Future Internet (FI) technologies and wanted to support emerging FI-enhanced applications of public and social relevance. As a use case project in Phase 2 of the FI-PPP, FIspace was leveraging on outcomes of the Phase 1 use case projects *Finest* and *SmartAgriFood.*

The goal of FIspace[7] was to pioneer toward fundamental changes on how collaborative business networks will work in the future. FIspace has developed a multidomain Business Collaboration Space (short: FIspace) that employs FI technologies for enabling seamless collaboration in open, cross-organizational business networks.

The use case trial *Meat Information on Provenance*, or MIP trial, one of eight use case trial experimentation sites in Europe, aimed at ensuring that consumers, authorities/regulators, and meat supply chain partners have reliable information concerning the origin of a meat product, eg, birth calf, breeding, slaughtering, deboning, processing, packaging, etc. (Fig. 2.13).

The overall goal of the MIP trial was to realize an efficient, scalable, and decentralized data provision in the meat supply chain that enables tracking (where does a specific meat item come from) and tracing (where are all related meat items).

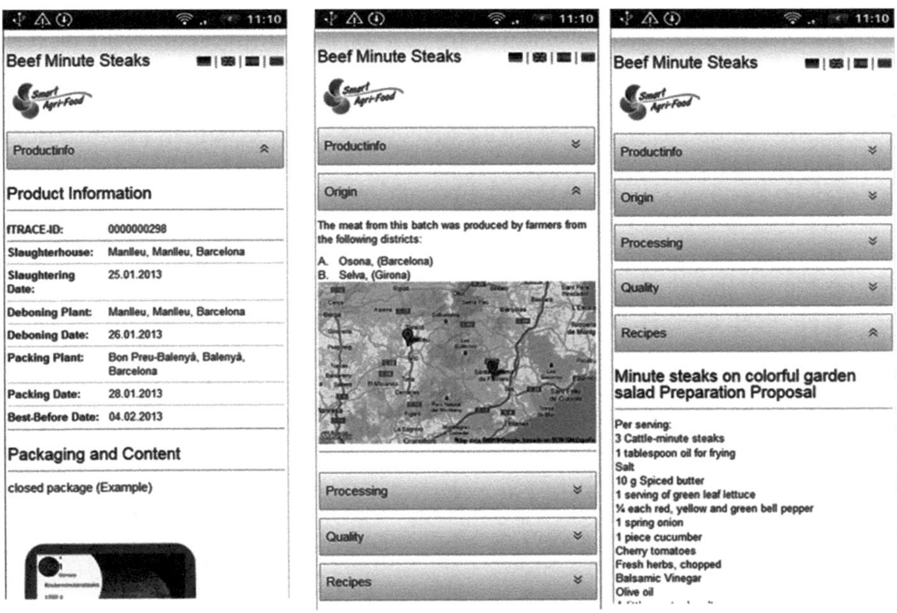

Figure 2.12 fTRACE mobile app (version of January 2013 tested in a Spanish supermarket BonPreu[6]).

[6]https://www.bonpreu.cat/.

[7]http://fispace.eu/.

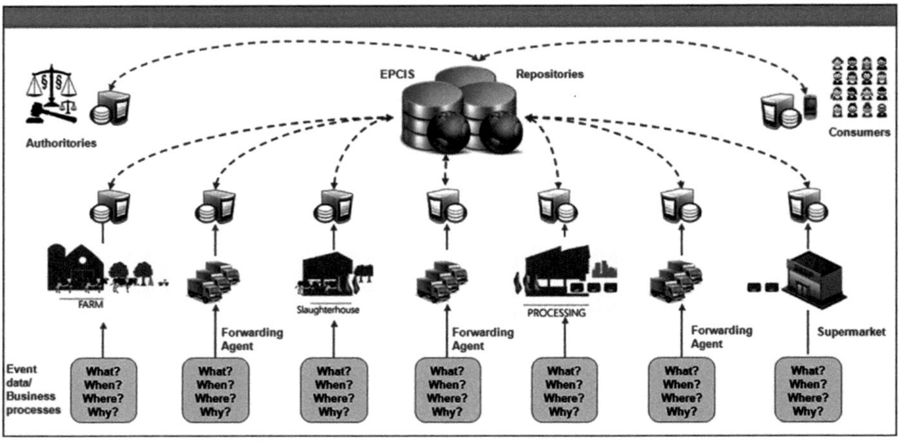

Figure 2.13 Overall view on FIspace's MIP trial transparency system for beef.

To achieve its goal the MIP trial has developed a system that enables meat supply chain stakeholders and authorities or regulators to track and trace and, at the same time, allows meat consumers to use their smartphone, other mobile device, or computer to get immediate and direct access to all information on the specific meat item they see at the butcher or in the supermarket. This meat transparency system follows meat through the supply chain and stores its history. Apps help supply chain partners to upload the data they want or have to share, and apps help them, authorities/regulators, and consumers to get the stored history in a format that fits the intended user.

Tracking and tracing is enabled by the MIP transparency system in two directions: from farm to fork and from fork back to the farm. This information will be stored in one or more EPCIS repositories to make it available for tracking and tracing by apps (see Section 4.2). These apps will inform consumers (what is the history of the meat item I am considering to buy?), meat supply chain partners (where are the meat items that went through my company?), and, in case of food alerts, authorities (where can we find all meat items that belong to some suspected lot?).

As the system is based on the EPCIS standard,[8] roll-out of the meat transparency system to other supply chains (other types of meat, other types of food, and other countries) is easy and straightforward. In FIspace a working prototype has been developed that serves three groups: (1) consumers, (2) meat supply chain partners, and (3) authorities and regulators (Fig. 2.13).

First, one or more EPCIS repositories have been implemented, which are based on Frequenz's Information Repository and Intelligence Server (IRIS),[9] one of only a few EPCIS implementations that can handle the new EPCIS standard 1.1 (released in May 2014). Next to this data infrastructure with functional interfaces for capturing and querying events, a series of apps has been developed based on the requirements by the MIP trial team. These will be discussed below.

[8]EPC Information Services, ie, Electronic Product Code Information Services, version 1.1, May 2014, http://www.gs1.org/sites/default/files/docs/epc/epcis_1_1-standard-20140520.pdf.
[9]http://frequentz.com/solutions/information-repository-intelligence-server/.

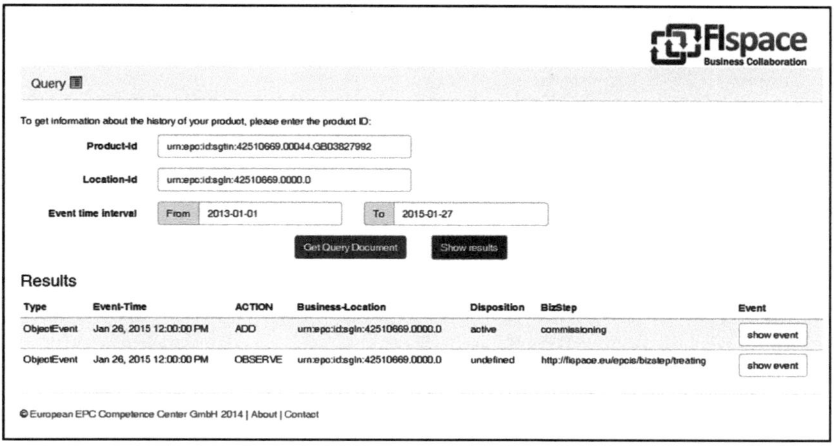

Figure 2.14 Results of a query to the EPCIS repository.

The apps can be used from two URLs: http://fispace.eu/apps.html#meat-information-provenance (FIspace platform) and http://31.204.118.29/(test site at EECC, username *meier* or *schulz*, password *pwd*). A video of the MIP apps can be found at http://www.eecc.info/files/mip/mip.html.

The *Farm Capture App* (not shown here) is a simple webpage-like app to enable farmers to copy animal passport data (and more) as EPCIS events and master data on other aspects to an EPCIS repository. Other supply chain partners should capture and upload their events from their ERP systems.

The *Query App* enables standard EPCIS query operations, ie, queries about raw, intermediate, or end-meat products for specific business locations, processes, time intervals, etc.

The screenshot in Fig. 2.14 shows the results of a query to a product by entering its ID. A more detailed overview of the Query App can be found in the Showcase of App 455.41.1 Meat Transparency System App Query EPCIS repositories.[10]

The *Discovery App* (not shown here) discovers which business party has information about a specific object, eg, a product identified via Global Trade Item Number (GTIN).

The *Aggregation App* collects and automatically aggregates traceability information about a given end or intermediate product and their presentation. This app is based on the Query App and the Discovery App.

The screenshot in Fig. 2.15 shows the results of a location search, aiming at finding all products that were in a specific location within a specific time frame. Users have to enter the GLN of a company, and the app will list all meat items and the products that were produced afterward from items that were at the specified location at the specified time. A more detailed overview of the Discovery App can be found in the Showcase of App 455.41.3 Aggregation App.[11]

[10]http://fispace.eu/query-app.html.

[11]http://fispace.eu/aggregation-app.html.

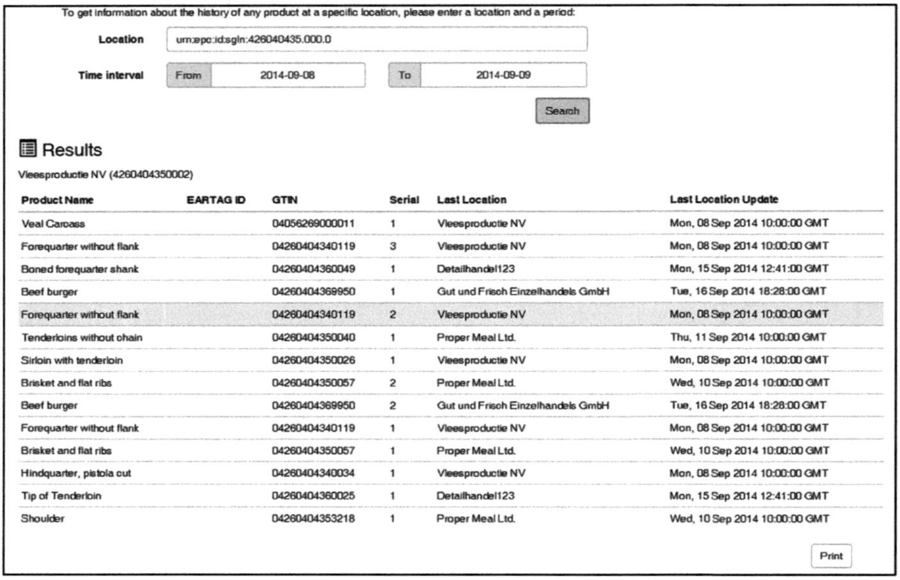

Product Name	EARTAG ID	GTIN	Serial	Last Location	Last Location Update
Veal Carcass		04056269000011	1	Vleesproductie NV	Mon, 08 Sep 2014 10:00:00 GMT
Forequarter without flank		04260404340119	3	Vleesproductie NV	Mon, 08 Sep 2014 10:00:00 GMT
Boned forequarter shank		04260404360049	1	Detailhandel123	Mon, 15 Sep 2014 12:41:00 GMT
Beef burger		04260404369950	1	Gut und Frisch Einzelhandels GmbH	Tue, 16 Sep 2014 18:28:00 GMT
Forequarter without flank		04260404340119	2	Vleesproductie NV	Mon, 08 Sep 2014 10:00:00 GMT
Tenderloins without chain		04260404350040	1	Proper Meal Ltd.	Thu, 11 Sep 2014 10:00:00 GMT
Sirloin with tenderloin		04260404350026	1	Vleesproductie NV	Mon, 08 Sep 2014 10:00:00 GMT
Brisket and flat ribs		04260404350057	2	Proper Meal Ltd.	Wed, 10 Sep 2014 10:00:00 GMT
Beef burger		04260404369950	2	Gut und Frisch Einzelhandels GmbH	Tue, 16 Sep 2014 18:28:00 GMT
Forequarter without flank		04260404340119	1	Vleesproductie NV	Mon, 08 Sep 2014 10:00:00 GMT
Brisket and flat ribs		04260404350057	1	Proper Meal Ltd.	Wed, 10 Sep 2014 10:00:00 GMT
Hindquarter, pistola cut		04260404340034	1	Vleesproductie NV	Mon, 08 Sep 2014 10:00:00 GMT
Tip of Tenderloin		04260404360025	1	Detailhandel123	Mon, 15 Sep 2014 12:41:00 GMT
Shoulder		04260404353218	1	Proper Meal Ltd.	Wed, 10 Sep 2014 10:00:00 GMT

Figure 2.15 Screenshot of the Aggregation App showing the results of a location search, aiming at finding all products that were in a specific location within a specific time frame.

The result list provides all the necessary information to take the appropriate measures. This includes the products' GTIN, the last location, and when the product arrived at the location.

A fifth app, the Tapia App,[12] is developed and codesigned in collaboration with FIspace's Tailored Information for Consumers (TIC) trial. This collaboration can build on previous experiences in FI-PPP Phase 1 project SmartAgriFood.

FIspace's MIP trial meat transparency system is able to run the following scenarios:

1. Farmer: *In case of problems with a processor: where are my products?*
   This use case needs first the Farm Capture App for entering data (eg, birth of a calf) and subsequently the Aggregation App to show the data. This Aggregation App uses the Query App to find the data and, in case the data are in several EPCIS repositories, the Discovery App is needed too. A farmer has to enter his GLN[13] only.
2. Slaughterhouse: *Which processors handled my products?*
   If the slaughterhouse wants to find a suspect meat processor that handles the carcasses of his slaughterhouse, the Query App will be used to determine it. This can be done by entering the GLN of the suspected processor.
3. Meat Processor: *Do all my beef burgers come from a certain farmer?*
   This use case uses the information from the Aggregation App, that in turn uses the Query App and, if needed, the Discovery App. The meat processor enters the SGTIN of his products.
4. Distributor: *When was the animal slaughtered and how old was the animal?*

[12]http://fispace.eu/consumer-app.html.

[13]Global Location Number (http://www.gs1.org/gln).

This use case also uses the Aggregation App and indirectly the Query App and, in case of more than one EPCIS repository, the Discovery App. This can be done by entering LGTIN of a product.

5. Authorities: *Where are all products now to which processor X had custody?*
   This use case also uses the Aggregation App and indirectly the Query App and, in case of more than one EPCIS repository, the Discovery App. Enter GLN and time frame to find all products of processor X.
6. Consumer: *What is the history of my piece of meat?*
   The Tapia App (TIC) for consumers is used to provide the history of the meat. Consumers scan the QR-code on the meat. The TIC will then invoke the Aggregation App to show the history of the piece of meat, from farm to shop.
7. Discovery App use case: *Where are all relevant event data located? Which business party has information about a specific object?*
   This use case uses only the Discovery App. Enter the (S)GTINs[14] to find the location of the object.

Although FIspace's MIP trial transparency system is a prototype, it has showed potential so far and will likely be implemented and tested in a real supply chain in Germany. More details on the implementation of the prototype can be found in Kassahun et al. (2015).

6. Discussion and Conclusion

Traceability is a vital issue for food supply chains in particular to guarantee food quality and provenance. In this chapter, we have proposed a method to define and analyze traceability systems in food supply chains. The main value of the method is that it helps to systematically map food supply chains and design the information architecture of food traceability systems. As such, it helps to manage the complexity of coordinated traceability systems in food supply chains, and consequently, it enhances the implementation of coordinated supply systems for traceability. Traceability is currently still often achieved by conventional systems, focusing on a single company or a specific part of the supply chain and using too basic technologies, eg, product labeling, EDI, email, and paper trails. As a consequence, most existing traceability systems follow a one-step-forward and one-step-back principle. This chapter also introduces a traceability system, based on EPCIS, for end-to-end visibility and real-time tracking and tracing, but such systems are rather immature at present, and many challenges have to be faced, including the inclusion of other data beyond whereabouts, ie, information about allergens, safety, nutritional value, place of origin or provenance, animal welfare, use of antibiotics, sustainability, and taste.

One of the major challenges in (fresh) food supply chains are the extremely high number of items that pass through them. Therefore transparency systems should be scalable, ie, extendable in numbers (eg, food items, shops, supply chain partners), types of food products (eg, meat, fruits and vegetables, dairy products), and information provided (eg, type of farming, classic, organic, animal friendly, sustainable,

[14]Global Trade Item Number (http://www.gs1.org/gtin).

etc.), type of slaughtering (classic, halal, kosher, etc.), and other information (quality indicators, nutritional content, allergenic substances, monitored temperature, veterinary and bacteriological tests). The wide diversity of information should be structured for its users and include (1) what is required by laws and regulations and (2) what is within the interest of consumers, supply chain partners, and authorities/regulators.

During the development of the EPCIS-based transparency system, it appeared that the previous version of EPCIS (version 1.01) did not allow irreversible processes, such as slaughtering, and did not support single specific product identification with SGTIN nor products in batches or lots identification with LGTIN. It took GS1 EPCglobal, representing a large part of the worldwide business community, only 18 months to extend the standard to EPCIS 1.1, which includes transformation processes and can handle LGTIN for lots or batches of products and SGTIN, which combines a GTIN product identifier with a unique serial number (GS1 EPCglobal, 2014). At present, tracking and tracing occurs with a granularity that is determined by the size of a lot/batch. For meat, that is typically what is slaughtered in half a day. If the batch sizes become smaller in the future, tracking and tracing becomes more accurate, finally tracking and tracing to individual animals and other food products.

One of the major obstacles experienced both in the development of fTRACE and in the EPCIS-based transparency system for meat was that supply chain partners were reluctant to share data with other supply chain partners and with authorities/regulators. This lack of trust can be mitigated along two paths: a technical path, ie, designing the system in such way that each partner can only get access to data of products he is involved in, and an organizational path, ie, persuading hesitant supply chain partners by showing the benefits of such a system, eg, optimizing his business processes. Authorities/regulators should have access to all data in case of food alerts that urge them to find all food items somehow connected to suspect ones. The lack of trust of supply chain partners to share data is strongly related to a centralized versus noncentralized architecture of the traceability/transparency system. In EPCIS-based systems, there is always a central repository, which does not contain EPCIS event data itself, but only its location. In this way, each supply chain partner can control sharing of its own data without the risk sharing data that may lead to more competition.

In conclusion, we have to emphasize that, at present, end-to-end traceability/transparency systems for fresh food hardly exist or are based on too basic technologies, but in this chapter, we discussed some interesting developments beyond. First, we see that traceability/transparency systems move from supporting a part of the supply chain to end-to-end visibility. Second, we see the adoption of existing standards such as EPCIS for capturing and analyzing business processes, but also to identify products (GTIN, LGTIN, SGTIN), companies (GLN), or to specify business processes (CBV). We expect that these initiatives will be further developed, implemented, and merged to a single approach for transparency infrastructures for fresh food and beyond.

Acknowledgments

The research leading to this paper received funding from the European Community's Seventh Framework Program (FP7/2007–2013) within the project FIspace under grant agreement No. 604

123. The authors acknowledge the individuals and companies involved for their support and the European Community for their funding and especially Tim Bartram, Sabine Kläser, Ralf Tröger, and Angela Schillings-Schmitz (GSI-Germany) for sharing their knowledge and experience, both on fTRACE and FIspace's MIP trial prototype, and Falk Nieder, Nicolas Becker, and Fabian Schenk (European EPC Competence Center) for developing FIspace's MIP trial prototype.

References

Albersmeier, F., Schulze, H., Jahn, G., Spiller, A., 2009. The reliability of third-party certification in the food chain: from checklists to risk-oriented auditing. Food Control 20 (10), 927–935. http://dx.doi.org/10.1016/j.foodcont.2009.01.010.

Beulens, A.J.M., Broens, D.F., Folstar, P., Hofstede, G.J., 2005. Food safety and transparency in food chains and networks – relationships and challenges. Food Control 16 (6), 481–486.

Christopher, M.G., 1998. Logistics and Supply Chain Management: Creating Value-adding Networks. Pitman Publishing, London.

Cooper, M.C., Ellram, L.M., 1993. Characteristics of Supply Chain Management and the implications for purchasing and logistics strategy. International Journal of Logistics Management 4 (2), 13–24.

EC, 2002. Regulation (EC) No 178/2002 of the European Parliament and of the Council of 28 January 2002 laying down the general principles and requirements of food law, establishing the European Food Safety Authority and laying down procedures in matters of food safety. Official Journal of the European Communities L 31, 1–24.

Folinas, D., Manikas, I., Manos, B., 2006. Traceability data management for food chains. British Food Journal 108 (8), 622–633.

Fritz, M., Schiefer, G., 2009. Tracking, tracing, and business process interests in food commodities: a multi-level decision complexity. International Journal of Production Economics 117 (2), 317–329. http://dx.doi.org/10.1016/j.ijpe.2008.10.015.

GS1, 2014. Core Business Vocabulary (CBV). Standard Version 1.1, May 2014. GS1 AISBL, Brussels, Belgium.

GS1 EPCglobal, 2014. EPC Information Services (EPCIS) Version 1.1 Specification. GS1 Standard Version 1.1. GS1 AISBL, Brussels, p. 169.

Kassahun, A., Becker, N., Scholten, H., Kläser, S., Bartram, T., Nieder, F., 2015. In: Realizing Chain-wide Transparency in Meat Supply Chains Using Web Applications Based on GS1 Standards Paper Presented at the EFITA/WCCA/CIGR 2015 Conference, June 29–July 2, 2015, Poznan, Poland.

Kassahun, A., Hartog, R.J.M., Sadowski, T., Scholten, H., Bartram, T., Wolfert, J., Beulens, A.J.M., 2014. Enabling chain-wide transparency in meat supply chains based on the EPCIS global standard and cloud-based services. Computers and Electronics in Agriculture 109, 179–190. http://dx.doi.org/10.1016/j.compag.2014.10.002.

Lambert, D.M., Cooper, M.C., 2000. Issues in supply chain management. Industrial Marketing Management 29 (1), 65–83.

Lambert, D.M., Garcia-Dastugue, S.J., Croxton, K.L., 2005. An evaluation of process-oriented Supply Chain Management frameworks. Journal of Business Logistics 26 (1), 25–51.

Lazzarini, S.G., Chaddad, F.R., Cook, M.L., 2001. Integrating supply chain and network analyses: the study of netchains. Chain and Network Science 1 (1), 7–22.

OMG, 2010. Business Process Modeling Notation (BPMN). Retrieved May 17, 2010, from http://www.bpmn.org/.

Porter, M.E., Millar, V.E., 1985. How Information Gives You Competitive Advantage. Harvard Business Review. 63(July–Aug).

Powell, W.W., 1990. Neither market nor hierarchy – network forms of organization. Research in Organizational Behavior 12, 295–336.

Ryu, J., Taillard, D., 2007. GS1 Global Traceability Standard: Business Process and System Requirements for Full Chain Traceability GS1 Standards Document. GS1.

SCC, 2008. SCOR Supply Chain Operations Reference Model: Version 9.0. Supply-Chain Council.

Schuster, E.W., Allen, S.J., Brock, D.L., 2007. Global RFID: The Value of the EPCglobal Network for Supply Chain Management. Springer, New York and Berlin.

Trienekens, J.H., Wognum, P.M., Beulens, A.J.M., van der Vorst, J.G.A.J., 2012a. Transparency in complex dynamic food supply chains. Advanced Engineering Informatics 26 (1), 55–65. http://dx.doi.org/10.1016/j.aei.2011.07.007.

Trienekens, J.H., Wognum, P.M., Beulens, A.J.M., van der Vorst, J.G.A.J., 2012b. Transparency in complex dynamic food supply chains. [Network and supply chain system integration for mass customization and sustainable behavior]. Advanced Engineering Informatics 26 (1), 55–65.

Trienekens, J., Vorst, J.V.D., Verdouw, C., 2014. Global food Supply chains. In: van Alfen, N.K. (Ed.), Encyclopedia of Agriculture and Food Systems, second ed. Academic Press, pp. 499–517.

van der Meulen, B., 2013. The structure of European food law. Laws 2 (2), 69.

Vorst van der, J.G.A.J., 2000. Effective Food Supply Chains; Generating, Modelling and Evaluating Supply Chain Scenarios, Doctoral Thesis. Wageningen University. 2000.

Vorst van der, J.G.A.J., April 2006. Product traceability in food-supply chains, accreditation and quality assurance. Journal for Quality, Comparability and Reliability in Chemical Measurement 11 (1–2), 33–37.

van der Vorst, J., Beulens, A., Beek, P.V., 2005. Innovations in logistics and ICT in food supply chain networks. In: Jongen, W.M.F., Meulenberg, M.T.G. (Eds.), Innovations in Agri-Food Systems: Product Quality and Consumer Acceptance 2005. Wageningen Academic Publishers.

Verdouw, C.N., 2010. Business Process Modelling in Demand-driven Agri-Food Supply Chains: A Reference Framework. (Ph.D. Doctoral Thesis). Wageningen University, Wageningen.

Verdouw, C.N., Beulens, A.J.M., Trienekens, J.H., van der Vorst, J.G.A.J., 2011. A framework for modelling business processes in demand-driven supply chains. [Production planning & control]. Production Planning & Control 22 (4), 365–388. http://dx.doi.org/10.1080/09537287.2010.486384.

Verdouw, C.N., Beulens, A.J.M., Trienekens, J.H., Wolfert, J., 2010a. Process modelling in demand-driven supply chains: a reference model for the fruit industry. Computers and Electronics in Agriculture 73 (2), 174–187.

Verdouw, C.N., Beulens, A.J.M., Trienekens, J.H., Wolfert, J., 2010b. Business process modelling in demand-driven agri-food supply chains. Paper Presented at the Food System Dynamics, Innsbruck-igls.

Verdouw, C.N., Bondt, N., Schmeitz, H., Zwinkels, H., 2014a. Towards a smarter greenport: public-private partnership to boost digital standardisation and innovation in the Dutch horticulture. International Journal on Food System Dynamics 5 (1), 44–52.

Verdouw, C.N., Vucic, N., Sundmaeker, H., Beulens, A., 2014b. Future internet as a driver for virtualization, connectivity and intelligence of agri-food Supply chain networks. International Journal on Food System Dynamics 4 (4), 261–272.

Verdouw, C.N., Robbemond, R.M., Verwaart, T., Wolfert, J., Beulens, A.J.M., 2015. A reference architecture for IoT-based logistic information systems in agri-food supply chains. Enterprise Information Systems (Available On-line).

Webster, F.E., 1992. The changing-role of marketing in the corporation. Journal of Marketing 56 (4), 1–17.

Wiersinga, R.C., Hiller, S.R.C.H., Zimmermann, K.L., Verdouw, C.N., Lans, I.A.V.D., 2010. Consumer driven and innovative fruit supply chains. In: Trienekens, J., Top, J., Vorst, J.V.D., Beulens, A. (Eds.), Towards Effective Food Chains – Models and Applications. Wageningen Academic Publishers, Wageningen, pp. 113–133.

Problems and Implementation Hurdles in Food Traceability

3

K.M. Karlsen, P. Olsen
Norwegian Institute of Food, Fisheries and Aquaculture Research (Nofima), Tromsø, Norway

1. Introduction

In the literature, traceability was first mentioned within the automobile industry in the 1970s. Traceability was used for product liability reasons. Then traceability was used to document software projects and enterprise modeling. The food scandals of the 1990s put the traceability of food on the agenda. The outcome of these scandals was that traceability was included in the regulations for food. Lately, other applications of traceability within the food industry have been identified, like better industrial statistics and the potential for storytelling. Today traceability is used in many different industries, especially production industries.

Several definitions of traceability exist. A practical and often used definition of traceability is found in the International Standardization Organization (ISO) 8402 (1994) where traceability is defined as: "The ability to trace the history, application or location of an entity by means of recorded identifications." The newer ISO 9000 definition (ISO, 2000) has a slightly less specific definition of traceability: "The ability to trace the history, application or location of that which is under consideration." For both these definitions, there is an additional clause which states that when relating to products, traceability specifically entails "the origin of materials and parts, the processing history, and the distribution and location of the product after delivery." The EU Common Food Law defines traceability as (EC-178/02, 2002) "The ability to trace and follow a food, feed, food-producing animal or substance intended to be, or expected to be incorporated into a food or feed, through all stages of production, processing and distribution." An attempt at unifying many of these definitions while removing the ambiguities and the circular specifications (using the word "trace" to define "traceability," but not defining what "trace" means) was made in Olsen and Borit (2013), which suggested "The ability to access any or all information relating to that which is under consideration, throughout its entire life cycle, by means of recorded identifications." Although the ambiguity and lack of definition relating to the word "trace" is a significant weakness in most of these definitions, the concept is clear: Each link records what it is doing, relates it to "that which is under consideration," and provides a mechanism for getting access to these recordings later on.

Advances in Food Traceability Techniques and Technologies. http://dx.doi.org/10.1016/B978-0-08-100310-7.00003-X

Table 3.1 Explanation of the Terms Batch, Trade Unit, and Logistic Unit

Batch	Trade Unit	Logistic Unit
"A batch unites products/items that have undergone the same transformation processes." (GS1, 2007)	"…any item (product or service) upon which there is a need to retrieve pre-defined information and that may be priced, or ordered, or invoiced at any point in any supply chain." (GS1, 2007)	"An item of any composition established for transport and/or storage that needs to be managed through the supply chain." (GS1, 2007)
"…a quantity that has gone through the same process at a specific place and time period before moving to another place." "A production batch is the traceable unit that raw materials and ingredients go into before transformed into products placed in new Trade Units and Logistic Units." (TraceFood, 2015)	"…the smallest traceable unit that is exchanged between two parties in the supply chain." (TraceFood, 2015)	"In practice it is made up by one or more separate TU's. In some cases, the trade unit and the logistic unit are the same." (TraceFood, 2015)

2. Principles of Traceability

2.1 Which Units to Trace?

A key concept in many articles, recommendations, and standards relating to product traceability is that of "traceable unit." GS1[1] (2007) divides the traceable units into (1) batch, (2) trade unit (TU), and (3) logistic unit (LU). The definitions of the terms batch, trade unit, and logistic unit are described in Table 3.1. Olsen and Aschan (2010) described the relationship between batches, TUs, and LUs in one link in a supply chain. TUs and LUs are external traceable units and batch is an internal traceable unit.

2.2 Which Information Is Relevant to Trace?

According to Storøy et al. (2008), information is divided into (1) "transformation information," and (2) "product information." "Transformation information" covers the identification of traceable units and of the transformation relationships that took place, and "product information" covers properties relating to origin, processing history, and location.

Transformations are points where the resources are mixed, transferred, added, and/or split up (Derrick and Dillon, 2004). The relationship of the traceable units can be one-to-one, many-to-one, one-to-many, or many-to-many.

[1] An international not-for-profit association designing and implementing global standards and solutions to improve supply chain management.

The identification of the traceable unit should be unique, at least in a given context, and the identification code can be used to access all the recorded product information. Recording transformation information can be seen as costly in a traceability system, as strictly speaking, the identification code in itself does not tell you anything you want to know about the product, and the same goes for the recording of the transformation relationships (exactly which raw material/ingredient (input) units were used to produce which product (output) units). Access to product information is the benefit of a traceability; it is the product properties you are interested in, but for those to be made available, you must either provide the identification code of the product in question, or (if the question relates to origin or a previous process) you must use the transformation recordings to find your way back to the recordings in the previous process(es) and access the product information recorded there.

2.3 How to Prevent Information Loss?

Several studies have highlighted the lack of unique identification of traceable units and transformation recordings (Frosch et al., 2008; Donnelly et al., 2009a; Karlsen et al., 2010). Transformation documentation is necessary in order to trace products (Donnelly et al., 2009a). The batch size of a product must be defined before any information can be linked to the product and thus be traceable (Bertolini et al., 2006).

A place where information loss occurs is called a critical traceability point (CTP) (Karlsen et al., 2010). Such points occur when information about a product or process is not linked to a traceable unit and recorded systematically. In other words, this information is not traceable and it is not possible to retrieve information again at a later point. CTP identification is necessary for traceability implementation, because certain recordings are necessary to prevent information loss. This includes recording the relationship between traceable units and the unique identification of the traceable units.

The following principles of traceability are applied: information is traceable by being linked to a unique identification of the traceable units. In addition, the relationship between these units must be recorded (the so-called transformation). These principles are illustrated with a simplified example in Fig. 3.1. A catch of wild-caught fish is identified as A. The catch information is linked to this identification. The landed fish is sorted into different containers on the basis of species, fish size, and quality. The fish in container A1 is used in the production of batch A1-1. In Fig. 3.1, this is illustrated with a box of fish. The identifications of the traceable units are linked together, and the measure of the yield of the fish from one landing can thus be linked to catch information, which in turn can be used to plan production and coordinate the activities of the fishing fleet to maximize profit for both fishing vessels and the production plant. In the real world, this is more complicated.

The overall recommendation to prevent information loss:

1. Start using unique and standard numbers on traceable units delivered.
2. Record the relationship between production batch and delivered trade unit explicitly: which production batch generated which tradable units?
3. Record all production information keyed to production batch number.

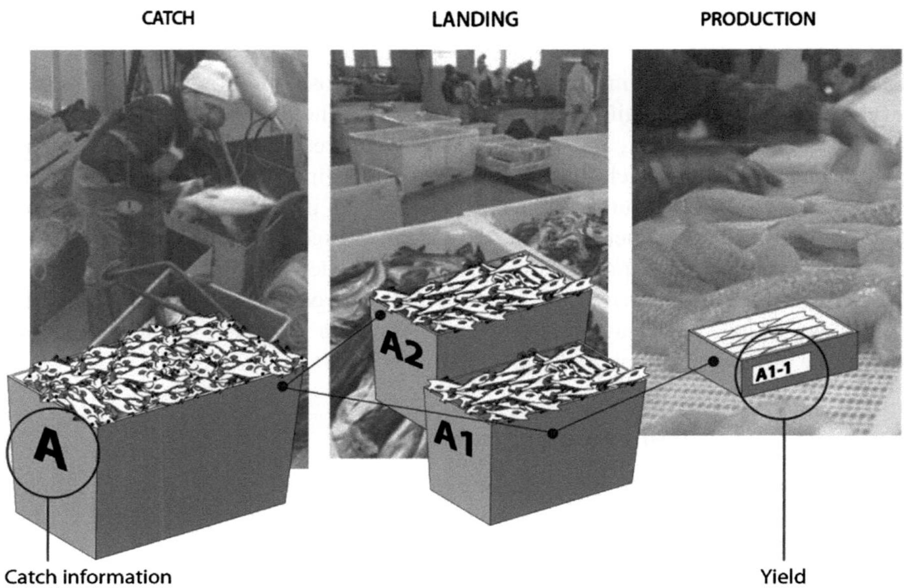

Figure 3.1 Simplified example of the principles of traceability.
Illustrator: Oddvar Dahl, photo: Frank Gregersen, Nofima.

4. Start assigning locally unique numbers to their own ingredient batches.
5. Record the relationship between ingredient batch and received trade unit explicitly: which ingredient batch incorporated which trade units? If there already are unique identifiers on the received trade units, they should be recorded and used; if not, a unique identifier should be assigned upon reception. Suppliers should be encouraged to use unique and standard numbers (GS1) to simplify this process.
6. Record the relationship between input batch and production batch explicitly whenever possible.
7. Disseminate information keyed to identity of delivered traceable unit (or production batch number) widely. Send information to customer or consumer upon request, or put it in searchable form on the company website so that anyone can trace the attributes, processes, constituents, and respective origin of the product and all its parts.

2.4 Internal and Chain Traceability

Previous studies have shown that information about food products and production processes can be lost internally within companies, as well as between companies in supply chains (Bertolini et al., 2006; Karlsen and Senneset, 2006; Randrup et al., 2008; Donnelly et al., 2012).

Internal traceability is the ability to trace the product information internally in a company and has the following characteristics:

- It is within one company.
- It is in one geographical location.

- It has interface to production management systems.
- There are few privacy issues.
- It is simpler to do.

Many companies have good routines and software systems for keeping track of internal traceability. This kind of software is often linked with dedicated production management software and general Enterprise Resource Planning (ERP) systems.

3. Chain Traceability

Chain traceability has the ability to trace the product information through links in a supply chain and has the following characteristics:

- It occurs between companies and between countries.
- It depends on internal traceability being present.
- There are major privacy issues.
- Standards are needed.

The enforcement of chain traceability implies the development of systems giving information on the entire life cycle of food products, "from the farm (or the sea) to the fork." As a company located in one specific link of a supply chain, you may very well have completely manual systems for the record keeping of traceability information. However, electronic solutions facilitate business partners to reconstruct the complete process history of any food efficiently and quickly. Software houses now offer sophisticated traceability systems that give you access to all aspects of a food's provenance at any time and from any level of the food chain.

3.1 Granularity Levels of Traceability

A traceability system can be simple (one-up-one-down traceability); costs would be low, and implementation would be easy. Traceability can also be complex. The chosen level will determine the complexity of the traceability system and can affect the practical solutions and specification of the IT systems in the implementation of traceability. The application of information for quality and process optimization purposes may demand finer traceability levels. Coarser traceability levels can be used when, eg, the risk of contamination is low, or when the requirements for controlling production processes are less stringent.

The chosen traceability level is dependent on a company's internal and external need for traceable information. A fine level will increase the complexity of the traceability system and will entail higher costs, because there is more information to record, a higher number of transactions, and new systems and procedures would possibly have to be introduced (Golan et al., 2004). It is clear that the implementation of food traceability is affected by the granularity level of the traceability system.

4. Standardization of Information and Information Exchange in Various Links in the Chain

4.1 Why Is Standardization Important?

Traceability is a concept in the realm of information logistics, which in turn deals with the flow of information within or between any number of organizations. The flow of information between organizations is invariably at least partly computerized: the traceable unit identification, the transformation recordings, and all related product properties are normally stored in a database on a network or a computer. Getting access to and exchanging product-related information thus in practice means generating, sending, receiving, and interpreting electronic messages.

In the old days, when product information was mainly on labels or in accompanying paper documentation, the format was less important. Humans were the senders and receivers of these messages, and humans could be trusted to decipher the label or document received and extract the information needed. Computers are less flexible than humans; when computers are the senders and receivers of messages, the format of these messages needs to be defined quite unambiguously.

In theory, each sender–receiver pair could agree on a message format and exchange messages using it. In practice, for a given sender, there are numerous, often hundreds, of potential recipients, including all the buyers of this product type, people further down the chain, and consumers, as well as authorities and other organizations. The same many-to-one relationship applies to recipients; there are numerous, often hundreds, of organizations that may send product-related messages to you, in particular about raw materials or ingredients you bought from them.

All this leads to a very strong relationship between the development and use of standards and efficient implementation of product traceability. For the electronic interchange of product-related messages to be efficient, standards are needed for the traceable unit ID (what number series and format to use), for representation of the unit ID (plain number/bar code/QR code/electronic tag/etc.), for naming of the product data elements (Donnelly et al., 2008, 2009b), and for the content/value of the product data itself.

4.2 Standards for Food Product Traceability

Going into a detailed discussion of all the existing standards is beyond the scope of this chapter, but the overview below illustrates that different standards relevant for food traceability exists:

- standards for number series, content of bar codes, and electronic tags, eg, GS1 (GS1, 2007)
- standards for exchanging electronic messages relating to (food) product properties, eg, Extensible Markup Language (XML), Universal Business Language (UBL)
- standard describing the principles and concepts of food traceability, eg, ISO 22005 "Traceability in the feed and food chain — General principles and basic requirements for system design and implementation" (ISO-22005, 2007)
- standards defining data element names and content in specific sectors; some of the most advanced examples of this type are probably ISO 12875 "Traceability of finfish

products — Specification on the information to be recorded in captured finfish distribution chains" and ISO 12877 "Traceability of finfish products — Specification on the information to be recorded in farmed finfish distribution chains" (ISO 12875, 2011; ISO 12877, 2011).

5. Different Drivers of Traceability

In the literature, 10 drivers for traceability have been identified (Karlsen et al., 2013): fulfillment of legislation, documentation of food safety, quality, welfare, sustainability, certification, competitive advantages, chain communication, bioterrorist threats, and optimize production.

According to Mai et al. (2010), the potential benefits to using traceability can be different in different links in a supply chain. Traceability drivers were studied in three different food supply chains. A summary of the identified drivers of traceability in the studied food supply chains is provided in Table 3.2. The identified drivers of traceability for the studied companies were legislation, food safety, quality, competitive advantages, chain communication, and production optimization (6 of 10). The most important driver of traceability was production optimization (five companies), followed by competitive advantages and quality (four companies), legislation (three companies), and food safety and chain communication (two companies).

Company six and seven had the highest numbers of drivers for traceability: legislation, food safety, competitive advantages, chain communication, and production optimization (5 of 10 drivers of traceability). Company one had no drivers of traceability.

A study carried out by Sparling et al. (2006) documented a difference between factors that motivated companies to use traceability before implementation and actually achieved benefits from using traceability after implementation. Company two had implemented traceability, while the other studied companies had not or had only partly implemented traceability. The identified drivers for the studied companies would probably have changed after an implementation of traceability.

6. Problems and Implementation Hurdles in Food Traceability

Implementation of food traceability can be complex and involves many different aspects that affect each other. A number of problems and implementation hurdles can occur before, during, and after implementation of food traceability. Several authors have identified a number of critical factors for the implementation of traceability.

Sohal (1997) identified six critical factors for developing and implementing traceability in an automobile manufacturer. Manufacturers must:

1. have top management understanding of **Computer-based/Computer-Integrated Manufacturing** (CIM),
2. communicate **benefits**, understand the benefits and overcome the **fear** of a new system,
3. **train** its employees, and
4. have a long-term **plan** for CIM.

Table 3.2 **Identified Traceability Drivers of Traceability in Three Food Supply Chains**

	Supply Chain 1		Supply Chain 2			Supply Chain 3	
Drivers	**Company 1**	**Company 2**	**Company 3**	**Company 4**	**Company 5**	**Company 5**	**Company 6**
1						*X*	*X*
2						*X*	*X*
3		*X*	*X*	*X*	*X*		
4							
5							
6							
7				*X*	*X*	*X*	*X*
8						*X*	*X*
9							
10		*X*	*X*	*X*		*X*	*X*

(1) Legislation, (2) food safety, (3) quality, (4) sustainability, (5) welfare, (6) certification, (7) competitive advantages, (8) chain communication, (9) terrorist threats, and (10) production optimization.

Senneset et al. (2007) pointed out eight essential criteria for the implementation of electronic chain traceability in a supply chain for seafood. The companies must:

1. have chain traceability **procedures** at point of reception,
2. have chain traceability **procedures** at point of **delivery**,
3. have internal traceability **procedures**,
4. have an internal traceability **software system**,
5. use electronic/automatic recording **technology** at reception,
6. use electronic/automatic recording **technology** at delivery,
7. use **standardized identification** for traceable units, and
8. have a chain traceability **software system**.

In addition, Karlsen et al. (2011) documented the following criteria for implementation of electronic chain traceability:

1. **motivation**,
2. identification of **benefits**,
3. identification of **investments**, and
4. development of **optimal solutions**.

Motivation has a central role when discussing the factors affecting the implementation of traceability (Karlsen et al., 2011), and an important question is how to motivate employees to do necessary actions to implement traceability?

Communicating and understanding the benefits of a traceability system is important for the successful implementation of traceability (Sohal, 1997). The willingness of the employees at all levels within a company to cooperate and their motivation to implement traceability has enormous impact on the implementation process.

The motivation can be different between the employees (top manager, middle managers, and shop-floor workers) in a company. For example, the top manager will implement an internal traceability system to be able to use the traceable information in decision support processes, while the shop-floor workers are not motivated to use the new system and change the practical routines in the production. Another example: the quality management in a company wants to implement a traceability system to increase documentation of the internal production processes, but the top manager does not see the benefits of having a traceability system and does not want to invest in necessary technology.

The motivation can also vary significantly between the different links of a supply chain. For example, a production plant is unsure of the benefits they can derive from a traceability solution, and the distributing company in the same supply chain does not put pressure on the production plant for improved documentation, so motivation for the production plant to implement traceability can be rather low.

Motivation for implementing traceability is strongly linked to the costs and benefits of using traceability. There are different costs and benefits to using traceability, and employees within a company and companies in supply chains apply traceable information differently. Any implementation of traceability in food supply chains should thus include an open discussion of the distribution of costs and benefits between companies in the supply chain (Mai et al., 2010) but also between the different levels within a company.

An evaluation of costs and benefits using traceability will determine the complexity of the traceability system and can affect practical solutions and IT system specifications in the implementation process. The development of optimal practical solutions to prevent information loss can be a challenge (Karlsen et al., 2011); thus involvement of the shop-floor workers in this process is essential. Another challenge in the implementation process is the development of user-friendly interfaces, where building relationships between the users and the Information and Communications Technology supplier is a key factor.

Implementing an efficient traceability solution may require big investments (Sohal, 1997). There are different types of costs associated with traceability implementation (eg, administrative, material, operational, equipment/technology, initial, and ongoing costs) and these investments are highly variable (Can-Trace, 2007; Mai et al., 2010). The investments necessary for successful traceability are dependent on several factors. These investments can be affected by which software solutions and electronic recording equipment are available in the company. Other factors affecting investments can be the degree of integration required in the software systems for successful internal traceability (simple or full integration), investments in new IT solutions, and necessary reengineering of current IT systems. Top managers will not make the investments necessary for better product documentation if they cannot identify the benefits they stand to gain by making these investments.

7. Future Trends

Identifying applications for traceability and benefits of traceable information in food supply chains and how this can motivate the employees within a company to implement traceability is a clear area for further studies. There is also a need to increase knowledge of the motivation for implementation of traceability between the different links of a supply chain. A central issue is how motivate the companies to implement traceability. Other interesting questions:

- Are the benefits and motivation for traceability implementation different depending on the companies' positions in a supply chain?
- Are there different benefits and motivation for traceability implementation of different foodstuffs ("high"-value products vs. "low"-value products)?
- How will the production concept and use of technology affect motivation?
- How can food traceability requirements be formulated and implemented in such a way as to be nondiscriminatory, facilitate trade, and provide a level playing field for international food business operators?
- How can we make diverse traceability systems interoperable, and how can we make sure that data recorded in one place will follow the product through the supply chain?
- What food traceability standards are needed on the different levels, and how do we ensure the uptake and harmonization of practices?
- How will changing consumer preferences and increased consumer demands for food information influence the traceability systems?
- How can traceability aid in the documentation of sustainability, ethical production, and animal welfare?

- How can traceability aid in combating product misdescription and food fraud?
- How can we trust the data recorded in a traceability system?

There are many ongoing and upcoming international and national research programs and projects addressing these issues, and food traceability is still an emerging field for R&D.

References

Bertolini, M., Bevilacqua, M., Massini, R., 2006. FMECA approach to product traceability in the food industry. Food Control 17 (2), 137–145.

Can-Trace, 2007. Cost of Traceability in Canada: Developing a Measurement Model. Report March 2007. Agriculture and Agri-Food Canada, Ottawa, Canada. ISBN: 978-0-662-46769-4.

Derrick, S., Dillon, M., 2004. A Guide to Traceability Within the Fish Industry. SIPPO, Eurofish and Humber Institute of Food & Fisheries, Copenhagen, Denmark. ISBN: 1-900134-18-7.

Donnelly, K.A.-M., Karlsen, K.M., Olsen, P., 2008. Creating standardised data lists for traceability: a study of honey processing. International Journal of Metadata, Semantics and Ontologies 3 (4), 283–291.

Donnelly, K.A.-M., Karlsen, K.M., Olsen, P., 2009a. The importance of transformations in traceability – a case study of lamb and lamb products. Meat Science 83 (1), 69–73.

Donnelly, K.A.-M., van der Roest, J., Höskuldsson, S.T., Olsen, P., Karlsen, K.M., 2009b. Improving information exchange in the chicken processing sector using standardised data lists. Special Issue of the Journal on Data Semantics of Springer 312–320.

Donnelly, K.A.-M., Karlsen, K.M., Dreyer, B., 2012. A simulated recall study in five major food sectors. British Food Journal 114 (7), 1016–1031.

EC-178/02, 2002. Regulation (EC) No 178/2002 of the European Parliament and of the Council of 28 January 2002 Laying Down the General Principles and Requirements of Food Law, Establish the European Food Safety Authority and Laying Down Procedures in Matters of Food Safety. The European Parliament and the Council of the European Union.

Frosch, S., Randrup, M., Frederiksen, M., 2008. Opportunities for the herring industry to optimize operations through information recording, effective traceability systems, and use of advanced data analysis. Journal of Aquatic Food Product Technology 17 (4), 387–403.

Golan, E., Krissoff, B., Kuchler, F., Calvin, L., Nelson, K., Price, G., 2004. Traceability in the US Food Supply: Economic Theory and Industries Studies Agricultural Economic Report Number 830. United States Department of Agriculture, Washington, DC.

GS1, 2007. The GS1 Traceability Standard: What You Need to Know. GS1, Brussels, Belgium.

ISO 12875, 2011. Traceability of Finfish Products – Specification on the Information to Be Recorded in Captured Finfish Distribution Chains. International Organization of Standardization.

ISO 12877, 2011. Traceability of Finfish Products – Specification on the Information to Be Recorded in Farmed Finfish Distribution Chains. International Organization of Standardization.

ISO, 1994. Quality Management and Quality Assurance Vocabulary – ISO 8402. International Organization of Standardization, p. 17.

ISO, 2000. Quality Management Systems – Fundamentals and vocabulary-ISO 9000:2000. International Organization of Standardization.

ISO-22005, 2007. Traceability in the Feed and Food Chain – General Principles and Basic Requirements for System Design and Implementation. International Organization of Standardization.

Karlsen, K.M., Sennesen, G., 2006. Traceability: simulated recall of fish products. In: Luten, J., Oehlenschläger, J., Jacobsen, C., Bekaert, K., Særbo, A.E. (Eds.), Seafood Research from Fish to Dish, Quality, Safety and Processing of Wild and Farmed Fish. Wageningen Academic publishers, Wageningen, The Netherlands, pp. 251–262.

Karlsen, K.M., Donnelly, K.A.-M., Olsen, P., 2010. Implementing traceability: practical challenges at a mineral water bottling plant. British Food Journal 112 (2), 187–197.

Karlsen, K.M., Sorensen, C.F., Foras, F., Olsen, P., 2011. Critical criteria when implementing electronic chain traceability in a fish supply chain. Food Control 22 (8), 1339–1347.

Karlsen, K.M., Dreyer, B., Olsen, P., Elvevoll, E.O., 2013. Literature review: does a common theoretical framework to implement food traceability exist? Food Control 32, 409–417.

Mai, N., Bogason, S.G., Arason, S., Arnason, S.V., Matthiasson, T.G., 2010. Benefits of traceability in fish supply chains – case studies. British Food Journal 112 (8–9), 976–1002.

Olsen, P., Aschan, M., 2010. Reference method for analyzing material flow, information flow and information loss in food supply chains. Trend in Food Science and Technology 21 (6), 313–320.

Olsen, P., Borit, M., 2013. How to define traceability. Trends in Food Science & Technology 29 (2), 142–150.

Randrup, M., Storøy, J., Lievonen, S., Margeirsson, S., Árnason, S.V., Ólavsstovu, D., Møller, S.F., Frederiksen, M.T., 2008. Simulated recalls of fish products in five Nordic countries. Food Control 19 (11), 1064–1069.

Senneset, G., Forås, E., Fremme, K.M., 2007. Challenges regarding implementation of electronic chain traceability. British Food Journal 109 (10), 805–818.

Sohal, A.S., 1997. Computerized parts traceability: an implementation case study. Technovation 17 (10), 583–591.

Sparling, D., Henson, S., Dessureault, S., Herath, D., 2006. Costs and benefits of traceability in the Canadian dairy-processing sector. Journal of Food Distribution Research 37 (1), 154–160.

Storøy, J., Senneset, G., Forås, E., Olsen, P., Karlsen, K.M., Frederiksen, M., 2008. Improving traceability in seafood production. In: Børresen, T.E. (Ed.), Improving Seafood Products for the Consumer, Part VI Seafood Traceability to Regain Consumer Confidence. Woodhead Publishing Limited, Cambridge, UK, pp. 516–538.

Tracefood, 2015. Tracefood wiki. EU-project TRACE. Available at: http://www.tracefood.org/index.php/Main_Page (downloaded 28.07.15.).

Part Two

Development of Methods, Devices, and Sensors for Food Traceability

Alphanumerical and Optical Coding Systems for Food Traceability

4

Zs. Kemény, E. Ilie-Zudor
Institute for Computer Science and Control, Hungarian Academy of Sciences, Budapest, Hungary

1. The State of the Art in Food Traceability

Many perspectives are taken in literature describing the technical and operational fundaments of food traceability (Dabbene et al., 2014; Chrysochou et al., 2009), yet they share nearly the same requirements which are also in line with functionality levels defined for track-and-trace networks in general (Monostori et al., 2010) (see also Fig. 4.1).

Traceability in food supply chains requires (1) units of material as well as process elements to be unambiguously identified (corresponding to *identifier infrastructure* and *identifier-based operations*), (2) events and associations with other entities connected to the unique unit of material to be recorded and kept available for retrieval (*tracking-based operations*), and (3) means to be provided for accessing information gathered at various points of the supply chain and synthesizing these to form a consistent and continuous chain of product history between two given points of the supply chain (*item-centric services*). Technologies supporting food traceability therefore include identification and data acquisition technologies as well as data storage, communication, and system integration technologies. Since identification technologies determine much of the limitations and possibilities of the overall system, these will first receive focus.

1.1 Technologies Supporting Traceability

1.1.1 Unique Material Properties

Most of the material forwarded in a food supply chain stems from organisms and bears an individual fingerprint in certain physical and biochemical properties. Several test methods are known to capture these features to a degree where unambiguous identity can be established. *Genetic fingerprinting* determines an identity based on DNA extracted from the material in question (Leal et al., 2015), while *DNA barcoding* refers to examining selected genetic markers, usually to determine the biological species of the sample (eg, to test purity of minced or ground products). *Chemical composition,* eg, the percentage of isotopes or trace elements, can also be unique for a given location or batch (Leal et al., 2015). *Mass spectrometry* or *radiation spectrometry* (typically infrared) can reveal the aforementioned composition-related characteristics. While the fingerprint-like uniqueness of the aforementioned properties *present in the material itself* makes them useful in establishing identities, many of the test and data acquisition

Advances in Food Traceability Techniques and Technologies. http://dx.doi.org/10.1016/B978-0-08-100310-7.00004-1

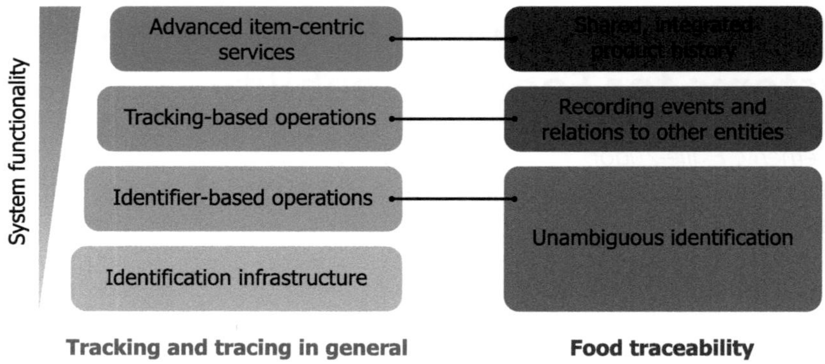

Figure 4.1 Functionality levels of track-and-trace networks matched with traceability measures in the food industry.

methods are slow, labor-intensive, and subject to an error margin that would make their transformation to computer-tractable identifiers—thought of as keys in databases—challenging, if not impossible. Therefore such test methods are normally relied on if authenticity or truth of assigned identifiers or credentials are challenged.

1.1.2 Optical Identifiers

Artificially assigned identifiers have a long history—livestock tags and human-interpretable labels with characters or symbols have long since been used, in conjunction with purely paper-based stock keeping, as forerunners of in-house traceability. Even today, some operations in traceable food supply chains are paper-based or require human intervention in capturing data and entering them into a computerized system. While text-only ID labels are on the retreat, the aspect of human readability continues to receive attention to keep identity or data content accessible to "low-tech" supply chain members or consumers, or to provide a last-resort fallback measure (Chrysochou et al., 2009). Human intervention is often associated with labor costs, error sources, and throughput limitations, suggesting the introduction of unique identifiers that can be captured (semi)automatically. Such solutions are referred to as *automatic identification, or AutoID techniques.*

Optical identifiers are one of the most common examples of automatic identification, and are, in essence, images consisting of regular fields of different colors that can be reliably recognized and mapped onto an interpreted message by a reading device (Palmer, 2007). The concept of optically perceptible patterns conveying information was proposed as early as the 1950s (Woodland and Bernard, 1952); however, practicable implementation of the concept had to wait for two to three decades until adequate equipment became available for acquiring and processing the information carried by optical labels. Widespread use, especially of bar codes, began in the 1980s, and since then, the most common type of bar code, UPC, which evolved into EAN13 (GS1, 2015a), became a symbol of commerce. Nowadays, the vast majority of optical codes uses two distinct colors only (ie, reflecting vs. nonreflecting), even though multicolor codes did occur in history and are currently in occasional use with devices (typically

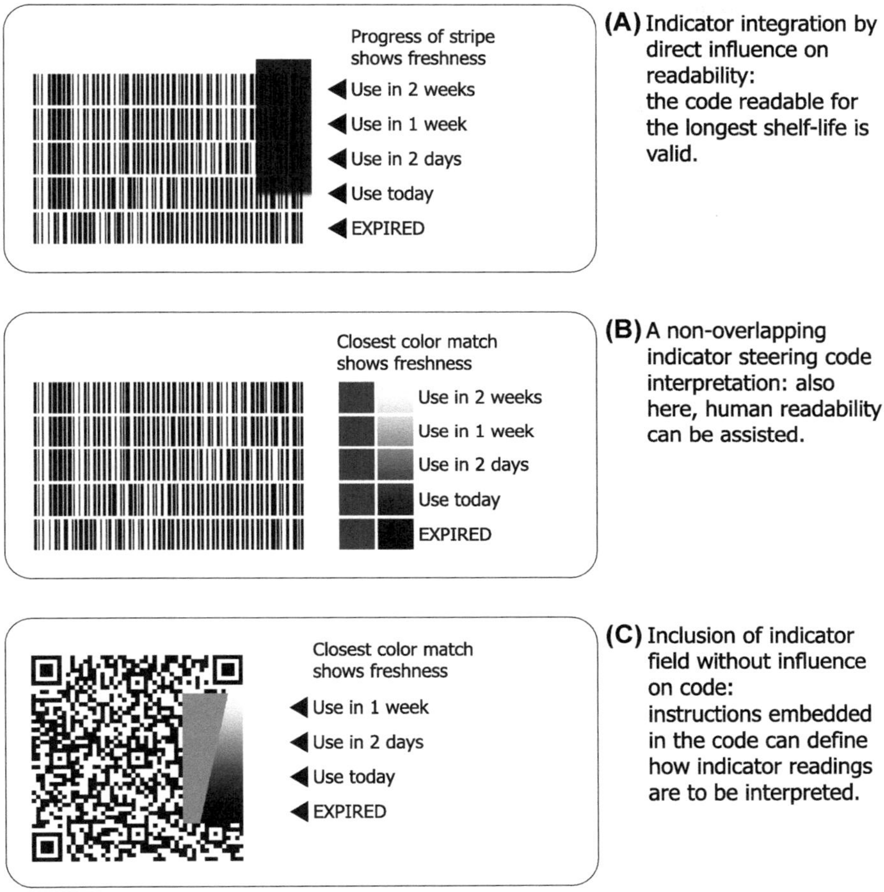

Figure 4.2 Possibilities of combining indicator fields with optical identifiers.

camera-equipped phones) that can compensate their low image resolution with good multicolor sensing capabilities (Zhou and Rong, 2011). Patterns used in optical identifiers can be of a one-dimensional (series of parallel bars), stacked 1D (multiple bar codes above each other), or true 2D (eg, the *Quick Response, or QR* code finding widespread use today (Denso Wave, 2015)). *Symbologies* (ie, optically coded data representations) can accommodate error detection and correction functionalities and can convey a relatively large amount of data, up to some kilobytes, at a relatively compact size. Optical codes can also be combined with optical indicators that change their shape or color when relevant changes in product composition (eg, decay, presence of microbes (Nopwinyuwong et al., 2010)), integrity (eg, package breach (Roberts et al., 2011)), or ambient conditions (eg, thermal abuse, exposure to humidity (Maschietti and Bianchini, 2010)), or expiry occur. Some commercial solutions are already available (CoolVu, 2010), but the possibilities of indicator integration, as shown in Fig. 4.2, are not yet fully exploited (Kemény et al., 2014) (note that optical indicators do already find widespread use in a stand-alone manner). The reading of optical codes is automatic but may require human assistance in adequately positioning the label or the reading device. Also, optical ID acquisition requires a line-of-sight free of opaque obstacles or even dirt contaminating the label surface.

1.1.3 Radio Frequency Identification

Radio Frequency Identification (RFID) is an AutoID technology deemed superior to optical codes in numerous ways. RFID uses induction (up to a few hundred kHz) or electromagnetic waves (in the several hundred MHz–GHz range, see also Fig. 4.3 for commonly used frequency bands) to transmit data between a reading/writing device (transceiver) and one or more RFID tags (transponder). The behavior of the transponder is initiated by the transceiver emitting at a given frequency, and in the majority of implementations, the transponder is also powered remotely by the transceiver's emission (the exception is active tags with their independent power source (Ilie-Zudor et al., 2006, 2011)). Since it is not visible light but much lower frequencies that must reach the tag with sufficiently little attenuation, tags can be placed inside the packaging of goods or shielded by a sturdy capsule from adverse environmental conditions. Moreover, information transmitted between transceiver and transponder is not a structured image but series of pulses; while optical codes are structured in space, the transmissions and responses in RFID technology are structured in time. Therefore it is not necessary to focus the signal from and to the reader into a narrow beam. The latter, along with the favorable propagation characteristics of the electromagnetic waves used, do not require placing the tag into a specific line of sight, thereby eliminating much of the human intervention needed for reading optical codes. Another key to the versatility of RFID technology is the application of a complex semiconductor device in the tags. While in most simple cases, these can only transmit a fixed identifier, transponder chips can also perform calculations themselves, thus enabling, eg, encrypted transmission, given the ease of physically eavesdropping wireless communication or collecting massive information in warehouse environments, encryption, and security altogether, continues to be a focal point of development and discourse (Mubarak et al., 2011). Transponder hardware can accommodate nonvolatile memory as well, allowing item-specific data to travel with the entity carrying the tag. RFID tags can be combined with sensors, even without their own independent power source (Jedermann et al.,

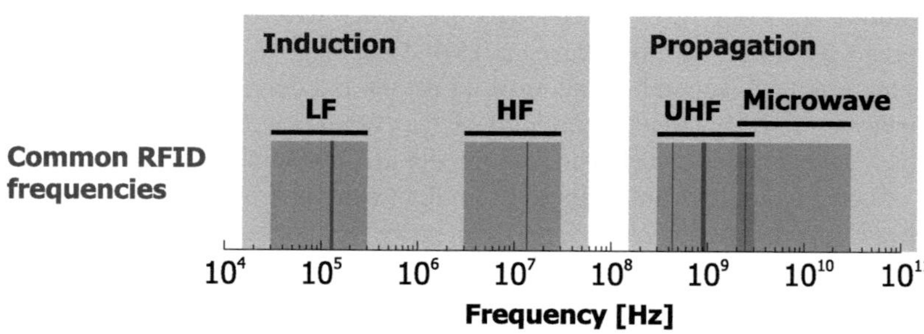

Figure 4.3 Frequency bands used by RFID; common frequencies are marked with *red* (gray in print versions) lines.

2009). Sensor-equipped RFID tags can perform preprocessing and in-tag recording of signals, and their operation can be initiated by an adequately powerful transceiver signal, eg, emitted in regular intervals for sensors distributed in a truckload of perishable goods, or health monitoring implants in livestock (Ruiz-Garcia & Lunadei, 2011). Sensor-equipped RFID is closely related to the concept of wireless sensor networks (WSN), even though a typical WSN may be different regarding sensor node autonomy and power management (Wang et al., 2006).

With all these advantages, there are limits to be reckoned with regarding technically and economically feasible use of RFID. Although costs have sunken dramatically, this still remains an issue for products with low profit margin and low per-unit value, eg, fresh vegetables. RFID requires a reading device, because exclusive application of RFID will exclude low-tech supply chain members and consumers from accessing or contributing to the information flow accompanying material movement. For this reason, RFID tags are often combined with barcode or human-readable labels. The material of RFID tags (nowadays, typically metal, silicon, and paper/plastic carrier) may not be compatible with some products for health/safety reasons, is not biodegradable, and is difficult to recycle, although the use of organic or degradable materials is already subject to research. Conductive materials (eg, metal cans, electrolytes) compromise RFID signals. Higher frequencies are less prone to such degradation, but limits do exist and must be observed when RFID deployment is considered.

RFID technology uses a variety of frequencies and air interfaces, but only the most common types and standards will be mentioned here. Low frequency (LF, mostly 125 kHz) tags, as in ISO/IEC 18000 Part 2 (International Organization for Standardization, 2009a), are read-only passive devices storing a unique identifier only and are typically applied in livestock identification. In the high frequency (HF, most are 13.56 MHz) range, two mutually incompatible types are widespread: "vicinity cards" (subject to ISO/IEC 15693 (International Organization for Standardization, 2010a)) for reading ranges around 1 m, and "proximity cards" (ISO/IEC 14443 (International Organization for Standardization, 2010b)) to be read from a few centimeters. Also of interest in the HF range, especially for consumers and small supply chain members, are "Mifare" tags (ISO/IEC 14443 Type A) that can be read by numerous smart phones equipped for near field communication (NFC). Many of the Mifare tags offer generous rewritable memory at affordable per-unit cost. Among UHF solutions, EPC Gen2 tags are of the highest commercial relevance, especially in a global trade context. Covered also in ISO/IEC 18000 Part 6C (International Organization for Standardization, 2009b), the EPC Gen2 standard has been elaborated by EPC global in an effort to harmonize protocols for worldwide use.

1.1.4 IT Infrastructure for Collection and Sharing

Infrastructural support for collecting, storing, and sharing traceability-related information is still undergoing important changes, and the transition to full product and process transparency is still far from complete. Enterprise information systems have gradually shifted from proprietary development of financially powerful, large pioneers to scalable frameworks that are offered as customizable collections of off-the-shelf products and services by IT enterprises (Moon, 2007). These solutions now accommodate a range of modules and functionalities that enable the automatic collection of information connected to individual entities (both products and production resources) and processes (Kemény et al., 2007). The need for efficient business-to-business (B2B) communication is clearly shifting toward more comprehensive cross-company transparency; as a result, past means of information exchange (such as "classical" EDI, Electronic Data Interchange) are gradually superseded by solutions built on a more modern basis, such as web services (Dorn et al., 2007). In response to growing complexity, security, and privacy concerns, as well as gradual penetration of transparency solutions in the small-business sector, analysis, data, and process modeling and access control techniques are undergoing considerable development (Dorn et al., 2007; van Blommestein, 2007).

While enterprise IT frameworks provide key architectural elements for implementing traceability solutions, it is the global or national traceability platforms and services that are decisive in establishing reliable traceability over an entire supply chain. In this regard, sector-specific initiatives with legislative support are one possibility—examples are the TRAde Control and Expert System (TRACES) in Europe (Council of the European Union, 2004), or Canadian Agri-Traceability Services (CATS) in Canada (Quigley and Bisset, 2014). Also, global organizations take a leading role in providing platforms and services for integration and supply chain transparency. The EPCglobal Network, created in a joint venture of GS1 and GS1 US, is the most important among implemented initiatives and provides a network for trade transparency in general, relying on GS1-maintained standards (Object Naming Services, ONS (GS1, 2013); EPC Information Services, EPCIS (GS1, 2014); as well as EPC Discovery and Security Services (GS1, 2015b)). GS1 US also has a leading role in building up whole-chain traceability in the US food industry.

1.2 Application Practice

Today's practice of food traceability still outlines a picture of transition. Drivers can, nevertheless, be identified and continue to push food supply chains toward improved transparency. On one hand, globalization of trade, large-scale productions and intensified use of new technologies often rushed into production with meager assessment of effects create higher risks of foodborne illness, while, on the other hand, growing awareness among consumers and lawmakers, as well as recognition of transparency as a source of market advantage by supply chain members, keeps food supply chain transparency in focus. Trade across organizational and administrative borders continues to be a driver for improving legal harmonization and technical interoperability within regulatory bodies, standards organizations, and solution providers. In this process, Europe is deemed leader, followed by North America, while transparency in Asia matches up to the state of the art in certain sectors of food business and specific countries and companies only. Much improvement is in preparation, nevertheless, in countries where technical preconditions and general awareness undergo rapid development.

However, even where national and regional regulations are already in place and are undergoing regular refinement, and national or global services and networks are being currently rolled out, it is still of varying success how much transparency is actually achieved for the supply chain as a whole. This can be attributed to varying degrees of ripeness in de facto use of underlying technologies and transparency and process practices. While major players in the industry have preceded national and international regulations by years or decades (and have enough capital to adjust their existing system to the legislative requirements, too), the small-business sector is either reluctantly yielding to the pressure of legislation and large companies or is just about to discover the potential benefits of transparency for their own business.

With automatic identification techniques still having not reached their peak of penetration, many elements in food supply chain practice remain paper-based or rely fundamentally on human intervention, with data being entered manually into electronic form where required (eg, reading of livestock ear tags or processing accompanying documents filled out by hand). While the per-unit costs of identification equipment have considerably sunken, partly due to growing penetration and large-scale production, there is still much confusion, especially in the small-business sector, about realistic costs and benefits of particular solutions, as well as the potential and expected return-on-investment, especially of RFID (Piramuthu et al., 2013). Many theoretical works have addressed the problem (some of these along with proper sizing of smallest traceable units (Donnelly et al., 2009)), and technology transfer or knowledge transfer continuously take place, mostly initiated by either government or technology stakeholders.

Communication across company borders currently leaves much room for improvement, not only regarding B2B communication per se, but also from the point of view of end-to-end traceability. A realistic legislative background of traceability cannot assume radical improvement is accomplished overnight. Partly for this reason, current regulations still allow simple "one-up, one-down" solutions (Crandall et al., 2013), where a given member of the supply chain is only required to specify its immediate predecessor and successor in records of material movement. Clearly, such policies leave the information chain vulnerable as one weak point can hamper tracing along the entire chain (Storøy et al., 2008) (see also Fig. 4.4). This is expected to improve with the introduction of global, regional, or national networks and services that support uniform access to all records relevant to reconstructing an entire traceability chain.

2. Perceived Problems and Demands

2.1 Industry-Specific Technical Problems

The food industry is one of the branches where identifiers—be it optical or RFID—may be exposed to extreme conditions that impair their readability or integrity. This is, in particular, the case when materials (both products and assets used in processes) undergo heat, radiation, freezing, or chemical treatment together with the identifiers. Although still highly experimental, tags screen-printed or prototyped using conductive materials outline promising temperature-tolerant alternatives to semiconductors employed in today's mass-produced RFID transponders (Janeczek et al., 2012). The protection (durable encapsulation) of tags is less of an issue here; in most cases,

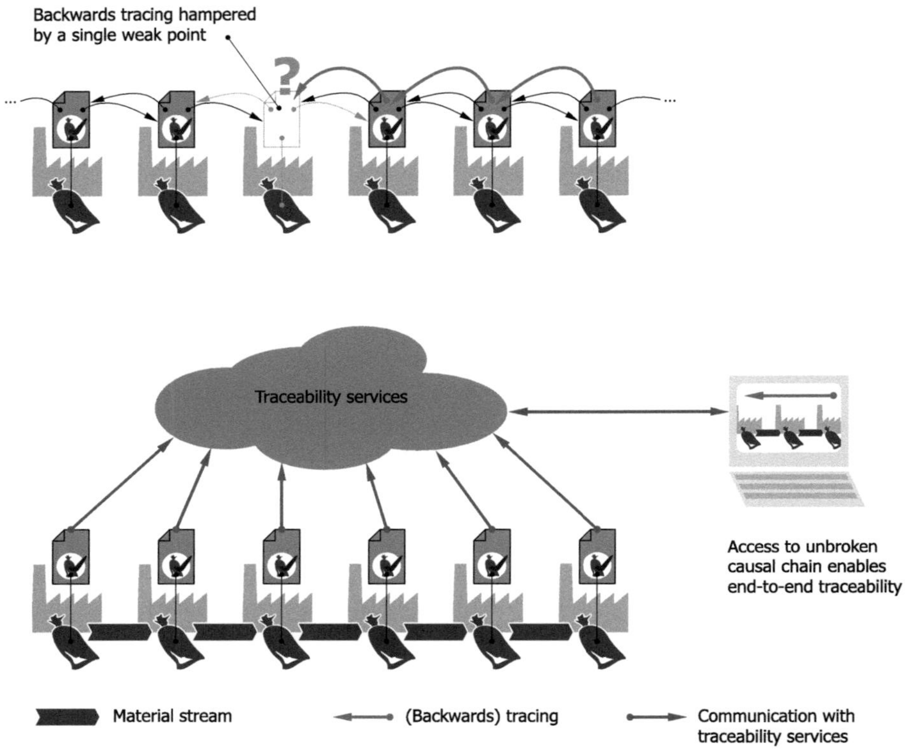

Figure 4.4 Discontinuities can hamper end-to-end traceability in "one-up, one-down" practice (top), while an independently operating traceability service can maintain all data for reconstruction of an entire causal chain (bottom).

food-tolerant durable materials are commercially offered. Optical codes can easily be printed on a more durable carrier, but possible problems arise here of impaired visibility, such as deposits or ice crystals, especially in cases where all wavelengths are affected that might be used for ID acquisition.

Another problem area inherent to certain branches of the food sector is continuous material flow. In some cases, the material flow cannot be strictly partitioned to disjointed volumes (eg, tankloads), and one must account for gradual dilution of residues, or material mixing processes that must be monitored in some quasicontinuous way. Gradual mixing has been addressed in literature by *fuzzy traceability* (Skoglund and Dejmek, 2007), along with estimations based on fluid dynamics or other process models. For bulk commodities, the use of trace pellets or similar means has been proposed, including food grade markers that do not compromise safety or quality upon consumption, yet these proposals are at best in the proof-of-concept phase (Dabbene et al., 2014). The characteristics of continuous material flow, and the difficulties of making bulk goods traceable, is one of the causes for the persistence of account-based perspectives in these branches of production, and it is likely that in such cases, a mixture of quantitative and identifier-based approaches become established in the long-term.

2.2 Small-Business Involvement and Demands

The involvement of small enterprises in traceability is still not much more than mere compliance with mandatory transparency regulations, similar to other branches of industrial production. The role of major players as driving forces of transparency has often been observed, and some cases report critical results regarding expected benefits of information transparency in the small-business sector (Kembro and Näslund, 2014). It is, nonetheless, likely that the persisting lag of small-business information sharing is also due to most of the small enterprises lacking the in-depth knowledge and resources for initial investment.

Aside from targeted knowledge transfer about confidence and security in B2B interaction (including the view of controlled information disclosure as an investment and not as a risk alone), the adoption of more affordable technologies can also present itself as a driver. Certain identifier technologies, for example, have spread remarkably, owing to easy and affordable smart phone support (eg, QR code, NFC). Inclusion of these technologies in transparency measures is expected to facilitate the involvement of small enterprises, as well as end consumers. Also noteworthy is the growing interest in traceability based on IoT (Internet of Things) technologies in Chinese research literature (Pang et al., 2015).

2.3 Changing Consumer Demands and Attitudes

End consumers are still left out of large parts of traceability systems, and are, in most cases, not able to access more than *credentials* regarding origin, safety, or specific attributes of food products (Dabbene et al., 2014). With consumers becoming more conscious and the food industry seeing consumer confidence more and more as a market advantage, the picture is expected to develop in the years to come.

Currently, however, the average consumer is far from ready for direct access to the transparency targeted within the supply chain. Recent surveys have shown that consumers' knowledge about critical properties of food products is still inaccurate (Chrysochou et al., 2009). Nevertheless, knowledge of traceability, per se, does have a positive association, and food safety is perceived in surveys as a "non-negotiable product attribute" in a consumer decision-making process (Verbeke et al., 2007). Demands for transparency and additional or more accurate product information can already be perceived in the less price-sensitive segments, and such tendencies are also recognized both by industrial stakeholders and regulatory bodies, moving from merely protecting consumers to presenting them an *informed choice* (Arienzo et al., 2008).

3. Potentials and Pitfalls

3.1 Tracking

Transparency measures enabling traceability are required by today's regulations as a "safety net" to localize and counteract foodborne hazards, and many participants of food supply chains are also voluntarily taking additional steps to strengthen consumer confidence and secure themselves competitive advantages. In addition to these, making provisions for traceability can also set the stage for product *tracking,* ie, current location, status, and other attributes of individual products can be followed with minimal time lag and high accuracy (Kemény et al., 2007; Alfaro and Rbade, 2009). The information gathered by tracking can be exploited to improve operational decisions in the supply chain (eg, dispatching with better utilization of assets, timely perception of resource-critical situations building up, and more time and more information for planning ahead). Also, collecting tracking information over longer periods of time can grant insight for better strategic decisions, such as long-term supply chain redesign.

Participation in tracking over company borders can be beneficial for smaller enterprises as well; with more accurate knowledge of processes across the supply chain, and today's planning methods at hand, the sharing of risks and benefits can be directly negotiated and formally defined in appropriate contracts (Egri and Váncza, 2013). Nevertheless, transparency across the supply chain also requires information to be seamlessly exchangeable and interpretable by the parties involved, ie, information *interoperability* has to be cared for, even for newcomers joining the network. This is assisted by adhering to common data models and semantics. Several top-level and domain ontologies have been built up, such as UN/CEFACT (van Blommestein, 2012), TraceFood XML (Storøy et al., 2013), FTTO (Pizzuti et al., 2014), and GS1 XML (GS1, 2015c). While these standards provide a good starting point, much expert work and development may be required before initial adoption in a supply chain.

As mentioned before, fear of disclosing sensitive information is still taming the involvement, especially of small enterprises, in information sharing in supply chains. In some regards, this apprehension is indeed justified. Security threats do exist, and improper application of technology can easily lead to backlash; particularly if the industry begins to use something for a purpose it was not meant for (as was the case with communication protocols in the past). Nowadays, formal methods on security assessment and design, as well as efficient detection of security breaches, do provide certain guarantees (Marucheck et al., 2011).

Another potential pitfall of inclusion in a tracking network is the overwhelming technical and financial power of large players in the supply chain which can lock smaller suppliers into certain solutions and make them dependent on specific technologies. The introduction of nationwide, regional, or global data-sharing platforms has the potential of foregoing such problems.

Yet another aspect that requires caution is the rate and depth of measures taken for improved transparency. Making material movement and other operations transparent will imply some degree of intervention into the processes themselves. In the optimal case, this is equal to merely tapping an information stream hitherto unexploited; however, even such simple measures, such as identifier acquisition introduced out-of-pace, may be detrimental to established routine. This is especially critical in industries (including complicated food supply chains and logistics) where the legacy of complicated interdependencies only allows incremental roll out, with carefully planned options of reverting to past practice if needed (Ilie-Zudor et al., 2013).

3.2 Consumer Involvement

The events of recent years have raised the importance of consumer trust, and industrial stakeholders are recognizing the growing importance of consumer acceptance as a competitive advantage. While the notion of traceability has, in general, positive associations among consumers, even with as little as credential level (Dabbene et al., 2014), care must be taken with presenting additional information in the right form and right context.

The key in assessing consumer reactions is a proper understanding of a layperson's psychological background of—bounded-rational—decision-making. As opposed to experts who have detailed information at hand and have a profound background knowledge of dependencies and theoretical models, the general public has to rely on partial knowledge and intuition while making judgments "along the way" with minimal time and effort spent. The perception of things partly unknown and being "beyond control" is then guaranteed to introduce bias (Kher et al., 2013), leading to amplified focus on certain risks and distorted interpretation of further information or reaction of the industry to detected phenomena. Since the psychology of a layperson's decision-making cannot be changed by design, it is the responsibility of industry and lawmakers to prevent escalation of false-risk perception (Veil et al., 2015) and responsibly support customer awareness in its development.

A step into this direction is, as mentioned before, the gradual shift "from protecting to labeling" ie, from shielding consumers assumed to be ignorant, toward presenting them an *informed choice* (Arienzo et al., 2008). The latter, however, must take into account the average consumer's simplified perception of food properties, risks, and assurances (Verbeke et al., 2007), overlapping concepts of quality and safety, or use-by-date and best-before-date (Rohr et al., 2005), or notions of probability or uncertainty differing from expert understanding. Another important finding is that most consumers will rarely engage in *actively* seeking information (eg, demanding information material or searching background information), rather, they "pick up" what they are readily presented on labels, in news or informal talk. In an age where social media shows increasing impact on community development, information delivered physically with the product (either human-readable or as links for smart phones) has to be understood as merely one component in a complex picture of risk and benefit communication (Rutsaert et al., 2013).

The linking of easily accessible consumer information can, nowadays, open up another potential: aside from assisting the consumer in the purchase decision, it can also provide guidance thereafter, possibly in a context-aware way. An easy-to-comprehend selection of serving suggestions, or clear, individual instructions on use or disposal of a perishable product can improve consumer satisfaction, forego health risks ensuing after a sale, and, generally, reassure consumers to be more in control of their decisions (Pennanen et al., 2015).

4. Current Trends

The transparency of food supply chains has undergone substantial improvement since the early 2000s in several regions of the world, and their strict import criteria are a driver for improvement in other regions, too (Mol, 2014). While "one-up, one-down" forms of transparency—implying potentially vulnerable traceability across the entire supply chain—still persist in regulations, support for farm-to-fork traceability is promised by national, regional, or global services (Council of the European Union, 2004; Quigley and Bisset, 2014; Crandall et al., 2013; GS1, 2014). Compliance with traceability requirements is often perceived, especially by small enterprises, as more of a burden than an opportunity, yet adequate introduction of material and process transparency enables tracking of goods and gives insight for better planning and steering of the supply chain from operational to strategic levels. Major industrial stakeholders have already pioneered in track-and-trace solutions, nonetheless, the involvement of smaller enterprises is still lagging behind due to a number of "inhibitors": (1) initial investment barriers (Schiefer et al., 2014), (2) disinterest in information sharing in view of perceived concerns and little assurance of expected benefits (Kembro and Näslund, 2014), (3) multitude of competing interchange standards (Dabbene et al., 2014), and (4) understanding of complex or long-range supply chain processes still being unripe and subject to research (Dabbene et al., 2014).

While some of the aforementioned obstacles are likely to persist in the future, technological development is predicting changes in some aspects of food supply chain practice. Certain identifier technologies have become widespread and affordable and accessible to many low-end supply chain members and end consumers. QR code (Denso Wave, 2015) and NFC working in the HF band (International Organization for Standardization, 2010b) are such solutions for optical codes and RFID, respectively: both can accommodate a relatively large amount of data, sufficient for embedded links to product-related services and other auxiliary information, and can be reliably scanned by a variety of affordable smart phones. The latter technologies are commonly relied on in IoT (Internet of Things) solutions (Badia-Melis et al., 2015), while their combination with EPC Gen2 UHF tags linked to item-level services in EPCIS (GS1, 2014) can enable versatile transparency for all members of the supply chain, including consumers.

Still marginally present in commerce, developments in indicator and sensor technologies are anticipated to grow in importance and become less costly to apply. Smart labels delivering accurate freshness information are likely to affect gradual refinement of views on safety and shelf life (Grunow and Piramuthu, 2013; Kemény et al., 2014), and associated distribution of responsibilities in food supply chains.

Technological progress and its impact on social interaction and perception are already showing effects in consumer behavior: certain consumer groups have become more critical of safety and quality, while social phenomena are not always working out to the benefit of either consumers or industrial stakeholders (Veil et al., 2015). Developments forecast that supply chain members, as well as governance, will have to count on closer consumer involvement and will have to respond to social and communication challenges in a more comprehensive and responsible way (Arienzo et al., 2008; Rutsaert et al., 2013).

5. Conclusion

Traceability in food supply chains has become routine in several regions of the world, with Europe consistently ranked as the leader in global comparison. Current regulations typically require no more than isolated records (in a "one-up, one-down" manner), leaving end-to-end traceability vulnerable to flaws at any single point in a chain of evidence; however, ongoing technological development and the emergence of networks and services, in particular, aim to overcome such drawbacks with their roll-out. While still not equally relied on by supply chain members of various sizes, automatic identification techniques, both optical and RFID, continue to spread with decreasing investment barriers. Certain optical codes, such as QR, and radio-frequency identifiers, such as NFC, are of particular interest to "low-end" supply chain members and consumers, while EPC Gen2 UHF tags and associated item-level services offer comprehensive technical and organizational support for cross-company transparency in any supply chain (ie, not restricted to the food industry alone). Data-sharing practices still have considerable development to master, especially in data and process interoperability and security/trust aspects. Much research has been conducted in supply chain integration, yet truly versatile solutions and methodologies are yet to emerge. A further important trend is the growing degree of consumer involvement and impact on decisions in the food supply chain, partly due to new forms of social interaction but also owing to increased insight into composition and handling of food products: the latter enabled, to a considerable part, by increased transparency. Recent events suggest that industrial stakeholders of food supply chains will have to deal with consumer reactions in a more efficient and responsible way, and the integration of consumers (both in form of communication and value-adding services) must be understood and guided in a complex socioeconomical context.

References

Arienzo, A., Coff, C., Barling, D., 2008. The European union and the regulation of food traceability: from risk management to informed choice? In: Coff, C., Barling, D., Korthals, M., Nielsen, T. (Eds.), Ethical Traceability and Communicating Food, Vol. 15 of the International Library of Environmental, Agricultural and Food Ethics, Springer Netherlands, pp. 23–42. http://dx.doi.org/10.1007/978-1-4020-8524-6_2.

Alfaro, J.A., Rbade, L.A., 2009. Traceability as a strategic tool to improve inventory management: a case study in the food industry. International Journal of Production Economics 118 (1), 104–110. http://dx.doi.org/10.1016/j.ijpe.2008.08.030. Special section on problems and models of inventories selected papers of the fourteenth international symposium on inventories. http://www.sciencedirect.com/science/article/pii/S0925527308002533.

van Blommestein, F., 2007. Decentralized Metadata Development for Open B2B Electronic Business. In: Enterprise Interoperability. Springer, pp. 355–364.

van Blommestein, F., 2012. Teaching business systems to agree. In: Bouma, G., Ittoo, A., Mtais, E., Wortmann, H. (Eds.), Natural Language Processing and Information Systems, Vol. 7337 of Lecture Notes in Computer Science, Springer Berlin Heidelberg, pp. 308–313. http://dx.doi.org/10.1007/978-3-642-31178-9_39.

Badia-Melis, R., Mishra, P., Ruiz-García, L., 2015. Food traceability: new trends and recent advances. a review. Food Control 57, 393–401.

Chrysochou, P., Chryssochoidis, G., Kehagia, O., 2009. Traceability information carriers. The technology backgrounds and consumers perceptions of the technological solutions. Appetite. 53 (3), 322–331. http://dx.doi.org/10.1016/j.appet.2009.07.011. http://www.sciencedirect.com/science/article/pii/S0195666309005777.

CoolVu, 2010. Time-temperature indication. Food & Beverage International 9, 39–40.

Council of the European Union, 2004. Commission Decision 2004/292/EC on the Introduction of the TRACES System and Amending Decision 92/486/EEC. text with EEA relevance http://eur-lex.europa.eu/LexUriServ/LexUriServ.do?uri=OJ: L: 2004:094:0063:0064:EN: PDF.

Crandall, P.G., O'Bryan, C.A., Babu, D., Jarvis, N., Davis, M.L., Buser, M., Adam, B., Marcy, J., Ricke, S.C., 2013. Whole-chain traceability, is it possible to trace your hamburger to a particular steer, a U. S. perspective. Meat Science. 95 (2), 137–144. http://dx.doi.org/10.1016/j.meatsci.2013.04.022. http://www.sciencedirect.com/science/article/pii/S0309174013001423.

Dabbene, F., Gay, P., Tortia, C., 2014. Traceability issues in food supply chain management: a review. Biosystems Engineering 120, 65–80. http://dx.doi.org/10.1016/j.biosystemseng.2013.09.006. Operations Management in Bio-production Systems. http://www.sciencedirect.com/science/article/pii/S1537511013001554.

Denso Wave, 2015. QR Code Official Website. http://www.qrcode.com/en/.

Dorn, J., Grun, C., Werthner, H., Zapletal, M., 2007. A survey of B2B methodologies and technologies: from business models towards deployment artifacts. In: HICSS 2007. 40th Annual Hawaii International Conference on System Sciences, 2007, p. 143a. http://dx.doi.org/10.1109/HICSS.2007.37.

Donnelly, K.A.-M., Karlsen, K.M., Olsen, P., 2009. The importance of transformations for traceability a case study of lamb and lamb products. Meat Science. 83 (1), 68–73. http://dx.doi.org/10.1016/j.meatsci.2009.04.006. http://www.sciencedirect.com/science/article/pii/S0309174009000928.

Egri, P., Váncza, J., 2013. A distributed coordination mechanism for supply networks with asymmetric information. European Journal of Operational Research 226 (3), 452–460.

GS1, 2015a. GS1 General Specifications. Version 15, issue 2. http://www.gs1.org/barcodes-epcrfid-id-keys/gs1-general-specifications.

GS1, 2013. GS1 Object Name Service (ONS) Version 2.0.1. Ratified Standard. http://www.gs1.org/sites/default/files/docs/epc/ons_2_0_ 1-standard-20130131.pdf.

GS1, 2014. EPC Information Services (EPCIS) Version 1.1 Specification. GS1 Standard. http://www.gs1.org/docs/epc/epcis_1_1-standard-20140520.pdf.

GS1, 2015b. GS1 System Architecture Document—how GS1 Standards Fit Together. release 4.0.1. http://www.gs1.org/sites/default/files/docs/architecture/GS1_ System_Architecture.pdf.

GS1, 2015c. GS1 Set of XML Standards. http://www.gs1.org/ecom-xml/technical-user-guide/gs1-set-xml-standards.

Grunow, M., Piramuthu, S., 2013. RFID in highly perishable food supply chains—remaining shelf life to supplant expiry date? International Journal of Production Economics. 146 (2), 717–727. http://dx.doi.org/10.1016/j.ijpe.2013.08.028. http://www.sciencedirect.com/science/article/pii/S0925527313003861.

Ilie-Zudor, E., Kemény, Z., Egri, P., Monostori, L., 2006. The RFID technology and its current applications. In: 8th International Conference on the Modern Information Technology in the Innovation Processes of Industrial Enterprises, Budapest, Hungary, pp. 29–36.

Ilie-Zudor, E., Kemény, Z., van Blommestein, F., Monostori, L., van der Meulen, A., 2011. A survey of applications and requirements of unique identification systems and RFID techniques. Computers in Industry. 62 (3), 227–252. http://dx.doi.org/10.1016/j.compind.2010.10.004. http://www.sciencedirect.com/science/article/pii/S0166361510001521.

International Organization for Standardization, 2009a. ISO/IEC 180002:2009 Information Technology—Radio Frequency Identification for Item Management—Part 2: Parameters for Air Interface Communications Below 135 Khz. http://www.iso.org/iso/home/store/catalogue_tc/catalogue_detail.htm?csnumber=46146.

International Organization for Standardization, 2010a. ISO/IEC 15693-1:2010 Identification Cards—Contactless Integrated Circuit Cards—vicinity Cards—Part 1: Physical Characteristics. http://www.iso.org/iso/home/store/catalogue_tc/catalogue_detail.htm?csnumber=39694.

International Organization for Standardization, 2010b. ISO/IEC 14443-1:2008 Identification Cards—Contactless Integrated Circuit Cards—proximity Cards—Part 1: Physical Characteristics. http://www.iso.org/iso/home/store/catalogue_tc/catalogue_detail.htm?csnumber=39693.

International Organization for Standardization, 2009b. ISO/IEC 180006:2013 Information Technology—Radio Frequency Identification for Item Management—Part 6: Parameters for Air Interface Communications at 860 Mhz to 960 Mhz General. http://www.iso.org/iso/home/store/catalogue_tc/catalogue_detail.htm?csnumber=59644.

Ilie-Zudor, E., Ekárt, A., Kemény, Z, Karnok, D., Buckingham, C., Jardim-Gonçalves, R., 2013. Information modeling and decision support in logistics networks. In: Proc. of the 5th Int. Conf. on Experiments, Process, System Modeling, Simulation, Optimization, Vol. 1, Athens, Greece, pp. 279–286.

Jedermann, R., Ruiz-Garcia, L., Lang, W., 2009. Spatial temperature profiling by semi-passive RFID loggers for perishable food transportation. Computers and Electronics in Agriculture 65 (2), 145–154.

Janeczek, K., Jakubowska, M., Moniak, A., Kozio, G., 2012. Thermal characterization of screen printed conductive pastes for RFID antennas. Materials Science and Engineering: B 177 (15), 1336–1342. http://dx.doi.org/10.1016/j.mseb.2012.01.014. MicroTherm2011 Microtechnology and Thermal Problems in Electronics. http://www.sciencedirect.com/science/article/pii/S0921510712000608.

Kemény, Z., Bozóki, S., Ilie-Zudor, E., Monostori, L., 2014. Low-cost extension of information transparency throughout the product life-cycle via optical identification and quality indication. Procedia CIRP 25, 106–113.

Kemény, Z., Ilie-Zudor, E., Kajosaari, R., Holmström, J., van Blommestein, F., 2007. State of the art in tracking-based business. Public Deliverable D3.1 of the EU FP6 Project "Identity-based Tracking and Web-services for SMEs (TraSer)", 53 pp.

Kembro, J., Näslund, D., 2014. Information sharing in supply chains, myth or reality? A critical analysis of empirical literature. International Journal of Physical Distribution & Logistics Management 44 (3), 179–200. http://dx.doi.org/10.1108/IJPDLM-09-2012-0287.

Kher, S.V., De Jonge, J., Wentholt, M.T., Deliza, R., de Andrade, J.C., Cnossen, H.J., Luijckx, N.B.L., Frewer, L.J., 2013. Consumer perceptions of risks of chemical and microbiological contaminants associated with food chains: a cross-national study. International Journal of Consumer Studies 37 (1), 73–83.

Leal, M.C., Pimentel, T., Ricardo, F., Rosa, R., Calado, R., 2015. Seafood trace- ability: current needs, available tools, and biotechnological challenges for origin certification. Trends in Biotechnology 33 (6), 331–336.

Monostori, L., Csáji, B.C., Kádár, B., Pfeiffer, A., Ilie-Zudor, E., Kemény, Z., Szathmári, M., 2010. Towards adaptive and digital manufacturing. Annual Reviews in Control 34 (1), 118–128. http://dx.doi.org/10.1016/j.arcontrol.2010.02.007. http://www.sciencedirect.com/science/article/pii/S1367578810000131.

Maschietti, M., Bianchini, M.L., 2010. A new time-temperature indicator (TTI) based on high-viscosity liquids. In: 4th Cold Chain Management Workshop, Bonn, Germany, pp. 1–22.

Mubarak, M., Manan, J., Yahya, S., 2011. A critical review on RFID system towards security, trust, and privacy (stp). In: Signal Processing and its Applications (CSPA), 2011 IEEE 7th International Colloquium on, pp. 39–44. http://dx.doi.org/10.1109/CSPA.2011.5759839.

Moon, Y.B., 2007. Enterprise Resource Planning (ERP): a review of the literature. International Journal of Management and Enterprise Development 4 (3), 235–264.

Marucheck, A., Greis, N., Mena, C., Cai, L., 2011. Product safety and security in the global supply chain: issues, challenges and research opportunities. Journal of Operations Management 29 (78), 707–720. http://dx.doi.org/10.1016/j.jom.2011.06.007. Special Issue: Product Safety and Security on the Global Supply Chain. http://www.sciencedirect.com/science/article/pii/S0272696311000945.

Mol, A.P., 2014. Governing china's food quality through transparency: a review. Food Control. 43, 49–56. http://dx.doi.org/10.1016/j. foodcont.2014.02.034. http://www.sciencedirect.com/science/article/pii/S0956713514001030.

Nopwinyuwong, A., Trevanich, S., Suppakul, P., 2010. Development of a novel colorimetric indicator label for monitoring freshness of intermediate-moisture dessert spoilage. Talanta 81, 1126–1132.

Palmer, R.C. (Ed.), 2007. The Bar Code Book: A Comprehensive Guide to Reading, Printing, Specifying, and Applying Bar Code and Other Machine Readable Symbols. Trafford Publishing.

Piramuthu, S., Farahani, P., Grunow, M., 2013. Rfid-generated traceability for contaminated product recall in perishable food supply networks. European Journal of Operational Research. 225 (2), 253–262. http://dx.doi.org/10.1016/j.ejor.2012.09.024. http://www.sciencedirect.com/science/article/pii/S0377221712006881.

Pang, Z., Chen, Q., Han, W., Zheng, L., 2015. Value-centric design of the internet- of-things solution for food supply chain: value creation, sensor portfolio and information fusion. Information Systems Frontiers 17 (2), 289–319. http://dx.doi.org/10.1007/s10796-012-9374-9.

Pizzuti, T., Mirabelli, G., Sanz-Bobi, M.A., Gomez-Gonzalez, F., 2014. Food track & trace ontology for helping the food traceability control. Journal of Food Engineering 120, 17–30.

Pennanen, K., Focas, C., Kumpusalo-Sanna, V., Keskitalo-Vuokko, K., Mat- ullat, I., Ellouze, M., Pentikäinen, S., Smolander, M., Korhonen, V., Ollila, M., 2015. European consumers' perceptions of time-temperature indicators in food packaging. Packaging Technology and Science 28 (4), 303–323. http://dx.doi.org/10.1002/pts. 2105.

Quigley, K., Bisset, B., 2014. Analysis of the Regulatory Regime for Controlling Risks Related to the Canadian Food Supply Chain. Tech. rep., Dalhousie University. http://cip.management.dal.ca/wp-content/uploads/2015/01/Agriculture-Paper_final.pdf.

Roberts, L., Lines, R., Reddy, S., Hay, J., 2011. Investigation of polyviologens as oxygen indicators in food packaging. Sensors and Actuators B: Chemical 152, 63–67.

Ruiz-Garcia, L., Lunadei, L., 2011. The role of RFID in agriculture: applications, limitations and challenges. Computers and Electronics in Agriculture 79 (1), 42–50. http://oa.upm.es/9621/1/LuisRG_2011_COMPAG_The_role_of_RFID_in_agriculture_Applications_limitations_and_challenges.pdf.

Röhr, A., Lüddecke, K., Drusch, S., Müller, M., Alvensleben, R., 2005. Food quality and safety—consumer perception and public health concern. Food Control 16 (8), 649–655. http://dx.doi.org/10.1016/j.foodcont.2004.06.001. 7th Karlsruhe Nutrition Congress on Food Safety. http://www.sciencedirect.com/science/article/pii/S0956713504001203.

Rutsaert, P., Regan, Á., Pieniak, Z., McConnon, Á., Moss, A., Wall, P., Verbeke, W., 2013. The use of social media in food risk and benefit communication. Trends in Food Science & Technology. 30 (1), 84–91. http://dx.doi.org/10.1016/j.tifs.2012.10.006. http://www.sciencedirect.com/science/article/pii/S0924224412002142.

Storøy, J., Senneset, G., Forås, E., Olsen, P., Karlsen, K., Frederiksen, M., 2008. Improving traceability in seafood production. In: Børresen, T. (Ed.), Improving Seafood Products for the Consumer, Part VI: Seafood Traceability to Regain Consumer Confidence. Woodhead Publishing in Food Science, Technology and Nutrition, Cambridge, UK, pp. 516–538.

Skoglund, T., Dejmek, P., 2007. Fuzzy traceability: a process simulation derived extension of the traceability concept in continuous food processing. Food and Bioproducts Processing 85 (4), 354–359.

Storøy, J., Thakur, M., Olsen, P., 2013. The tracefood framework-principles and guidelines for implementing traceability in food value chains. Journal of Food Engineering 115 (1), 41–48.

Schiefer, G., Reiche, R., Deiters, J., 2014. Transparency in food networks-where to go. International Journal on Food System Dynamics 4 (4), 283–293.

Verbeke, W., Frewer, L.J., Scholderer, J., Brabander, H.F.D., 2007. Why consumers behave as they do with respect to food safety and risk information. Analytica Chimica Acta 586 (12), 2–7. http://dx.doi.org/10.1016/j.aca.2006.07.065. Papers Presented at the 5th International Symposium on Hormone and Veterinary Drug Residue Analysis Drud Residue Analysis 2006. http://www.sciencedirect.com/science/article/pii/S0003267006016424.

Veil, S.R., Reno, J., Freihaut, R., Oldham, J., 2015. Online activists vs. Kraft Foods: a case of social media hijacking. Public Relations Review. 41 (1), 103–108. http://dx.doi.org/10.1016/j.pubrev.2014.11.017. http://www.sciencedirect.com/science/article/pii/S0363811114001891.

Woodland, N., Bernard, S., 1952. Classifying Apparatus and Method. U.S. Patent 2,612,994.

Wang, N., Zhang, N., Wang, M., 2006. Wireless sensors in agriculture and food industry—recent development and future perspective. Computers and Electronics in Agriculture 50 (1), 1–14.

Zhou, P., Rong, X., 2011. Applications of 2D barcode for mobile tagging. Advanced Materials Research 174, 171–174.

Radio-Frequency Identification Usage in Food Traceability

5

F. Dabbene
IEIIT–CNR, Politecnico di Torino, Turin, Italy

P. Gay
IEIIT–CNR, Politecnico di Torino, Turin, Italy; Università degli Studi di Torino, Grugliasco, Italy

C. Tortia
Università degli Studi di Torino, Grugliasco, Italy

1. Introduction

The use of Radio-Frequency Identification (RFID) has a great potential for traceability, logistics, supply chain management, and quick response systems. RFID is a mean of product automatic identification (Auto-ID) where the product can even identify itself without human intervention, independently of line-of-sight, in motion, and simultaneously with other items.

The automated monitoring of processing, displacement, and storage of food items or lots of bulk materials, defined in traceable units, has a precise state-of-the art manufacturing process which is not only considered merely an internal traceability system, but an extraordinary tool to control, manage, and model the process, as well as to share traceability data with the other stakeholders, improving the performances of the supply chain (Ruiz-Garcia and Lunadei, 2011).

Traceability systems have been put in place mainly to enhance food safety, limiting the wide spreading of foodborne diseases consequently to a series of dangerous sanitary outbreaks which constantly arise and must be controlled on a large scale due to market expansion. Using rapid procedures to recall food products is very important for the social and economic impacts on our society. While institutions and legislators are mainly considering social aspects, firms have to also consider the economic costs that a recall could arise as direct costs for the retrieval process and indirect costs due to, for example, brand image losses.

By automated Information and Communication Technology (ICT) enabled systems, the granularity of traceability could be enhanced, as the available human resources limit traditional traceability systems, which require additional operations.

The online, automated, and shared inventories on the Electronic Product Code (EPC) network trace back the food product location in the food supply chain and trace forward other lots that could be contaminated. This aspect is very important for food safety in case of recall, when defective products must be removed from the market as fast as possible to limit potential hazards to human health.

Advances in Food Traceability Techniques and Technologies. http://dx.doi.org/10.1016/B978-0-08-100310-7.00005-3

Each stakeholder of the food supply chain has to face the market pressure and define strategies to identify areas of possible improvement to put in place strategies to obtain savings as, for example, lean manufacturing. The segregation and identification of bulk and packed food products, for example, can generate unexpected revenues due to waste reduction and quality assessment and define the price grading of production. Moreover, waste reduction positively influences sustainability and respect of the environment of the overall supply chain.

Besides, certain RFID applications play a role in continuously sensing and monitoring parameters which are highly involved in value creation in perishable product food chains.

2. Technology and Equipment for Radio-Frequency Identification

An RFID system consists basically in the communication between an *interrogator* (or reader) and a *tag* (or *transponder*) (Fig. 5.1). Both interrogator and tag contain an antenna, which transmits and receives in a communication channel where the reader sends information to the tag (*forward link*) and the tag answers to the reader (*reverse link*).

Tags can be categorized as *passive*, *semipassive*, or *active*:

- passive tags do not contain a battery and are powered by the energy harvested from the electromagnetic field emitted by the interrogator;
- semipassive tags or Battery Assisted Passive (BAP) contain a battery which is used only to power the tag IC and is not for communication; and
- active tags, which have a battery that supplies power to all functions.

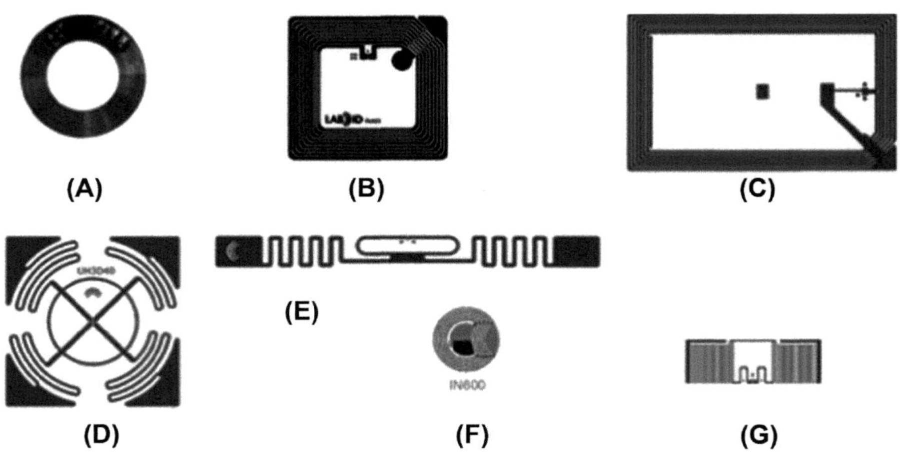

Figure 5.1 Passive tags at different frequencies. Type (A) is a Low-Frequency tag copper coil of an half duplex (HDX) transponder used in RF animal ear tag identifiers (134 kHz). Tag models (B) and (C) have HF square coils (13.56 MHz), while those identified by the letters (D), (E), (F), and (G) are UHF tag antennas of different shapes.
Courtesy of Lab-ID, Castel Maggiore, Bologna, Italy.

The simplest RFID tracking system is based on the attaching of a passive tag to the item which is then identified and acknowledged in fixed reading stations positioned in strategic logistic points where the product auto-identifies while it is displaced through RFID system installations and records its position during processing and logistic phases. RF identification can be performed also by mobile devices. On the market are available logistic handhelds equipped with internal RFID modules or external tools which facilitate the reader approach to the tag (eg, wand antennas for animal identification or external loop antennas to read through tagged food items arranged on shelves).

When RFID systems accessed industrial exploitation, frequency band standards have been defined and adopted. Most of the systems operate in Low-Frequency (LF), High-Frequency (HF), Ultra-High-Frequency (UHF), or microwave. Hereafter, we describe the technological aspects of systems operating at different frequency bands, explaining potential benefits and disadvantages in the food sector. As a rule, given that food matrices are typically characterized by high moisture, it has to be mentioned that, at a higher frequency, radio waves reduce their penetration and are more subject to reflection, adsorption, and detuning phenomena that can negatively influence the data transmission. Indeed, LF (125–134 kHz) systems have already found a number of fields of application when, eg, the tag is applied to organic tissues like in animal identification, according to the standards ISO 11784 and 11785, 1996 (International Standard Organization, 1996a,b), where the tag is applied directly to the ears, neck, or leg of the animal or is even injected or swallowed. Another element that causes electromagnetic field interference is metal: both HF and UHF radiation are shielded by even thin films of metal. For these reasons, LF systems are adopted in industrial environments when the tag has to be applied to (or is surrounded by) metal containers or machines. This is the case, for example, of the identification of metal containers, structures, or machines where food is handled, eg, in bulk state (bins, tins, boxes, etc.), metallic shelves, and other machines.

Both LF and HF tags employ inductive (magnetic) coupling: the reader/tag systems act as a magnetic transformer providing coupling between the current flowing in the reader antenna and the voltage across the tag. The reader electrically charges the tag, which demodulates reader signal and responds, backscattering an altered signal on the basis of data stored in the microchip.

LF tag antennas use coils, typically in copper. The voltage induced on an LF tag is proportional to the size and the number of turns of the coil. The reading range of an LF system depends on an interrogator and tag antenna size and their alignment. The reading zone is roughly comparable to the size of the reader antenna.

Tags at a High-Frequency band (13.56 MHz) are widely employed in ticketing, smart cards and labels, credit cards, antitheft, passports, asset tracking, and supply management applications. In HF bandwidth, several standards were developed, as, eg, ISO 15693 (International Standard Organization, 2010) for tracking, ISO/IEC 14443 (International Standard Organization, 2008) for smart and proximity cards, and ISO/IEC 18092 (International Standard Organization, 2013b) for Near Field Communication (NFC) among devices, as well as other standards for electronic money transfer. The reading range is comparable to that of LF systems.

In comparison with LF tags, HF tags are cheaper, as copper coil turns are fewer and, in some cases, laid down on a surface, using a production technique similar to that of a printed circuit board. In HF bandwidth, higher data rates allow a rapid interchange of data, which can support more complex protocols and elaborations as necessary, for example, for anticollision, data encryption, and high-capacity memory storage. In some cases, HF tags can carry enough data to maintain traceability without accessing databases.

UHF frequency covers the range from 300 MHz to 3 GHz. The majority of the passive logistic UHF tag is respondent to the Electronic Product Code (EPC) Gen 2 (GS1, 2015) standard, further integrated in the ISO Standard 18000–6 (last version: International Standard Organization, 2013a). These tags work in the frequency range 860–950 MHz, but three bands in this range are adopted in different geographical regions: 865–868 MHz in Europe, India, the Middle East, and Africa (ETSI); 902–928 MHz in the United States, South America, and some regions of Asia; and 950–956 in Japan.

On the tag antenna is attached and connected the microchip which contains the control logic and the memory of the tag. Transponder antenna shapes are often dipoles, which are very efficient but are more sensitive to orientation or have different designs to favor object identification at different operational distance, orientation, and sizes.

The EPCglobal Gen2 protocol regulates the interaction between the RFID reader and tags defining the physical and logical requirements for a passive-backscatter, interrogator-talks-first system. To read multiple tags simultaneously in a dense population, an anticollision algorithm, based on a dynamic framed slotted Aloha method, has been developed. As soon as the tag enters the RF field domain, it is enabled to accept the interrogator request. A selection of a group of tags can be made by a query command which communicates to the tag one among four sessions: S0, S1, S2, and S3. Each session is provided with a two-state inventoried flag (A or B). When the tag is energized, the persistence time of the inventoried flag value is indefinite for S0, S2, and S3, while it is in the range of 0.5–5 s for the S1 session. The algorithm implements a number of slotted frames equal to 2^{Q-1}, where Q is an integer number, which defines the slot dimension. Each of the selected tags randomly choose one among the available time slot (in the range 0, 2^{Q-1}) and will backscatter a 16-bit random number (RN16) when the interrogator calls for the specified time slot. Upon receiving a tag RN16, the interrogator acknowledges the tag with the same RN16.

If the RN16 is correct, the tag will finally backscatter its 96-bits ID. At this point, the tag commutates to the acknowledged state and also inverts the inventoried flag specified by the "Query" command.

Tags contain four memory banks: one is reserved, which contains kill and access passwords that allow locking and unlocking of the memory tag. The EPC bank (96 bit) contains the unique product code assigned by EPCglobal, the TID bank (which is permanently blocked and contains the unique tag ID number assigned by the tag producer), and a user memory (usually of 512 bits or more), which can store information that can also be rewritten.

At the ultra-high frequency band, the tag and reader communicate by radiative coupling. Each tag type has a minimum threshold power value required to activate the chip and guarantee the transmission and then can be activated only in a reading zone where the power is beyond this minimum threshold. A further requirement for a successful tag detection is the capability of the reader to detect the backscattered signal emitted by the tag, which is very weak with respect to the signal emitted by the reader.

The consequence is that the reading range could be limited both by the forward as well as by the backscattered signal, even if the sensitivity threshold is very low (−60 ÷ −80 dbm): in most cases the limits of the reading range is due to the forward link. UHF systems reading zone is very wide, but its shape can be very complex as, due to reflections, part of the field can propagate in other directions and back to the reader with the result that the reading zone could be not homogeneous. It is quite usual that the tag can disappear or reappear while it is moved away from the reader. As tag-reading efficiency is influenced by tag-reader mutual orientation, the type and shape of the tag should be carefully considered when integrating an RFID identification system. Each tag has a typical radiation pattern which gives information about best radiation directions of the tag, both in harvesting and backscattering energy. Passive tags, which are often dipoles when subject to a planar antenna interrogation, where the direction of electrical and magnetic field are constant (linear polarization), give optimal response if the electric field impinges on the vertical line between tag extremities. If the tag assumes different orientation, reading efficiency decreases and in some orientations is null. To cope with this problem, another antenna type is used, which electrical field alignment rotates (circular polarization) to also give the opportunity to be detected to tags in suboptimal orientations. Tag constructors also created tag antennas less sensitive to orientation as, eg, double orthogonal dipoles or other shapes.

In free space, reading distances in the ultra-high frequency domain are very high: in passive systems the operating reading range is of about 5 m (from 2 to 7 m), while active tags can communicate with the reader up to 18 m (Rizzi et al., 2011). In UHF data, transfer rate and memory capacity are high (up to 640 kbits^{-1} and some kbit, respectively).

UHF RFID technologies can be introduced for food traceability for item-level contactless automatic inventory management in processing plants, warehouses, distribution centers, and point of sales. As UHF tags have a low cost, they are widely used in supply chain management. Due to capabilities in solving multiaccess problems using the EPC Gen-2 interrogation protocol based on slotted Aloha algorithm, as well as improved performance in far-field identification at affordable costs, UHF is expected to grow and will be introduced in the near future in markets where RF technologies were limited in the past. Main criticalities in the use of UHF systems in food supply chains are due to the food chemical and physical properties as well as the harsh working environment (Barge et al., 2013b), as will be treated in a following Section 4. Research is continuously improving performances of tags and readers, as well as their configuration to obtain a reliable identification in different contexts.

In semipassive tags (Battery Assisted Passive or BAP) the internal circuit is autonomously powered, so they do not rely on their efficiency in harvesting power from the reader field. For this reason the signal emitted by the tag is much stronger if compared to a passive tag. BAPs can also be active in absence of an interrogator, their reading zone is very large, and they are more easily detectable by the readers in a critical environment.

Both semipassive and active tags can be powered even if they are not in a reader reading zone.

Active tags have an onboard power supply, an active transmitter, and a receiver (which are not present in semipassive tags) and can begin a communication session with a reader when certain conditions are sensed as, for example, a fixed temperature has been reached. The autonomous communications hardware enables active tags to be used in places with large amounts of metal, which is typically very unfriendly to passive and semipassive transponders. One of the criticalities of battery-assisted tags is the limited life of the power systems.

An active tag can also communicate with a network of other active tags and a variety of sensors (eg, temperature, humidity, light exposure, pressure, pH) and microcontrollers.

Even if the price per unit is very high in comparison with other RFID tags, active tags offer the possibility to be part of Wireless Sensors Networks (WSN), with significant cost reduction for wires due to permanent installations (Badia-Melis et al., 2015a).

3. Advantages of Radio-Frequency Identification in Food Traceability

3.1 Process Control

By means of RFID the information about the status of the products in the flow from processing to delivery can be continuously updated and a complete management of the process can be put in place, including aspects of traceability, quality, and safety of food products.

When traceability information flow is maintained inside the boundaries of the single firm, we refer to the term *internal traceability*, by which we obtain, at different level of detail, the information about the state of units of raw materials and products being processed and stored.

A well-structured information system, which can track the physical flow during the production process, can be used to define management objectives and dynamic production planning (Dong et al., 2006) to improve the overall performance of the food supply chain, enhancing value and reducing waste in terms of labor, space, materials, and asset allocation. The advantages are extended to the whole supply chain if collected information is shared in common ICT platforms among all the stakeholders accessible by different users at the appropriate privacy level (often referred as *external traceability*).

The individual choosing the tag at processing must consider if it will be reused in external processes or if it leaves the food item when it leaves the processing plant. In the first case, to reduce costs, the choice is oriented to disposable cheap transponders (eg, UHF labels). In the case of recycling, more rugged tags coated with plastic or resins materials which can be sanitized and reused attached directly to food items (eg, matured cheese wheels, Barge et al., 2014) can be used.

In some cases, RFID systems can be used to sort lots of raw or finished products at different sites inside the production plants, manufacturing products at the desired quality grade in function of external factors (destination, price, taste of the customers), such as, for example, for coffee or cocoa lots.

RFID-based supply chain will fit the concept of "smart factory" where internal manufacturing is dynamically planned and driven by external factors (eg, taste of the customer, environmental respect, ethical issues, seasonality, etc.).

Production planning strategies, especially in the case of perishable and seasonal products, can also benefit from RFID adopting strategies as first-expiry-first-selling or pricing in function of quality parameters as freshness (Herbon et al., 2014).

3.2 Warehouse and Logistics

Due to the capabilities to face multiaccess problems using the EPC Gen-2 (GS1, 2015) interrogation protocol based on slotted Aloha algorithm, as well as improved performance in far-field identification at affordable costs, UHF systems have high potentialities in warehouse management. This standard is nowadays the reference for tags used in food logistics.

It has been reported by Lim et al., 2013, who presented the vast range of possibilities of ameliorating warehouse management, that the inventory and storage management is the largest domain treated by RFID literature.

Common UHF logistic passive tags are composed of an antenna usually made of aluminum or copper on an inlay which could be a polyethylene terephthalate (PET) or Polyvinyl chloride label and is usually adhesive, or embedded in other materials.

RFID labels can be attached to food items at different level of traceable unit size: a whole pallet containing several units, on secondary packaging, or on single items, at different costs. The use of EPCglobal standard allows supply chain stakeholders to track information, at a detailed level, about processed, shipped, and globally distributed goods in a unified protocol.

In warehousing, the checking of identity and quantity of items unloaded from the transport carriers can be fully automated. Items can be depalletized, packed, and repacked and put away for storage until they are picked for the next delivery without losing their identity.

Bigger semipassive tags with enhanced reading performance could be placed on bins, logistic pallets, and other containers even in metal where food is processed, stored, or transferred inside the warehouse. Other small containers (eg, plastic crates for fresh food such as fruit, vegetables, cheese, etc.) can be identified by recyclable, cheaper transponders.

Reading systems are fixed or mobile and are placed in strategic points of the logistic process. Fixed reading stations are often configured in portals (or gates) equipped with several antennas which are connected to the same reader. Antenna positioning and configuration is crucial to obtain a complete coverage and accuracy in inventorying. These portals are placed, for example, in supplying points for production machines, in the entry of storage cells, in loading/unloading platforms, in packaging lines, etc. In warehousing, RFID systems can be connected to warehouse management middleware to give correct and real-time information about item positioning to find optimal handling solutions and load planning considering the available resources.

To detect where and when items are stored and register vehicles (eg, forklifts) paths, the facility can be conceptually divided by referenced grids where readers are placed in strategic points to monitor the item's displacement using antenna buried in the floor and placed in the ceiling and racks which can be used as a location sensors. Measurements of material handling are a powerful instrument to define key performance indicators to improve Material Handling Engineering (MHE).

Picking orders for delivery collection can be established and dispatched to forklifts or to operators which wear or handle an RFID Personal Digital Assistant (PDA). The capability of identifying each single item during picking operations also controls deliveries of mixed loads carrying various types of products, reducing errors and possible protests.

3.3 Retail

To establish a good position on the market, retailers are required to provide high product availability, the best quality at a competitive cost. One strategy to increase sales is to correctly replenish shelves to satisfy customers buying food.

On-Shelf Availability (OSA) is composed by shelf, store, and warehouse availability, which means that a product can be not present on the shelf but is in the store (in a wrong place) or is not on the shelf nor in the store but in the warehouse, or, in the worst case, is not on the shelf, nor in the store, nor in the warehouse (Thiesse and Buckel, 2015). One of the strategies to respect a maximum degree of on-shelf availability is to enhance the safety stock level in the back room and inventory, but in the case of fresh food and expiry products, this results in a decrease of quality and freshness. A system preventing the out of shelf by self-order on the basis of recurrent automatic inventory and alerts in case of shrinkage can add value to fresh food product supply chains (Bertolini et al., 2013). Auto-ID technologies can contribute to collaborative replenishment policies in food cold chain among suppliers and retailers, which, as was demonstrated, can contribute to reach an equilibrium where benefits are distributed among the supply chain stakeholders (Lan et al., 2014). RFID hardware is capable of detecting product movements between the backroom and the sales floor in a store at item or case level. The tag can be applied in shops as well as directly in processing. Recently we have assisted a huge growth of RFID adoption in very competitive apparel players (eg, Decathlon, 2015, and Walmart experiences, Vowels, 2006), where it is used, among other scopes, to guarantee shelf replenishment in order to fulfill the request of fast-buying customers.

3.4 Cold-Chain Monitoring

As the maintaining of the correct temperature determines the value, the overall quality (including sensorial), and the safety of perishable refrigerated or frozen food, the temperature monitoring during the time the food spends in cold-chain phases (processing, precooling, storage, transport) is precisely controlled. Temperature in excess of critical thresholds can seriously affect shelf life, even if the time of exposure is limited. In safety control by RFID active systems the worst case, in which the spoilage time-temperature level is reached, can be monitored on the whole supply chain, with benefit to distributors, retailers, and consumers (Grunow and Piramuthu, 2013). In the logistic industry, costly systems are traditionally mounted in rooms and vehicles where the food is maintained and transported with limits due to the lack of data or reliability. One single logistic node can disrupt the reliability and usefulness of the cold chain respecting the whole supply chain. Active or semipassive HF or UHF (Gen 2) RFID sensors can replace traditional data loggers in the monitoring of temperatures and gradients in refrigerated packaging or rooms for perishable product logistics. By means of sensor-enabled data loggers, which are physically displaced in food proximity, the temperature and information collection is independent from handling and can be registered in any checkpoint wirelessly. RF data loggers are disposable or recycled on the supply chain. As the dimensions and weight of RFID temperature sensors allow the insertion of transponders attached or inserted in the products, monitoring is continuous during the handling along the whole supply chain. Passive tags can be used as well, with some limitations, but with the advantage of a lower cost and the absence of a battery which reduces tag life; they do not have an internal clock and have to communicate with a reader to log time. Another example of disposable tags are blister pack-type semipassive transponders manufactured very cheaply on a flexible substrate; they can be activated and begin to log.

The availability of cheap wireless miniaturized temperature sensors have facilitated studies on the temperature gradients in trucks loaded with perishable goods (Jedermann et al., 2009) or in cases containing food items (Narayana et al., 2014). The same technologies are applied in the pharma industry.

Predictive models on food spoilage on the basis of temperature data determine the remaining shelf life and manage the logistic strategy moving from the traditional first-in-first-out to the first-expired-first-out selling.

Advances in the domain of RF sensors have led to measuring and acquiring other physical and chemical parameters, such as moisture, and other chemical compounds produced during storage and logistic phases (Bhadra et al., 2015).

3.5 Smart Packaging

A new generation of buyers, the so-called "millennial", is facing the food market: this generation is accustomed to buying online, is connected to social media, is aware of environmental and ethical issues, and is accustomed to making decisions based on feedback and information on the web. Food packaging can be used as a communication medium to fulfill the expected information of this new category of buyers. As it has been demonstrated that buyers usually do not pay attention to labels with an excess of text, other communication technologies, such as automatic identification and electronic media, can vehicle additional information to the consumer. QR code, as well as RFID, can be used to access supplementary information by means of cell phones and/or internet connections referred to the item identified by means of the packaging, with the aim to encourage the purchasing intention.

Smart packaging can be divided in two major categories: active packaging and intelligent packaging. In a definition accepted by several authors in literature, *intelligent packaging* is a system able to monitor the condition of packed foods to give information about the food quality or its environment during transport and storage (López-Gómez et al., 2015).

Conversely, *active packaging* can be defined as a system in which the product, package, and the environment interact in a positive way to extend the shelf life or maintain or improve the condition of packaged goods (Biji et al., 2015; Restuccia et al., 2010). RFID are mostly used in intelligent packaging systems to provide the consumer with information on parameters such as relative humidity, light exposure, pressure, pH, and temperature. Semipassive temperature loggers eventually equipped with external probes can be used.

3.6 Anticounterfeit

Traceability can be exploited to prevent, deter, and eliminate illegal, unreported, and unregulated (IUU) productions. The potential of traceability systems to prevent fraud in a supply chain derives from its ability to trace the history, process, and location of an entity by means of recorded identification and to provide unique and robust identification of items and/or lots (Dabbene et al., 2014). Among other technological systems (eg, security printing, watermark, holograms, tattoo, etc.), track-and-trace technologies such as bar code and RFID are considered anticounterfeit devices (Li, 2012). The position of the item, uniquely identified by a serial number stored in a tag, is registered in a strategic supply chain node, generating an electronic pedigree (e-pedigree) that contains data about product authentication from manufacturers to the point of sales (Choi et al., 2015). In the food sector, for instance, these systems are applied to certify Protected Designation of Origin (PDO), Protected Geographical Indication (PGI), and other high-quality supply chains protected by brands (wine, cheese, meat, ham, etc.). Special RFID labels are provided which have to be expressively destroyed at purchasing (eg, at the counter) or when the product is consumed (eg, opening the box or the bottle) to avoid multiple uses of the same label. In some cases, only the energy harvesting part of the tag is altered, leaving the internal loop intact to keep the stored information but avoiding the further detection in the far field.

There are different available techniques to check RFID authentication: accessing a centralized database, using an offline object authentication, or by track-and-trace methods. Centralized database checking, which is a reliable technique used, eg, in ticketing applications, relies on online product authentication in real-time by a plausibility check of the unique code performed over the Internet. These systems are secure and very efficient, but require a permanent Internet connection and the need of maintaining a (reliable) back-end database. Offline systems include encrypted tags where authentication is performed locally, for instance, by cryptographic algorithms embedded in handheld devices, without the need of network or remote connections. There are still some open problems, as, eg, how to prevent label (RFID) cloning in offline authentication procedures and how to permanently link RFID labels to food products to impede the repeated application of the same identifier to many products.

In any case, the security level of most of the anticounterfeit systems can be enhanced using and precisely checking information collected along the supply chain. As the cost for traceability is already included for other purposes, the use of track-and-trace information can strengthen anticounterfeit systems for protecting from fakes without increasing prices or compromising competitiveness.

4. Constraints of Radio-Frequency Usage in Food Traceability

4.1 Effect of Food Composition and Packaging on RFID Reading Performance

One obstacle to the introduction of UHF in the traceability of food products is the technical difficulty to obtain exhaustive and reliable identification of food items in critical production environments. This has led some end users to have a low confidence about the reliability of the system.

The reliability in UHF RFID system object identification in manufacturing and logistics depends on the tag performances when it is attached to objects. In fact, the presence of products near the tag can modify the overall tag antenna impedance altering the reader-to-tag matching. This can negatively affect the minimum power to activate the tag, influence signal modulation, detune and narrow the operating bandwidth, and modify the backscattered signal.

To successfully read a passive RFID tag the circuitry must be energized by enough power. In free space, the power received by the tag antenna can be expressed using the following Friis equation:

$$P_{RX} = P_{TX} G_{TX} G_{RX} \left(\frac{\lambda}{4\pi r} \right)^2 \tau \rho_T \qquad [5.1]$$

where P_{RX} is the power received by the tag, P_{TX} is the power transmitted by the reader, G_{TX} and G_{RX} are the gains of the transmitting (reader) and receiving (tag) antennas, respectively, r is the distance among transmitter and receiver, λ is the wavelength of the RF signal, τ is the polarization factor, and ϱ_T accounts for the proper antenna-to-chip coupling.

Each tag type has a minimum threshold power value P_{min} required to activate the integrated circuit (IC) and to guarantee the transmission. The reader must detect the backscattered signal emitted by the tag, which is modulated according to the Gen 2 protocol. The same type of equation basically governs this reverse path. The presence of food products near the tag could modify the overall impedance of the tag antenna, altering the antenna-to-tag matching, represented in (1) by ϱ_T. An improvement in readability has been obtained by tag producers by continuous improvement of tag IC sensitivity.

The material, size, and shape of food items to which the tag is attached strongly influence the energy harvesting capability of the tag, as the presence of mass of materials near the tag could modify the overall tag antenna impedance altering the antenna-to-tag matching. RFID identification of dry and porous food products such as biscuits and snacks is not critical, as these are very permeable to electromagnetic waves (Barge et al., 2013b); in these cases, tags can be identified even if they are hidden into the food carton containing the packed products (secondary packaging). Materials containing a high quantity of lipids like oil or sweet chocolate spreads are less critical than liquid food products containing a high quantity of water that have strong negative effects on the tag capability of energy harvesting (juices or beverage) (Barge et al., 2013b). Even food packaging can limit the energy delivered to the IC of the transponder; for example, RF radiation is especially affected by the presence in the packaging of metal, such as in the case of aluminum foils, but also glass has a negative influence on readability of the attached UHF labels for some ions contained within.

The performance results in terms of reading distance, minimum power of activation, and readability of tags of different shapes and sizes on the empty packaging of food can be dramatically different (Bogataj, 2012).

As identification of food products can be very critical, it is then very important to find, for each type of food and packaging as well as their combinations, optimal configurations which can provide high inventory accuracy.

The success in object identification in manufacturing and logistics by RFID systems depends on the correct choice and coupling of the transmitting and receiving devices. The correct configuration of the reader regarding tuning of the parameters governed by the C1G2 protocol, which influences signal decoding as well as the query commands through reading cycles during inventory, is also strongly influencing the overall precision of identification, reducing the error rates. One problem that happens in RFID systems integration is collision. There are different types of collision problems that can cause errors: tag-to-tag collision and reader-to-reader-collision. The first occurs when a dense population of tags is present in the coverage area of one reader. In fact, a vast number of tags originate simultaneous broadcasting of signals which have to be arbitrated by anticollision algorithms (eg, Aloha, see Section 2). Especially in dynamic conditions at high speeds, the time a tag spends in the reading area is limited and arbitrating could enhance errors.

On the contrary, reader-to-reader collision occurs when two or more readers (and antennas) are active in the same work space. Methods for avoiding reader collisions are based either on time sharing of reader transmission on the different antennas, on setting spatial distance among readers, on the assignment of different frequencies, and on the use of electromagnetic shield materials (Carbunar et al., 2009). As readers are very sensitive in receiving backscatter signals (up to −70 db or more), if a reader electrically activates a tag, another reader (even at a large distance) can capture the emitted signal.

4.2 Traceable Unit Size

Another important aspect is the size of the unit of product that is associated with a tag and has to be individually tracked. This amount of product has been defined by Moe (1998) as Traceable Resource Unit (TRU), where it indicates a "unique unit, meaning that no other unit can have exactly the same, or comparable, characteristics from the point of view of traceability." This concept was later formalized in the ISO Standard 22005/2007 (International Organization for Standardization, 2007), where the notion of lot is defined as a "set of units of a product which have been produced and/or processed or packaged under similar circumstances." The size of TRU determines the granularity of the traceability system: a finer granularity allows for acting at a more detailed and range-limited level in the case of a possible recall, but also a higher operational cost in terms of amount of data to be collected and processed. Among operation costs, the identification and monitoring costs (eg, labels, tags, readers, etc.) can constitute an important issue in relation to the type of technology employed for the unique identification of products (Aiello et al., 2015).

The optimal granularity level is very difficult to determine, since it depends on product type and customer. Granularity is strictly related to the different measures and criteria for the performance evaluation of traceability systems (Dabbene and Gay, 2011). Performance criteria are aimed to quantify the cost and the impact of a possible product recall, which is affected, among others, by the size of the TRU, the degree of dispersion of raw materials and intermediates, and the speed of the traceability system to supply information and manage the crisis.

Unfortunately, in most parts of current supply chains, the granularity traceability level is not the result of a formal analysis and optimization study, but it is principally the consequence of a combination of tradition, short-term convenience, and the use of available facilities (Dabbene et al., 2014). Typical traceability levels are pallet, case, or item. In general, coarse granularity can be used when the risk of contamination is low or when requirements for controlling production processes are less stringent, while finer granularity is recommended otherwise (Karlsen et al., 2012).

There are food products, which are implicitly managed as single items, as, eg, cheese wheels, hams, and wine bottles, whose cost is rather high and justifies the adoption of an RFID tag to each piece (Badia-Melis et al., 2015b). In some cases, tags are applied to the container or package (eg, bottles of wine), while in other cases, they have to be hooked directly to the food product (see eg, Barge et al., 2014; for the case of cheese wheels). Meyer et al. (2009) pointed out that the gradual reduction in the cost of RFID technology would made it possible to track and trace not only expensive products but also small and cheap ones. According to Piramuthu et al. (2014), the choice of tagging or not tagging single items should not be established on the basis of the price of the individual items and the relative cost of RFID tags. The benefits of incorporating item-level RFID tags at the store level cannot be fully achieved unless there is an appropriate and consistent level of information generated for/by these items. The fraction of not-tagged RFID items will threaten the overall efficiency of the store by an amount that exceeds the ratio of tagged/not-tagged items. However, costs, in some cases, remain one of the major obstacles in the diffusion of the RFID technology in agriculture and the food industry. In many applications the size of the TRU, and therefore the granularity, then is chosen to make tag cost sustainable with respect to the food item value. This is, for example, the case of carton boxes and secondary packaging units. These aspects have been recently analyzed considering advantages and disadvantages of using RFID at increasing traceability granularity (item, case, or pallet level) statistically considering the errors due to incomplete reading at different levels (Thiesse and Buckel, 2015).

4.3 Food Item Inventory in Static and Dynamic Conditions

Using fixed readers, the detection of a tag or of a population of tags can happen in static scenarios, when, during interrogation, all tags are still, or dynamic conditions, while objects are moving along a path in the reader coverage area. Most traceability systems (LF, HF, and UHF), if correctly configured, can detect both static and moving tags. Reading rates (% of tag correctly detected/tag present in the reading zone) depend on different factors and their interactions, as, eg, the reading range, the size and position of antennas, the mutual orientation, the reader RF power output, the transit speed, and the number, size, and type of tags. In several cases, the difficulties in introducing RFID were due to incorrect applications of RF systems which, early on, caused the loss of reputation and reliability.

Several studies were conducted to evaluate the correct configuration of readers and tags to obtain enough precision in reading tags. At all the considered frequencies, one important aspect to consider implementing in RFID systems is reader and tag mutual orientation. In LF and HF shapes and reading distance of reading area is dependent on the alignment of the electromagnetic coupling of copper loops. When an LF or LF gate is composed by two antennas the direction of the electric current can be inverted in the two loops, obtaining a different shape of magnetic field. In any case a "safe reading zone" has to be found inside which the tag is likely to be read with a high degree of confidence (Barge et al., 2013a). In LF transmission, speed is very low and the application of any anticollision algorithm is very difficult because of the time employed to perform subsequent queries. LF gate mostly detect one tag at a time, which can be also an advantage when a queue of objects has to be detected in its original order. In HF frequency, multiple detection could be also performed, but in dynamic conditions the speed of transit in front of the antenna should be very low. As the majority of UHF tags are essentially long and thin, they behave like a dipole and receive the maximum possible signal if oriented along that field, while they receive no signal at all if it is perpendicular. While it transits through portals, the tag on a food item should be positioned in a way that it is correctly oriented for its identification. Dynamic conditions are not always negative as tag orientation during motion changes continuously, recovering tags which were missing as they were not first optimally positioned. Therefore high-speed transit can reduce reading efficiency.

4.4 Recycling Issues

The presence of different materials composing tags, as pieces of metal from antennas, electronic material (microchip), adhesives, PET, and high-density polyethylene (HDPE) from labels, complicate the recycling process. In the case of steel recycling, copper contamination can affect the quality of the obtained steel (Kumar et al, 2009).

In animal identification, slaughterhouses have to be equipped with systems that allow separating RFID rumen boluses (which are made of very strong ceramic, glass, and copper) from animals' entrails. When tags are directly attached to food material (eg, cheese) the tag must be safely detached, eventually facilitating using tags in contrasting colors. All these aspects must be carefully considered before integrating RFID systems in the food sector.

5. Examples and Applications of RFID in Food Supply Chains

5.1 Livestock and Meat Supply Chain

Electronic Animal Identification (EID) was first introduced in the mid-1970s (Eradus and Jansen, 1999). The advantages of using RFID for individual identification were early understood and the spreading of this technology in the animal farming was very fast. Both active and passive tags eventually coupled to sensors recorded data for monitoring performance, health, reproduction status, feeding, and milk production and also interacted with automated devices of the farm as feeding stations, milking robots, etc.

As individual identification became very important for disease eradication, electronic identification was mandatorily introduced for some species (beef calves, sheep and goats, and pigs) by the safety authorities of many countries (eg, Europe, Australia, New Zealand, Canada, and the United States). In 1996, two standards (ISO 11874 for the 64-bit code structure and ISO 11785 for the combined FDX-HDX interrogation protocol) working at 134.2 kHz were published and widely adopted.

LF RFID tags can be applied both externally and internally to the animal. Ear tags are the more popular among the external applied tags; other types can be tied to the neck, legs, and wings. For ruminants, a tag inserted in a ceramic device, which can be swallowed and further internally retained in the reticulum of the animal, was developed by a Spanish research group in the 1990s (Caja et al., 1999). Other transponder models, used, eg, in the identification of horses and pet animals, are injectable.

The choice of the identifier model should consider the length of the animal's life, the animal behavior, and the layout of barns and structures where the animal has to be identified. Static readers are usually placed in gates or corridors, while mobile handhelds, eventually added as an extension of the antenna to more easily approach the animal, are portable.

Due to past food scandals (eg, Bovine Spongiform Encephalopathy or BSE), meat requires the collection and registration of a huge amount of data, which are stored in different databases, generally not linked.

At breeding farms, animal data are input in official national databases following national standards; this allows an easy interconnection within official databases at the local level. Using auto-identification by RFID systems and real-time data interchange, registration about birth, transport, and slaughter, is continuously updated. In meat supply chains, ICT frameworks, where mandatory, as well as voluntary (eg, for quality labeling), traceability data are stored in distributed architecture, can improve the actual traceability information system increasing speed, accessibility, reliability, and safety (Barge et al., 2013a).

As in the worldwide-accepted RFID standards the animal identifier contains only the National ID code, animal information can be retrieved only accessing a shared database. Communication between remote hosts based on eXtensible Markup Language/Simple Object Access Protocol framework can be put in place for automatic animal registration and real-time data interchange with the national bovine register database when a traced event occur (eg, animal entering/leaving a farm or being slaughtered). Nevertheless, at the international level the identification code numbering is different, and the communication between the national official databases is difficult due to lack of homogeneity. In spite of the many efforts that have been made, European official databases cannot actually be used for business electronic transactions in the case of imported calves. A common standard for different worldwide data sharing is necessary, as well as data communication among the different national registers.

The availability of a web-shared information system, in private, shared platforms, will allow the adoption of a common traceability path involving small producers and slaughterhouses, which can maintain traditional data recording, even if the animal is identified with both an electronic and a visual device.

Due to the technical limits of systems operating in LF, the use of HF, as well as UHF, systems have been investigated. This is the case, for example, when the breeder or researchers want to identify the behavior of a group of animals simultaneously, such as to monitor the feeding habits of piglets (Hessel et al., 2008).

5.2 Fish

Fish quality is highly dependent on freshness and therefore on optimal condition in cooling during transport, especially in cases of long trips. In the fish industry, which has rapidly internationalized and expanded international trade, the consumer demands high safety and quality levels, as well as transparency in the supply chain.

Positioning GPS/GPRS systems coupled with a network of passive RF identifiers and sensors can be used to control chill or frozen fish supply chain, allowing intervention in case of interruption in the cold chain, avoiding losing product value for deterioration.

Active or semipassive RFID tags can be applied inside refrigerated containers and storage rooms to monitor temperature. These types of monitoring are cheaper than traditional onboard sensors, and data can be accessed wirelessly, eg, without opening trucks in the loading/unloading area.

RFID HF (Abad et al., 2009), as well as UHF, have also been envisaged to monitor temperature and humidity sensors in the packed fish supply chain with tags inserted inside the polystyrene transport box systems (Trebar et al., 2013).

5.3 Cheese and Dairy Products

In farm dairies, radio-frequency systems for livestock traceability are used coupled to automatic milking systems to trace milk lots delivered to cheese manufacturing plants, as described in the previous chapter. Milking yields, oestrus detection, rumination, and individual grain feeding can be monitored individually using RFID coupled to other sensing devices and actuators.

The traceability of milk presents the same criticalities as other liquid or bulk products, which are usually stored in tanks and progressively merged during the production process (Comba et al., 2013). Farm milk records (farm number, size of the lot, date, and hour of delivery) as well as analytical determined quality parameters (eg, pH, presence of antibiotic residues, protein and fat content, somatic cell number, and total microbial count) have to be recorded. De Las Morenas et al. (2014), proposed an RFID solution to trace and monitor the temperature of milk sample vials collected in farms during the transport, which improves reliability in milk analysis for high-quality milk bonus payment.

During the steps of merging milk stored in silos to the first operations in cheese production, traceability can be maintained at the processed milk lot level since the cheese reaches its solid state. Then, Auto-ID technologies can be put in place to identify smaller cheese lots or single cheese wheels.

Cheese identification is very critical due to the environmental conditions during the manufacturing phases (curd molding, pressing, dry or brine salting, ripening) and frequent product handling (eg, daily turning, brushing, and scraping during ripening). For these reasons, traditional identification methods cannot be used for item-level identification. Research studies have demonstrated (Regattieri et al., 2007; Pérez-Aloe et al., 2010; Papetti et al., 2012; Barge et al., 2014) that, if properly coated by special resins or plastic materials approved for food contact, RFID passive (LF or HF) transponders could be directly inserted in the cheese rind (Fig. 5.2). In this case the tag can be reutilized after a proper sanitization, reducing costs.

In the case of fresh, small, portioned, and packed cheese the RFID tracking can be applied identifying assets and packaging by RFID labels.

In seasoned cheese, by monitoring the stock level in storage rooms, the scheduled processing can be driven by the demand and on the basis of cheese shelf life avoiding overproduction, which causes a decrease in cheese quality, the failure in respecting established rules for certified cheese such as PGI and PDO, the loss of value, and avoiding costs due to overtime labor.

Figure 5.2 Cheese wheels identified by means of circular LF or HF tags. Transponders are applied under a casein disk during curd molding on the side (up, left), or on the cheese wheel face (up, right). During ripening the tag remains included in the cheese rind and can be detected by different handled models when placed on shelves in the storage room (down, left, and right figures).

5.4 Fruit and Vegetables

For fruit, RF traceability can be performed in the field as well as plants where calibration, washing, selection, and storage take place by tracking assets as, for example, bins used to transport fruit lots. Afterward, plastic recyclable boxes where tags are incorporated during molding can be used in storage, handling, and transport, controlling the inventory and the state of delivery (Mainetti et al., 2013). Perishable food traceability can be coupled to storage cell RFID sensors, which acquire continuously ambient parameters like, eg, temperature, humidity, and atmosphere composition (Amador and Emond, 2010). Sensors able to monitor fruit quality can be embedded in RFID readers to assess the conservation stage of fruit, eg, apples, directly linking this information to a traceability platform via RFID tags (Vergara et al., 2007).

Other applications are proposed (Ampatzidis and Vogioukas, 2009) which use on-field identifiers in orchards for yield mapping and fruit weight recording by each tree with the associated harvested fruit bin.

5.5 Potted Plants and Nursery

Especially in northern regions, the market of aromatic plants is growing, and sales of potted plants are usual in main distribution chains. Pot identification by RFID allows tracking in greenhouse operation and automation, logistics, inventory and order management, and counterfeit limitation.

The tag can be hidden in the pot or tied to the plant or the soil as a visible RFID label which also carries printed information (Tortia et al., 2010) (Fig. 5.3).

Some authors have prospected the adoption of an electronic identity card for grapevine clones (Luvisi et al., 2012) and other plants' propagation material, inserting a transponder directly in plant tissues (Luvisi et al., 2011) with the aim to guarantee disease-free plants.

6. Future Trends

As it is difficult to separate and attribute benefits to each upstream and downstream partner, the opportunity of RFID-integrated information systems adoption is often felt as an unbalanced cost afforded only by some stakeholders without taking advantages. Future economic models centralized on the whole supply chain will allow supply chain actors to evaluate advantages and disadvantages of RFID identification and make their decision considering the opportunities this technology can offer.

One of the challenges for RFID systems producers and integrators in the food sector is to enhance RFID efficiency and accuracy in food product identification, which is very hard due to their composition and the critical environment where they are often processed, stored, and transported. This aspect is crucial, as RFID identification should cover a wide range of food products to justify the adoption in large-scale applications, such as in logistic chains. The cost of the tag, the battery life length in active tags, the improved hardware and software,

Figure 5.3 Example of RF identification of a potted plant. The UHF dipole tag is inserted in a customized label (up, left) tied to the soil by a holed lower part that allows the roots of the plant, during its growth, to penetrate and surround the label which results in strong links to the plant and soil (up, right). The detection can be performed during trolley rotation on wrapping platforms (down, right) by four stationary UHF antennas.

and the harmonization of standards are research subjects still in progress and worth future engagement. Moreover, strategies limiting potential errors due to missing tag reading have to be stochastically considered and included in implementation frameworks to make the traceability system robust and reliable. Other domains and applications which could benefit from this technology have to be discovered to evaluate its full potential.

References

Abad, E., Palacio, F., Nuin, M., Zárate, A.G., de, Juarros, A., Gómez, J.M., Marco, S., 2009. RFID smart tag for traceability and cold chain monitoring of foods: demonstration in an intercontinental fresh fish logistic chain. Journal of Food Engineering 93 (4), 394–399. http://dx.doi.org/10.1016/j.jfoodeng.2009.02.004.

Aiello, G., Enea, M., Muriana, C., 2015. The expected value of traceability information. European Journal of Operational Research 244, 176–186.

Amador, C., Emond, J.P., 2010. Evaluation of sensor readability and thermal relevance for RFID temperature tracking. Computers and Electronics in Agriculture 73 (1), 84–90. http://dx.doi.org/10.1016/j.compag.2010.04.006.

Ampatzidis, Y.G., Vougioukas, S.G., 2009. Field experiments for evaluating the incorporation of RFID and barcode registration and digital weighing technologies in manual fruit harvesting. Computers and Electronics in Agriculture 66 (2), 166–172. http://dx.doi.org/10.1016/j.compag.2009.01.008.

Badia-Melis, R., Ruiz-Garcia, L., Garcia-Hierro, J., Villalba, J., 2015a. Refrigerated fruit storage monitoring combining two different wireless sensing technologies: RFID and WSN. Sensors 15 (3), 4781–4795. http://dx.doi.org/10.3390/s150304781.

Badia-Melis, R., Mishra, P., Ruiz-Garcia, L., 2015b. Food traceability: new trends and recent advances. A review. Food Control 57, 393–401.

Barge, P., Gay, P., Merlino, V., Tortia, C., 2013a. Radio frequency identification technologies for livestock management and meat supply chain traceability. Canadian Journal of Animal Science 93, 23–33. http://dx.doi.org/10.4141/cjas2012-029.

Barge, P., Gay, P., Merlino, V., Tortia, C., 2013b. Effect of packaging on radio frequency identification of food products. International Commission of Agricultural and Biological Engineers, section V. In: CIOSTA XXXV Conference "From Effective to Intelligent Agriculture and Forestry", Billund, Denmark, July 3–5, 2013.

Barge, P., Gay, P., Merlino, V., Tortia, C., 2014. Item-level radio-frequency identification for the traceability of food products: application on a dairy product. Journal of Food Engineering 125, 119–130. http://dx.doi.org/10.1016/j.jfoodeng.2013.10.019.

Bertolini, M., Ferretti, G., Vignali, G., Volpi, A., 2013. Reducing out of stock, shrinkage and overstock through RFID in the fresh food Supply chain: evidence from an Italian retail pilot. International Journal of RF Technologies 4, 107–125.

Bhadra, S., Narvaez, C., Thomson, D.J., Bridges, G.E., 2015. Non-destructive detection of fish spoilage using a wireless basic volatile sensor. Talanta 134, 718–723.

Biji, K.B., Ravishankar, C.N., Mohan, C.O., Srinivasa Gopal, T.K., 2015. Smart packaging systems for food applications: a review. Journal of Food Science and Technology 1–11. http://dx.doi.org/10.1007/s13197-015-1766-7.

Bogataj, U., Maček, M., Muck, T., Klanjšek Gunde, M., 2012. Readability and modulated signal strength of two different ultra-high frequency radio frequency identification tags on different packaging. Packaging Technology and Science 25 (7), 373–384.

Caja, G., Conill, C., Nehring, R., Ribó, O., 1999. Development of a ceramic bolus for the permanent electronic identification of sheep, goat and cattle. Computers and Electronics in Agriculture 24 (1–2), 45–63.

Carbunar, B., Ramanathan, M.K., Koyutürk, M., Jagannathan, S., Grama, A., 2009. Efficient tag detection RFID. Journal of Parallel and Distributed Computing 69, 180–196. http://dx.doi.org/10.1016/j.jpdc.2008.06.013.

Choi, S.H., Yang, B., Cheung, H.H., Yang, Y.X., 2015. RFID tag data processing in manufacturing for track-and-trace anti-counterfeiting. Computers in Industry 68, 148–161. http://dx.doi.org/10.1016/j.compind.2015.01.004.

Comba, L., Belforte, G., Dabbene, F., Gay, P., 2013. Methods for traceability in food production processes involving bulk products. Biosystems Engineering 116 (1), 51–63. http://dx.doi.org/10.1016/j.biosystemseng.2013.06.006.

Dabbene, F., Gay, P., 2011. Food traceability systems: performance evaluation and optimization. Computers and Electronics in Agriculture 75, 139–146. http://dx.doi.org/10.1016/j.compag.2010.10.009.

Dabbene, F., Gay, P., Tortia, C., 2014. Traceability issues in food supply chain management: a review. Biosystems Engineering 120, 65–80. http://dx.doi.org/10.1016/j.biosystemseng.2013.09.006.

De Las Morenas, J., García, A., Blanco, J., 2014. Prototype traceability system for the dairy industry. Computers and Electronics in Agriculture 101, 34–41. http://dx.doi.org/10.1016/j.compag.2013.12.011.

Decathlon, 2015. L'innovation au service de la relation. web site: http://media.decathlon.fr/communique-presse/innovation-services/landing/landing.html.

Dong, L., Kehoe, D., Drake, P., 2006. Dynamic planning with a wireless product identification technology in food supply chains. The International Journal of Advanced Manufacturing Technology 30 (9–10), 938–944.

Eradus, W.J., Jansen, M.B., 1999. Animal identification and monitoring. Computers and Electronics in Agriculture 24 (9), 1–98.

Grunow, M., Piramuthu, S., 2013. RFID in highly perishable food supply chains – remaining shelf life to supplant expiry date? International Journal of Production Economics 146, 717–727.

GS1, 20/7/2015. EPC Radio-Frequency Identity Protocols Generation 2 UHF RFID – Specification for RFID Air Interface Protocol for Communications at 860 MHz-960 MHz, Version 2.0.1. Accessed Online: http://www.gs1.org/sites/default/files/docs/epc/Gen2_Protocol_Standard.pdf.

Herbon, A., Levner, E., Cheng, T.C.E., 2014. Perishable inventory management with dynamic pricing using time–temperature indicators linked to automatic detecting devices. International Journal of Production Economics 147, 605–613.

Hessel, E.F., Reiners, K., Böck, S., Wendl, G., van den Weghe, H.F.A., 2008. Application of high frequency transponders for simultaneous identification of weaned piglets. In: AgEng 2008 Congress: Agricultural & Biosystems Engineering for a Sustainable World. Hersonissos, Crete, June 23–25, 2008, Greece.

International Standard Organization, 1996a. ISO Standard 11784:1996 Radio Frequency Identification of Animals – Code Structure. Available at: http://www.iso.org/iso/catalogue_detail?csnumber=25881.

International Standard Organization, 1996b. ISO Standard 11785:1996: Radio Frequency Identification of Animals – Technical Concept. Available at: http://www.iso.org/iso/home/store/catalogue_ics/catalogue_detail_ics.htm?ics1=65&ics2=040&ics3=99&csnumber=19982.

International Standard Organization, 2007. ISO Standard 2 2005/2007. Traceability in the Feed and Food Chain: General Principles and Basic requirements for System Design and Implementation. Available at: http://www.iso.org/iso/catalogue_detail?csnumber=36297.

International Standard Organization, 2008. ISO/IEC 14443 Identification Cards – Contactless Integrated Circuit Cards – Proximity Cards. Available at: http://www.iso.org/iso/home/search.htm?qt=14443&sort=rel&type=simple&published=on.

International Standard Organization, 2010. ISO/IEC 15693-1:2010 Identification Cards – Contactless Integrated Circuit Cards – Vicinity Cards. Available at http://www.iso.org/iso/home/search.htm?qt=15693+&sort=rel&type=simple&published=on.

International Standard Organization, 2013a. ISO/IEC 18000 – 6: 2010. Information Technology – Radio Frequency Identification for Item Management – Part 6: Parameters for Air Interface Communications at 860 MHz to 960 MHz. Available at: http://www.iso.org/iso/home/store/catalogue_ics/catalogue_detail_ics.htm?csnumber=59644.

International Standard Organization, 2013b. ISO/IEC 18092:2013 Information Technology – Telecommunications and Information Exchange between Systems – Near Field Communication – Interface and Protocol (NFCIP-1).

Jedermann, R., Ruiz-García, L., Lang, W., 2009. Spatial temperature profiling by semi-passive RFID loggers for perishable food transportation. Computers and Electronics in Agriculture 65 (2), 145–154. http://dx.doi.org/10.1016/j.compag.2008.08.006.

Karlsen, K.M., Dreyer, B., Olsen, P., Elvevoll, E.O., 2012. Granularity and its role in implementation of seafood traceability. Journal of Food Engineering 112, 78–85.

Kumar, P., Reinitz, H.W., Simunovic, J., Sandeep, K.P., Franzon, P.D., 2009. Overview of RFID technology and its applications in the food Industry. Journal of Food Science 74 (8), R101–R106.

Lan, H.J., Zhao, L., Su, L., Liu, Z.G., 2014. Food cold chain equilibrium based on collaborative replenishment. Journal of Applied Research and Technology 12 (2), 201–211.

Li, L., 2012. Technology designed to combat fakes in the global supply chain. Business Horizons 56 (2), 167–177. http://dx.doi.org/10.1016/j.bushor.2012.11.010.

Lim, M.K., Bahr, W., Stephen, C., Leung, H., 2013. RFID in the warehouse: a literature analysis (1995–2010) of its applications, benefits, challenges and future trends. International Journal of Production Economics 145 (1), 409–430.

López-Gómez, A., Cerdán-Cartagena, F., Suardíaz-Muro, J., Boluda-Aguilar, M., Hernández-Hernández, M.H., López-Serrano, M.A., López-Coronado, J., 2015. Radiofrequency Identification and Surface Acoustic Wave Technologies for Developing the Food Intelligent Packaging Concept. Food Engineering Reviews 7 (1), 11–32.

Luvisi, A., Panattoni, A., Bandinelli, R., Rinaldelli, E., Pagano, M., Triolo, E., 2011. Implanting RFIDs into Prunus to facilitate electronic identification in support of sanitary certification. Biosystems Engineering 109 (2), 167–173.

Luvisi, A., Panattoni, A., Bandinelli, R., Rinaldelli, E., Pagano, M., Triolo, E., 2012. Ultra-high frequency transponders in grapevine: a tool for traceability of plants and treatments in viticulture. Biosystems Engineering 113 (2), 129–139. http://dx.doi.org/10.1016/j.biosystemseng.2012.06.015.

Mainetti, L., Patrono, L., Stefanizzi, M.L., Vergallo, R., 2013. An innovative and low-cost gapless traceability system of fresh vegetable products using RF technologies and EPCglobal standard. Computers and Electronics in Agriculture 98, 146–157.

Meyer, G.G., Främling, K., Holmström, J., 2009. Intelligent products: a survey. Computers in Industry 60, 137–148.

Moe, T., 1998. Perspectives on traceability in food manufacture. Trends in Food Science & Technology 9 (5), 211–214.

Narayana, S.A., Pati, R.K., Vrat, P., 2014. Managerial research on the pharmaceutical supply chain – a critical review and some insights for future directions. Journal of Purchasing & Supply Management 20, 18–40.

Papetti, P., Costa, C., Antonucci, F., Frigorilli, S., Solaini, S., Menesatti, P., 2012. RFID web-based info tracing system for the artisanal Italian cheese quality traceability. Food Control 27 (1), 234–241.

Pérez-Aloe, R., Valverde, J.M., Lara, A., Castaño, F., Carrillo, J.M., Gonzalez, J., Roa, I., 2010. Use of RFID tags for data storage on quality control in cheese industries. In: Balasa, F. (Ed.), Data Storage. InTech. ISBN: 978-953-307-063-6. http://dx.doi.org/10.5772/8867 (Chapter 13). Available at: http://www.intechopen.com/books/data-storage/use-of-rfid-tags-for-data-storage-on-quality-control-in-cheese-industries.

Piramuthu, S., Wochner, S., Grunow, M., 2014. Should retail stores also RFID-tag 'cheap' items? European Journal of Operational Research 233, 281–291.

Regattieri, A., Gamberi, M., Manzini, R., 2007. Traceability of food products: general framework and experimental evidence. Journal of Food Engineering 81 (2), 347–356.

Restuccia, D., Spizzirri, U.G., Parisi, O.I., Cirillo, G., Curcio, M., Iemma, F., Puoci, F., Vinci, G., Picci, N., 2010. New EU regulation aspects and global market of active and intelligent packaging for food industry applications. Food Control 21 (11), 1425–1435. http://dx.doi.org/10.1016/j.foodcont.2010.04.028.

Rizzi, A., Montanari, R., Bertolini, M., Bottani, E., Volpi, A., 2011. Logistica e tecnologia RFID – Creare valore nella filiera alimentare e nel largo consumo. Springer-Verlag Italia, Milano, Italy. http://dx.doi.org/10.1007/978-88-470-1929-4 (In Italian).

Ruiz-Garcia, L., Lunadei, L., 2011. The role of RFID in agriculture: applications, limitations and challenges. Computers and Electronics in Agriculture 79 (1), 42–50. http://dx.doi.org/10.1016/j.compag.2011.08.010.

Thiesse, F., Buckel, T., 2015. A comparison of RFID-based shelf replenishment policies in retail stores under suboptimal read rates. International Journal of Production Economics 159, 126–136.

Tortia, C., Barge, P., Gay, P., Merlino, V., Piccarolo, P., 2010. Radio frequency identification technologies for traceability of potted flowering plants. In: XVIIth World Congress of the International Commission of Agricultural Engineering (CIGR), Quebec City, Canada, 13–17 June.

Trebar, M., Lotrič, M., Fonda, I., Pleteršek, A., Kovačič, K., 2013. RFID data loggers in fish supply chain traceability. International Journal of Antennas and Propagation 2013 (3–4), 1–9. http://dx.doi.org/10.1155/2013/875973.

Vergara, E., Llobet, E., Ramírez, J.L., Ivanov, P., Fonseca, L., Zampolli, S., Scorzoni, A., Beckerd, T., Marco, S., Wöllenstein, J., 2007. An RFID reader with onboard sensing capability for monitoring fruit quality. Sensors and Actuators B: Chemical 127 (1(20)), 143–149. http://dx.doi.org/10.1016/j.snb.2007.07.107.

Vowels, A., 2006. A strategic case for RFID: an examination of Wal-Mart and its supply chain. In: Proceedings of the 2006 Southern Association for Information Systems Conference, March 11–12, Jacksonville, FL.

The Use of Molecular Biology Techniques in Food Traceability

6

M. Espiñeira
ANFACO-CECOPESCA, Vigo, Spain

F.J. Santaclara
IDIS CHUS, Santiago de Compostela, Spain

1. A Brief Introduction to Molecular Biology Techniques

Traceability has become an essential requirement to ensure the quality of food products that reach the market. Their implementation in the food industry involves the development of control systems of raw materials, from their entry into the chain of production to their marketing, ensuring the quality and reliability of food for both the producer and the consumer.

Into traceability systems of the food industry, the control of the authenticity of the raw material or processed product is a fundamental aspect. For this, it is necessary to verify that food is marketed under the commercial denomination to which it really belongs, and that it proceeds from a defined origin, as well as contain the raw material and the percentages of ingredients that it is declaring on the label. The growth of the international trade and the increase of the number of potentially marketable species require reliable and rapid methods to verify the authenticity of the products and their origin.

In products where the manipulation is minimal, usually those which are sold whole and without transformation processes, both fresh as chilled or frozen, the species identification based on morphological characters is possible and even relatively easy. However, in other cases where external morphological characteristics are eliminated during the processing phase, identification is not possible (Lago et al., 2014). In these cases, methodologies based on protein analysis or DNA (deoxyribonucleic acid) must be applied. They enable the identification regardless of the presence or absence of external body characters.

Genetic methods are the most commonly used because of their advantages over the morphological characters or protein-based methods. Although DNA may be altered with various food processes, this molecule is far more resistant and heat-stable than proteins, allowing the amplification by PCR of small DNA regions which are sufficient to enable identification even in the case of DNA fragmentation. Besides, given the degeneration of the genetic code and the presence of noncoding regions, this molecule provides more information than proteins (Lago et al., 2014). Another advantage of DNA is that this marker can potentially be extracted from any species in virtually any kind of organic substrate, such as muscle, fin, or blood, because it is present in all cells of an organism (Lockley and Bardsley, 2000; Teletchea, 2009).

Advances in Food Traceability Techniques and Technologies. http://dx.doi.org/10.1016/B978-0-08-100310-7.00006-5

The use of molecular biology techniques to the verification of the traceability makes it necessary to differentiate between the terms identification, authentication, and detection of species. The identification determines what species is present in a sample, and it answers the question: what species is it? The authentication checks if the species present in a sample is, or not, the declared species, and it answers the question: is it or is not this species? Finally, the detection confirms the presence or absence of the species of interest, and it answers the question: does it contain this species? The verification of traceability in food products by genetic techniques consists in checking the identity of the species that make those products. This gives a response to one or more of the above questions (Espiñeira et al., 2016).

Numerous genetic methods are currently applied to the **identification of species**. The main types of molecular techniques for species identification, detection, and authentication are Polymerase Chain Reaction (PCR), PCR-Restriction Fragment Length Polymorphism (PCR-RFLP), Forensically Informative Nucleotide Sequencing (FINS), Real-Time PCR (RT-PCR), digital PCR (dPCR), and Next-Generation Sequencing (NGS). Other techniques applied to traceability are the PCR-Length Polymorphism (PCR-LP), used, for instance, in the case of the microsatellites analysis and the Single Nucleotide Polymorphism (SNP) used for large-scale genotyping using high-throughput technologies. These last two techniques are used to determine at the **population**, **stock**, **variety**, or **cultivar level** and are particularly relevant when it comes to authenticate the origin or identity of products included in the European marks of Protected Designation of Origin (PDO) or Protected Geographical Indication (PGI).

In the following paragraphs these techniques are described with more detail.

The first of them is **PCR**, which is a technique based on the amplification and detection of specific DNA fragments by means of primers. The primers are a pair of small nucleotide sequences which limit the region to be amplified. The PCR reaction is a succession of cycles where each cycle has a DNA denaturation step to separate the chains, a primer alignment with a template DNA step, and a polymerization step to synthesize a new DNA between the two primers. Specific primers allow the generation of fragments that, after being separated and visualized by agarose gel electrophoresis, allow identification at the species level (Lago et al., 2014).

Another technique is the **PCR-RFLP**, which involves the amplification of a preselected DNA fragment with universal primers, followed by digestion with restriction endonucleases, which recognize specific short sequences (four to six nucleotides) of the amplified fragment and cut the DNA at those sites. These fragments can then be separated and visualized with gel electrophoresis. The development of a PCR-RFLP method requires sequence information for the DNA fragment of interest to select appropriate restriction endonucleases that produce species-specific DNA profiles after an enzymatic digestion (Rasmussen and Morrissey, 2011).

The **DNA Sequencing and phylogenetic analysis**, and in particular **Forensically informative nucleotide sequencing (FINS)**, involves PCR amplification and sequencing of a specific DNA fragment, followed by analysis of nucleotide variation between the target sequence and reference sequences of patterns of known species (Bartlett and Davidson, 1992). From the sequences, a distance matrix is constructed, through algorithmic calculations and analysis a phylogenetic tree is obtained, allowing the assignment of the unknown species to a group according to the genetic distance (Espiñeira et al., 2009a). This is a very robust and powerful technique for determining interspecific variability, suitable for the identification of closely related species, even in highly processed products (Lago et al., 2014).

RT-PCR (also known as quantitative PCR, Real-Time quantitative PCR, or qPCR) is a method of simultaneous DNA amplification and quantification. In Real-Time PCR a fluorescent reporter molecule is included in the assay, and this enables the products of the PCR reaction to be measured after each cycle once a threshold has been passed. The amount of fluorescence produced is proportional to the amount of amplicon produced during PCR, and it is used to calculate the amount of target DNA present at the beginning of the reaction. The Real-Time PCR is a technique that covers any need for rapid and accurate detection or quantification of small amounts of genetic material. Also, due to its sensitivity, it can be applied to highly processed samples, allowing the amplification of small DNA fragments, less than 200 bp. These features make it a highly effective tool for the detection of species in highly processed products, where the raw material has been subjected to high pressures and temperatures, and where DNA is highly degraded (Bustin, 2004).

The **Digital PCR** (dPCR) is a new method for accurate quantification of nucleic acids. It uses a limiting dilution analysis together with Poisson distribution analysis to allow the absolute quantification of the number of copies of target DNA. Digital PCR uses an amplification reaction system similar to a system of standard dPCR. The use of a nanofluidic chip provides a quick and easy mechanism which allows running thousands of parallel PCR reactions. As a refinement of the conventional qPCR, dPCR has the potential to allow more accurate and sensitive measurements of the number of copies of target DNA, especially in low concentration samples and complex samples. As result, dPCR has already been applied among others for pathogens and for the detection of different species in food products. Its potential is based on the ability to analyze samples containing mixtures of species with high sensitivity and in a single trial, performing multiple reactions in parallel (Espiñeira et al., 2016).

Next-Generation Sequencing (NGS) or massive sequencing is the latest trend in sequencing. Compared with traditional DNA sequencing based on the dideoxy chain termination technique, NGS relies on the immobilization of fragmented DNA templates on a solid support system. The spatially separated, immobilized fragments can then be amplified simultaneously by PCR and subjected to massively parallel DNA sequencing. This advance enables rapid sequencing of large stretches of DNA base pairs spanning entire genomes, with the latest instruments capable of producing hundreds of gigabases of data on a single sequencing run. The result is a sequencing technology that is simpler, faster, and more cost-effective and scalable (Rasmussen and Morrissey, 2011).

Other technique is the **PCR-Length Polymorphisms**. This method is used for studying the microsatellites, a tract of repetitive DNA in which certain motifs (ranging in length from two to five base pairs) are repeated at thousands of locations in the genome. These genomic regions have a large mutation rate in comparison with other DNA regions. Due to its high level of polymorphism, microsatellite markers are useful in the determination of individual profiles of single species and for population genetic studies, measuring gene flow and hybridization between populations, determining paternity, assigning individuals to their population of origin and determining population structure (González, 2003; Lago et al., 2014). Another of these techniques is **Single Nucleotide Polymorphisms (SNP)** analysis, which is a methodology used to compare genome variation between members of a species or in the same individual. An SNP is a single nucleotide variation in a DNA sequence, and it represents the most widespread type of sequence variation in genomes. This makes of them the ideal tool for population studies, making the comparisons of genomic diversities more straightforward than has been possible with microsatellites (Brumfield et al., 2003). An SNP that is common in one geographical group may be much rarer in another; it assigns an allele frequency by means of statistical analyses and estimates the presence of this SNP in each population (Lago et al., 2013a).

At present, most commonly used genetic methodologies for the traceability of the food industry are FINS and RT-PCR. This is due to their robustness, reliability, and sensitivity. On the other hand, dPCR and NGS are techniques of the latest trend which continue developing to satisfy market needs.

Molecular methodologies allow the competent authorities to perform regulatory control of processed products and to check the documents accompanying the products and their regulatory status in terms of security, traceability, authenticity verification, certification, and labeling, with the objective of ensuring the final quality of the product which reaches the consumer (Lago et al., 2013a).

2. Fish and Seafood Applications

The globalization of markets has led to an increase in the number of fish species that reach the consumer. Even for the same species, their different choices have increased, such as the selection of different origins or capture zones. This diversity has generated great variability in the prices of fish species, as well as in the quality of products made from them. These differences in market values have resulted in the emergence of fraudulent practices, which consist in the substitution of higher value or quality species for other similar species with inferior value or quality.

All this makes the introduction of traceability systems necessary for fisheries, as well as monitoring tools that enable the identification and authentication of species but can also come to distinguish the area of origin of the raw material, the stock, or town that it belongs to.

In this sense, methodologies based on molecular techniques are the most used because of their robustness and reliability. Several molecular methods for the detection, identification, and authentication of fish species of the main taxonomic groups have been developed (Table 6.1).

Often, species identification determines the origin or capture area of a product because a particular species is associated with a particular zoning. An example of this is reflected in the work of Espiñeira et al. (2008a), where the anglerfish species identification is related with its distribution range. The importance of this work lies not only

Table 6.1 **More Recent Studies Related to Traceability in the Most Important Taxa Ranked by Genetic Methodology Applied**

	FINS	RT-PCR	PCR-RFLP	Microsatellites
Scombroids	Infante et al. (2007) Espiñeira et al. (2009)	Chuang et al. (2012) Dalmasso et al. (2007) Liu et al. (2015)	Lin and Hwang (2008)	Abedi et al. (2012) Cheng et al. (2014)
Anglerfish	Espiñeira et al. (2008a)	Castigliego et al. (2015) Herrero et al. (2011d)	Armani et al. (2012a)	Blanco et al. (2006) Garoia et al. (2003)
Anchovies	Jerome et al. (2008) Santaclara et al. (2006)	Albaina et al. (2015)	Pappalardo and Ferrito (2015) Rea et al. (2009) Santaclara et al. (2006)	Borrell et al. (2012) Chiu et al. (2002) Kathirvelpandian et al. (2014) Sun et al. (2014)
Salmonids	Espiñeira et al. (2009b)	Herrero et al. (2011c) Hird et al. (2012) Li et al. (2013)	Espiñeira et al. (2009)	Ensing et al. (2013) Zhivotovsky et al. (2013)
Sardines	Jerome et al. (2003) Lago et al. (2011c)	Armani et al. (2012b) Herrero et al. (2011b)		Kasapidis et al. (2011)
Horse mackerels	Lago et al. (2011b)	Prado et al. (2013)	Karaiskou et al. (2007) Turan et al. (2009)	Abaunza et al. (2008) Kasapidis and Magoulas (2008)
Flatfish	Espiñeira et al. (2008b)	Herrero et al. (2012)	Comesaña et al. (2003)	Danancher and Garcia-Vazquez (2011) Molina-Luzán et al. (2012)
Swordfish	Herrero et al. (2011a)	Herrero et al. (2011a)	Hsieh et al. (2007)	Bradman et al. (2011) Muths et al. (2009)
Gadoids	Lago et al. (2013b) Pérez and Presa (2008)	Herrero et al. (2010) Hird et al. (2012) Sánchez et al. (2009)	Di Finizio et al. (2007)	Kijewska et al. (2011) Stroganov et al. (2010)

in the capacity of differentiation of the species, but also in the possibility of linking the identity of the species with its area of origin and the quality of their meat. Also, depending on the area of origin, there are different aspects such as the strictness of hygienic sanitary conditions of transformation plants, which are worse in third-world countries than in developed ones, or the amount of heavy metals accumulated, which are considered linked to the fish quality and can affect different species in different ways (Espiñeira et al., 2008a).

In cases where the identification of the species does not provide information about its origin, it is necessary to develop methods of population differentiation. These methodologies provide added value to some populations over others within the same species and regulate the PDO of products. Cases of population differentiation, which reflect the valorizations of a particular population, are the salmon from Norway and the anchovy from the Bay of Biscay. McConnell et al. (1995) used microsatellites to show a clear discrimination between Canadian and European populations of Atlantic Salmon. In anchovy, Borrel et al. (2012) used SNP to show the variation in populations of European anchovy from the Bay of Biscay and the Mediterranean. In addition, population genetic studies assess parameters as the genetic diversity and population structure, which can be useful for the management of resources focused to the sustainable exploitation (Lago et al., 2013a). Table 6.1 contains the most recent studies of population differentiation of the major taxonomic groups of fish through microsatellites.

The application of SNP for species identification and differentiation of stocks has mainly been used in salmonids and gadoids due to the commercial importance of the species that compose these groups. Its application is mainly focused on population studies for the conservation of genetic variability. These studies are focused on sustainable exploitation by the adequacy of fisheries' management plans. Among the studies focused on salmonids, one can find the work of Larson et al. (2014), who distinguished five salmon populations present in western of Alaska. There are also studies focused on conserving genetic variability of trout populations, as the study of Drywa et al. (2014), which differentiates *Salmo trutta* populations in the south of the Baltic Sea. In gadoid, SNP studies pursue both species identification and differentiation of stocks (Maretto et al., 2010; Lago et al., 2013b; Pocwierz-Kotus et al., 2015). An example of this is the work of Pocwierz-Kotus et al. (2015), that differentiates between eastern and western Baltic populations of *Gadus morhua*.

Exists a continuous progress in technical, and a constant development of studies that are focused on the traceability of fish species and products derived from them. These two aspects produce an increase in interest of producers, consumers, and authorities to increase the transparency of products that reach the market, thereby ensuring their safety and quality.

3. Meat Applications

Meat and meat products are one of the main nutritional components of the diet. Its production, processing, and marketing is an important part of the food industry. The consumer demands healthy food with distinctive qualities and high added-value meats.

Due to this, the meat industry has enforced strong measures toward the establishment of effective traceability systems to preserve food safety and quality from farm to fork (Shackell, 2008).

The traceability of meat and meat products is a major issue in the meat industry, with the two main drivers being food safety management and species verification. Meat speciation is substantially addressed to avoid of unfair competition practices of producers, processors, and sellers who would gain an economic advantage from food misrepresentation, given the profit that results from selling cheaper meats as premium-quality meats. It is also addressed to prevent overexploitation and illegal trafficking, as well as the consumption of endangered populations of wildlife species (Fajardo et al., 2010).

But meat traceability not only relates to the industrial economic profit resulting from the illegal trading, handling, or substitution of species but also to public health risks such as zoonosis or even allergies to particular meat proteins. The meat industry has dealt with crises such as Bovine Spongiform Encephalopathy (BSE), labeling scandals, chemical contamination scares, and microbiological poisoning, that have driven increased concern about the quality, origin, and integrity of meat (Cagney et al., 2004; Collee et al., 2006; Shackell, 2008; Tlustos, 2009).

The differentiation in the quality of the meat, consumer acceptance of it, and its market value do not only depend on the species, but also varies according to race or place of origin, and in many cases, depending on the sex of the animal (Abdulmawjood et al., 2012; Hersleth et al., 2012; Herrero et al., 2013; Realini et al., 2014). Because of this, traceability systems need to have accurate and efficient methodologies able to determine these aspects regardless of the degree of processing which the meat was undergone. The molecular methodologies answer these needs rapidly, sensitively, and efficiently, and provide an effective tool for traceability systems in meat products (Fajardo et al., 2013).

Several molecular methodologies focused toward the detection, identification, and authentication of species in meat products have been developed. The majority of genetic applications published to date for meat identification are focused on domestic animal species like cattle, sheep, goat, pig, turkey, or chicken (Ballin, 2010; Farrag et al., 2010; Fajardo et al., 2013). However, the increasing importance and high commercial value of game and exotic meats is driving the development of appropriate tools for the authentication of a growing number of wild or farmed exotic species (Fajardo et al., 2010).

The main molecular methodologies applied to meat species identification which should be noted are Polymerase Chain Reaction (PCR), Polymerase Chain Reaction-Restriction Fragment Length Polymorphism (PCR-RFLP), Forensically Informative Nucleotide Sequencing (FINS), and Real-Time PCR (RT-PCR). Table 6.2 shows some of the more recent works about application and development of molecular methods for meat species identification.

Other molecular techniques have also been introduced. This is the case of digital PCR in the field of meat product traceability. Until now, few studies have used this methodology, which is focused on the exact quantification of different species in meat and processed meat products. But among them, it is important to note the work of Cai

Table 6.2 Recent Studies Related to Implementation and Development of Molecular Tools for the Identification of Meat Species

Molecular Technique for Meat	Target Species	References
PCR	• Horse, cattle, mutton, pig, dog, chicken, and mice • Pig, cow, hare, red deer, rabbit, and wild boar • Pig • Chicken, duck, pigeon, and pig • Badger, cat, cow, dog, donkey, fox, goat, guinea pig, harvest mouse, hedgehog, horse, house mouse, human, pig, rabbit, rat, red deer, and sheep	• Dai et al. (2015) • Amaral et al. (2014) • Kumar et al. (2012) • Haunshi et al. (2009) • Tobe and Linacre (2008)
FINS	• Cattle, buffalo, sheep, and goat • Cattle, sheep, goat, domestic pig, horse, buffalo, chicken, turkey, ostrich, rabbit, kangaroo…(40 species) • Chicken, turkey, geese, quail, guinea fowl, ostrich, and emu • Cat, dog, lynx, cow, horse, red fox, red deer, roe deer, sheep…(28 species of mammals)	• Mane et al. (2013) • Lago et al. (2011a) • Girish et al. (2009) • Karlsson and Holmlund (2007)
PCR-RFLP	• Camel, buffalo, and sheep • Cattle, carabeef, chevon, mutton, and pig • Cattle, horse, goat, and pig • Cow, buffalo, sheep, camel, turkey, chicken, and donkey	• Farag et al. (2015) • Kumar et al. (2014) • Han et al. (2013) • Haider et al. (2012)
RT-PCR	• Cattle, goat, lamb, chicken, goose, turkey, pig, and horse • Duck, pig, and chicken • Deer, cattle, goat, horse, donkey, pig, and chicken • Seagull • Duck, goose, chicken, turkey, and pig • Pig, cattle, turkey, sheep, and chicken • Common pigeon, wood pigeon, and stock pigeon	• Okuma and Hellberg (2015) • Cheng et al. (2014) • You et al. (2014) • Kesmen et al. (2013) • Köppel et al. (2013) • Cammá et al. (2012) • Rojas et al. (2012)

et al. (2014), who applied digital PCR for the detection and quantification of pig and chicken meat and verified the reliability of this technique using meat mixtures with known compositions. It is also important to highlight the work of Floren et al. (2015). These authors applied the digital PCR in the routine use in laboratories for the detection and quantification of cattle, horse, and pig. The limit of quantification for DNA and meat mixtures was 0.01% and the limit of detection was 0.001%. Also, authors demonstrated that the quantification of traces of an unwarranted species cannot be achieved by using mtDNA, due to an at least five-fold variability between different tissues, but rather has to be based on a nuclear gene (Floren et al., 2015).

Table 6.3 **Studies Related to the Application and Development of Microsatellites and SNP in the Differentiation of Pig and Cattle Breeds**

Cattle		Pig	
Molecular Technique	**References**	**Molecular Technique**	**References**
Microsatellite	Orrú et al. (2006) Filippini et al. (2006) Dalvit et al. (2008) Rodríguez-Ramírez et al. (2011) Delgado et al. (2012) Acosta et al. (2013) Ema et al. (2014) Suh et al. (2014) Sharma et al. (2015) Mateus and Russo-Almeida (2015)	Microsatellite	• Alves et al. (2009) • Scali et al. (2012) • Conyers et al. (2011) • Oh et al. (2014) • Wang et al. (2015)
SNP	Negrini et al. (2008) Orrú et al. (2009) Allen et al. (2010) Mullen et al. (2013) Edea et al. (2013) Nishimura et al. (2013) Cheong et al. (2013) Dimauro et al. (2013) Bertolini et al. (2015)	SNP	• Kim et al. (2010) • Ramos et al. (2011) • Wilkinson et al. (2012) • Yang et al. (2014) • Srivastava et al. (2015) • Choi et al. (2015)

Moreover of the species identification, into traceability of meat products also has interest the breed differentiation or the geographical origin of the meat, as these factors also influence over the quality of meat. Methodologies based on Microsatellite and SNP have been applied for this purpose. Microsatellite methodologies have been mainly focused on cattle and pigs (Table 6.3), but also in sheep and chickens (Arranz et al., 2001; Baumung et al., 2006; Peter et al., 2007; Oka and Tsudzuki, 2014; Abebe et al., 2015). Examples of the works focused on beef are the work of Dalvit et al. (2008), who validated a set of 12 microsatellites for the assessment of a genetic traceability system in six cattle breeds (Dalvit et al., 2008) and the work of Rodríguez-Ramírez et al. (2011), who achieved breed differentiation of Brangus and Charolais/Brahman cattle through the identification of seven microsatellite markers. These molecular markers differentiated Iberian and Duroc pigs and European wild boar lineages, as well as wild boars and domestic pigs (Alves et al., 2009; Conyers et al., 2011).

SNP are rapidly replacing microsatellites due to a more robust genotyping and data interpretation, as well as a strong potential for automation (Nicoloso et al., 2013). Its application in the traceability of meat focused mainly on pork and beef, due to their commercial importance (Table 6.3) but also are focused on chicken and sheep (Twito et al., 2007; Groenen et al., 2011; Pariset et al., 2011). Some examples are the works of Negrini et al. (2008) and Allen et al. (2010) in European cattle breeds traceability or the work of Choi et al. (2015), where five pig breeds are differentiated.

Another factor influencing the quality of the meat and its market value is the gender of the animal. For example, male beef is regarded to be of a higher quality than cow or heifer meat (Price, 1995) and therefore yields higher prices (Zeleny and Schimmel, 2002). This fact leads control laboratories and research groups to develop reliable methods for gender determination to avoid the possibility of mislabeling and fraud. Several molecular methods for sex differentiation have been implemented, including PCR, PCR-RFLP, and more recently, the RT-PCR (Pollevick et al., 1992; Fontanesi et al., 2008; Abdulmawjood et al., 2012; Ballester et al., 2013; Gokulakrishnan et al., 2013; Herrero et al., 2013). Among recent works, we highlight one of Herrero et al. (2013), who developed a methodology based on Real-Time PCR assay using specific primer sets and probes for the detection of male beef. The method was validated for all kinds of beef products, including those subjected to intensive processing treatments, and obtained a robust methodology with high specificity, sensitivity, and rapidity. This methodology allows the authentication of male beef to avoid possible substitutions for meat of lower commercial value (Herrero et al., 2013).

Introduction of new species, development of new products, and the increasing demand for transparency in beef have promoted the development and implementation of molecular tools in the traceability systems. Constant studies, the development of new methodologies, and technological advances will improve traceability systems, encouraging transparency and ensuring the quality of meat products coming to market, thereby benefiting both producers and consumers.

4. Milk Products

The transparency in dairy products is an important issue regarding interests of the consumer due not only to the economic point of view, but also sanitary requirements, food allergies, or religious practices. Bovine, ovine, caprine, and buffalo milk are the main kinds used to make cheese or other dairy products. Common adulterations of dairy products are the partial or total substitution of higher value milk by other of less value, or the omission of a declared milk species. There is a larger quantity of bovine milk available, and its price is usually lower; also, the caprine milk yield is higher than ovine milk yield and its price is lower (Mafra et al., 2008; Nollet and Toldrá, 2009; Guerreiro et al., 2012). Also the absence of proper labeling, indicating the possibility of even traces of determinate milk in any dairy products, can be a risk for allergic persons, becoming a safety issue (Agrimonti et al., 2015).

The detection of milk species is important in cheese making too, especially those made from one pure species and with PDO, such as pure sheep or pure goat cheeses. In addition, some cheeses are manufactured with defined proportions of each type of milk, making the quantification in traceability system important (Bottero et al., 2003; Ulberth and Lees, 2003).

Several analytical approaches have been developed to apply in the traceability of milk products, the principal ones being electrophoretic, immunological, chromatographic, and molecular (Mafra et al., 2008). Among these analytical strategies the molecular techniques stand out for their robustness, reliability, sensitivity, and specificity, and their applicability, regardless of the treatment to which the product has been submitted (thermally-treated milks such as pasteurized milk, ultra-pasteurized, and powder milks). Among the molecular methods applied to the traceability of dairy products are mainly PCR, PCR-RFLP, and RT-PCR (Lanzilao et al., 2005; López-Calleja, 2005; Abdel-Rahman and Ahmed, 2007; López-Calleja et al., 2007; Mafra et al., 2007; Dalmasso, et al., 2011; Drummond, et al., 2013). Many works focused on differentiation of the species in dairy products are based on primers developed for the differentiation of species of meat. These ones are combined with new primers, and the conditions for validating the methodology and its application are optimized. The most recent works combine the simultaneous identification of various species and quantification by RT-PCR (López-Calleja et al., 2007; Dalmasso et al., 2011; Drummond et al., 2013).

Among the methodologies developed for the simultaneous identification of species in dairy products are the work of Gonçalves et al. (2012), who developed a method for the identification of cows, sheep, goats, and water buffalo in dairy products by multiplex PCR followed by fragment size analysis by capillary electrophoresis. Other works comprise the identification and quantification of species in dairy products. Drummond et al. (2013) describe a Real-Time PCR methodology for calculating the bovine and buffalo content in milk and meat-derived food products. It also highlights the work of Agrimonti et al. (2015), who developed a quadruplex quantitative Real-Time PCR (qxPCR) methodology for the rapid identification of DNA of cows, goats, sheep and buffalo in dairy products, as well as quantification of cow DNA.

The evolution of technology and the introduction of new techniques, such as digital PCR, will allow the development of increasingly sensitive and specific methodologies, promoting its implementation in the dairy sector and routine control tools as an indispensable part of the traceability system.

5. Cereals

Cereals are very common among the ingredients of many processed food products, and therefore the development of assays for their traceability is necessary, through specific identification and quantification methodologies. Molecular tools focused on the traceability of cereal products based on PCR and RT-PCR have been developed (Alary et al., 2002; Hernández et al., 2005; Terzi et al., 2005; Alary et al., 2007; Sonnante et al., 2009). Among the methodologies based on RT-PCR, it is important to note the work of Hernandez et al. (2005), who reported the development of four independent assays suitable for the identification and quantification of barley, rice, sunflower, and wheat. Also, Alary et al. (2007) developed a methodology based on PCR for detecting adulteration in chestnut flour by barley, bread, and durum wheat, as well as oat, rye, maize, and rice.

Molecular methodologies based on microsatellites or SNP are also used for the traceability of varieties within a particular species. It is especially important in the case of durum wheat-based products included as the European marks of PDO or PGI (Pasqualone, 2011). An example of this is the work of Pascualone et al. (2010), who designed a microsatellite-based method for verifying the presence of the four required durum wheat cultivars in PDO Altamura bread. Also Prins et al. (2010) reported a DNA-based method that can specifically and sensitively detect a number of different cereal species in samples of specialty varieties. For the specialty products, the PGI wheat variety Farro della Garfagnana determines the purity of the product by eliminating the possibility that one had been mixed with other cereals with a lower economical profile (Prins et al., 2010).

A very important aspect associated to cereals (wheat, kamut, spelt, rye, triticale, and barley) are respiratory allergies, dietary allergies, and intolerance to wheat, which produce celiac disease (gluten enteropathy), a common disorder that causes damage of the small bowell mucosa, affecting up to 1% of the population in Europe, North and South America, North Africa, and the Indian subcontinent (Hischenhuber et al., 2006; Tatham and Shewry, 2008). This adverse reaction is a lifelong illness, so persons with gluten enteropathy have to avoid the gluten-containing cereals by applying a gluten-free diet (Mustalahti et al., 2002; Zeltner et al., 2009). Adequate labeling and control is very important to avoid the inadvertent ingestion of products containing gluten (EU, 2011).

The most-used methods for gluten detection rely on direct detection of gluten proteins using an immunological test. As an alternative, molecular methods have emerged since, indirectly detecting the presence of corn that produces the protein highlighting PCR and RT-PCR (Allmann et al., 1993). One of the first PCR methodologies was developed by Köppel et al. (1998), who used a species-specific PCR technique to detect contaminations below 0.1% (w/w), about ten-fold more sensitive than ELISA. Other PCR methodologies have been developed to detect presence of traces of material derived from gluten-containing cereals (Dahinden et al., 2001; Olexová et al., 2006; Debnath et al., 2009). Also, RT-PCR methods were developed, performing quantitative assays (Sandberg et al., 2003; Piknová et al., 2008; Zeltner et al., 2009; Mujico et al., 2011). Among the most recent works, it is important to note the work of Mujico et al. (2011), who developed a new qPCR system which is capable of quantifying wheat contaminations in raw materials and other composite food types, with a sensitivity similar to the ELISA.

All applications mentioned show the integration of molecular tools in traceability systems for cereal products, promoting the implementation of these systems in production processes.

6. Fruit and Vegetable Foodstuffs

The fruits and its pulps are widespread and essential ingredients in variety products, as juices, jams, baby food, snacks, and yogurts. In the majority of these products, the percentage of fruit contained is declared. The widening market of these products has led to speculation that they may contain artificial aromas and be adulterated and mislabeled. Similarly, vegetable components are included in foodstuffs as raw or processed materials. Authentication testing and adulterant detection of these components are essential for value assessment, to prohibit unfair competition, and to ensure consumer protection against fraudulent practices. Furthermore, consumption of products containing undeclared constituents may cause problems such as allergy in sensitized individuals (Madesis et al., 2014).

The applications of molecular methods in the traceability of fruits and vegetables cover both the detection and identification of species, as well as the differentiation of varieties. The most-used molecular methodologies for the detection and identification of species of fruits and vegetables are PCR and RT-PCR (Ortola-Vidal et al., 2007; Palmieri et al., 2009; Scott and Knight, 2009; Han et al., 2012; Madesis et al., 2014). A reflection of this is the work of Han et al. (2012), who detected ingredients from seven fruits including the apple, pear, peach, grape, strawberry, mandarin, and orange by means of DNA-based technology. Palmieri et al. (2009) used RT-PCR to discriminate between five different berry genera and species and between these fruits and other fruit species mixed together in different types of fruit-based food products. It is especially important in the detection of vegetables species which are known for their allergenicity, like for instance celery, whose detection has been the basis for development of several molecular detection methodologies (Köppel et al., 2010; Pafundo et al., 2011; Fuchs et al., 2012). Other methodologies based on SNP and microsatellite have been applied to the differentiation of varieties in vegetables, as for instance tomatoes, allowing their characterization and favoring their traceability along the entire tomato food chain (Tedeschi et al., 2011; Sardaro et al., 2013).

Genetic traceability is indispensable to distinguish traditional varieties with specific and high-quality characteristics, and consequently, it is important to protect PDO and PGI indications and local economies (Caramante et al., 2011). An example of their application in fruits and vegetables is the work of Hernández et al. (2012), who developed a PCR method for the rapid and accurate screening for adulteration in smoked paprika, recognized as PDO "Pimentón de la Vera", used in making chorizo sausage. Also, Serradilla et al. (2013) developed an RT-PCR method for the authentication of the sweet cherry, marketed under the registry of the PDO "Cereza del Jerte."

The increasing variety of products whose composition includes fruits and vegetables, and new food trends to healthier and high value-added products, make it necessary to adapt the systems of traceability in this sector. Therefore progress in the development of molecular methodologies is continuous, favoring their application throughout the entire food chain, from farm to fork.

7. Wine

Wine quality depends on the vinification process and the geographical origin of the grapes but also highly relies on the varietal composition of the grape must; for this reason, wine traceability is important in relation to quality control and consumer information. The need of a traceability system is more evident in the case of wines identified by a PDO, in which only a limited number of cultivars can be used in order to guarantee a standardized quality (Siret et al., 2002).

Several molecular methodologies based on PCR have proven useful in differentiating varieties of grapes used for winemaking (Bowers et al., 1993; Thomas and Scott, 1993; Faria et al., 2000). But the application of microsatellite markers has provided better results for the differentiation of grape varieties in wine and musts (Sefc et al., 2001). Many works have focused their studies in the selection and application of microsatellites for the characterization and authentication of wine from the grape-residual DNA remaining after the manufacturing process (Faria et al., 2000; Siret et al., 2000; García-Beneytez et al., 2002; Siret et al., 2002; Ibáñez et al., 2003; Baleiras-Couto and Eiras-Dias, 2006; Rodríguez-Plaza et al., 2006; Savazzini and Martinelli, 2006; Faria et al., 2008). Among these works is the work of Faria et al. (2008), who described a quantitative microsatellite DNA-based method to determine the percentage of each of the varieties present in a must.

The main limiting factors for the molecular analysis of wine is to have an efficient DNA extraction method. Several studies reported the successful extraction of DNA, both from plant tissues (leaves, seeds, and stems) and from different grapevine products (juice or must) (Faria et al., 2000). But in wines, it is difficult to obtain a good quantity and quality of DNA. It contains a low quantity of DNA due to the removing of the grape in the steps of winemaking (decanting, clarification, and filtration process) and to the degradation of DNA during the fermentation (Siret et al., 2000; García-Beneytez et al., 2002). Studies have been made to overcome this difficulty, among them stand out the works of Pereira et al. (2011) and Recupero et al. (2012), who optimized the extraction of DNA from wine, in addition to developing molecular methodologies for the differentiation of wine varieties.

The importance acquired by the authentication of grape varieties in traceability systems in the wine sector is each time greater, so that molecular methodologies have become one of the tools used for this purpose. Its implementation in traceability systems in this sector will ensure the quality of the wines, both for producers and consumers.

8. Oils

Olive oil is a food product particularly prone to fraudulent practices, since it commands a higher price than other vegetable oils due to their nutritional and organoleptic properties (Giugliano and Esposito, 2005). Oil adulteration can involve blending premium oil with others produced from poor quality fruit or with other vegetable oils (almond, maize, palm, sunflower, or hazelnut). In cases of potentially allergenic substitutes, such as hazelnut, its use might represent a risk for sensitized individuals (Arlorio et al., 2010). Another fraudulent practice is the mislabeling regarding the information about the geographical origin, the cultivars, and/or the production methodology, especially important in the case of monovarietal oils and PDO olive oils (Consolandi et al., 2008). These factors highlight the need for an efficient traceability system.

DNA-based analysis methods are widely used to control olive oil traceability and authenticity, allowing identification of varietal composition and detecting any addition of seed oils. However, several difficulties have been reported to obtain amplifiable DNA from oil matrices (Gryson et al., 2002; Bazakos et al., 2012; Costa et al., 2012). During the oil-refining steps, the applied heat treatments and the use of activated clays and pH variations may cause DNA degradation, affecting its integrity and therefore its quality. Moreover, the presence of PCR inhibitors in the oils might severely decrease the ability to amplify DNA fragments (Costa et al., 2010; He et al., 2013; Ramos-Gómez et al., 2014). Numerous methods have been attempted for DNA extraction from vegetable oils (Costa et al., 2010; Giménez et al., 2010; Pafundo et al., 2010; Agrimonti et al., 2011; Ramos-Gómez et al., 2014; Muzzalupo et al., 2015; Raieta et al., 2015). Among the most recent studies stand out the work of Raieta et al. (2015), who describe a novel and optimized protocol based on the CTAB–hexane–chloroform method for DNA extraction from extra virgin olive oil.

Several molecular techniques have been applied to the traceability of the oils. PCR and RT-PCR are the most common methodologies used for species identification in vegetable oils (Spaniolas et al., 2008; Zhang et al., 2009; Giménez et al., 2010; Bai et al., 2011; Kumar et al., 2011; Wu et al., 2011). One of the most recent works about identifying species in different types of oils is that of Zhang et al. (2012), which distinguishes edible oils from olive, soybean, sunflower, peanut, sesame, and maize.

Microsatellites and SNPs are the molecular markers most used in the methods for the discrimination of varietal olive oils, ensuring PDO olive oil authentication regarding the cultivar (Alba et al., 2009; Ayed et al., 2009; Vietina et al., 2011; Bazakos et al., 2012; Kalogianni et al., 2015; Montemurro et al., 2015). Among the most recent applications developed is the work of Kalogianni et al. (2015), who designed a multiplex SNP genotyping assay for olive oil that enabled the identification of five common Greek olive cultivars.

The incorporation of molecular methodologies in oil traceability systems will allow their adaptation to the current needs, providing efficiency and speed in checking the authenticity of these products, reducing the risk of adulteration, and guaranteeing the quality of the oils present in the market.

9. Conclusions and Perspectives

The application of molecular methods is widely implemented in the traceability systems in the food sector, covering many types of food, as it is shown in this chapter. The rapid implementation of traceability systems in the food system are due to these techniques with high specificity, sensitivity, efficiency, and speed. In addition, compared with protein-based techniques, DNA-based techniques can be applied to any type of product, regardless of the treatment of processing to which it has been subjected.

Future trends in the development of molecular or genetic tools for food traceability are focused on the search of techniques to obtain more information in the shortest time possible. Techniques that allow the identification, both at the species and population level, also facilitate the determination of the geographic origin. At present, one of the most promising techniques is the digital PCR (dPCR), which allows the analysis of samples containing mixtures of species with high sensitivity and in a single assay, making multiple reactions in parallel through a nanofluidic chip. The application of emerging technologies, microfluidics, and nanotechnology in the development of molecular methods will obtain greater sensitivity, discriminatory power, reproducibility, and speed, thus increasing its potential in the traceability of the food sector.

References

Abaunza, P., Murta, A.G., et al., 2008. Stock identity of horse mackerel (*Trachurus trachurus*) in the Northeast Atlantic and Mediterranean sea: integrating the results from different stock identification approaches. Fisheries Research 89 (2), 196–209.

Abdel-Rahman, S.M., Ahmed, M.M.M., 2007. Rapid and sensitive identification of buffalo's, cattle's and sheep's milk using species-specific PCR and PCR-RFLP techniques. Food Control 18 (10), 1246–1249.

Abdulmawjood, A., Krischek, C., et al., 2012. Determination of pig sex in meat and meat products using multiplex real time-PCR. Meat Science 91 (3), 272–276.

Abebe, A.S., Mikko, S., et al., 2015. Genetic diversity of five local Swedish chicken breeds detected by microsatellite markers. PloS One 10 (4).

Abedi, E., Zolgharnein, H., et al., 2012. Genetic differentiation of narrow-barred Spanish mackerel (*Scomberomorus commerson*) stocks using microsatellite markers in Persian Gulf. American-Eurasian Journal of Agricultural & Environmental Sciences 12 (10), 1305–1310.

Acosta, A.C., Uffo, O., et al., 2013. Genetic diversity and differentiation of five Cuban cattle breeds using 30 microsatellite loci. Journal of Animal Breeding and Genetics 130 (1), 79–86.

Agrimonti, C., Pirondini, A., et al., 2015. A quadruplex PCR (qxPCR) assay for adulteration in dairy products. Food Chemistry 187, 58–64.

Agrimonti, C., Vietina, M., et al., 2011. The use of food genomics to ensure the traceability of olive oil. Trends in Food Science & Technology 22 (5), 237–244.

Alary, R., Buissonade, C., et al., 2007. Detection and discrimination of cereal and leguminous species in chestnut flour by duplex PCR. European Food Research and Technology 225 (3–4), 427–434.

Alary, R., Serin, A., et al., 2002. Quantification of common wheat adulteration of durum wheat pasta using real-time quantitative polymerase chain reaction (PCR). Cereal Chemistry 79 (4), 553–558.

Alba, V., Montemurro, C., et al., 2009. SSR-based identification key of cultivars of *Olea europaea* L. diffused in Southern-Italy. Scientia Horticulturae 123 (1), 11–16.

Albaina, A., Irigoien, X., et al., 2015. A real-time PCR assay to estimate invertebrate and fish predation on anchovy eggs in the Bay of Biscay. Progress in Oceanography 131, 82–99.

Alves, E., Fernández, A.I., et al., 2009. Identification of mitochondrial markers for genetic traceability of European wild boars and Iberian and Duroc pigs. Animal 3 (09), 1216–1223.

Allen, A.R., Taylor, M., et al., 2010. Compilation of a panel of informative single nucleotide polymorphisms for bovine identification in the Northern Irish cattle population. BMC Genetics 11 (1), 5.

Allmann, M., Candrian, U., et al., 1993. Polymerase chain reaction (PCR): a possible alternative to immunochemical methods assuring safety and quality of food detection of wheat contamination in non-wheat food products. Zeitschrift für Lebensmittel-Untersuchung und Forschung 196 (3), 248–251.

Amaral, J.S., Santos, C.G., et al., 2014. Authentication of a traditional game meat sausage (Alheira) by species-specific PCR assays to detect hare, rabbit, red deer, pork, and cow meats. Food Research International 60, 140–145.

Arlorio, M., Coisson, J.D., et al., 2010. Olive oil adulterated with hazelnut oils: simulation to identify possible risks to allergic consumers. Food Additives and Contaminants 27 (1), 11–18.

Armani, A., Castigliego, L., et al., 2012a. A rapid PCR-RFLP method for the identification of Lophius species. European Food Research and Technology 235 (2), 253–263.

Armani, A., Castigliego, L., et al., 2012b. Multiplex conventional and real-time PCR for fish species identification of Bianchetto (juvenile form of *Sardina pilchardus*), Rossetto (*Aphia minuta*), and Icefish in fresh, marinated and cooked products. Food Chemistry 133 (1), 184–192.

Arranz, J., Bayón, Y., et al., 2001. Differentiation among Spanish sheep breeds using microsatellites. Genetics Selection Evolution 33 (5), 529–542.

Ayed, R.B., Grati-Kamoun, N., et al., 2009. Comparative study of microsatellite profiles of DNA from oil and leaves of two Tunisian olive cultivars. European Food Research and Technology 229 (5), 757–762.

Bai, S., Li, S., et al., 2011. Rapid detection of eight vegetable oils on optical thin-film biosensor chips. Food Control 22 (10), 1624–1628.

Baleiras-Couto, M.M., Eiras-Dias, J.E., 2006. Detection and identification of grape varieties in must and wine using nuclear and chloroplast microsatellite markers. Analytica Chimica Acta 563 (1), 283–291.

Ballester, M., Castelló, A., et al., 2013. A quantitative real-time PCR method using an X-linked gene for sex typing in pigs. Molecular Biotechnology 54 (2), 493–496.

Ballin, N.Z., 2010. Authentication of meat and meat products. Meat Science 86 (3), 577–587.

Bartlett, S.E., Davidson, W.S., 1992. FINS (Forensically informative nucleotide sequencing): a procedure for identifying the animal origin of biological specimens. Biotechniques 12 (3), 408–411.

Baumung, R., Cubricá, C.V., et al., 2006. Genetic characterisation and breed assignment in Austrian sheep breeds using microsatellite marker information. Journal of Animal Breeding and Genetics 123 (4), 265–271.

Bazakos, C., Dulger, A.O., et al., 2012. A SNP-based PCR-RFLP capillary electrophoresis analysis for the identification of the varietal origin of olive oils. Food Chemistry 134 (4), 2411–2418.

Bertolini, F., Galimberti, G., et al., 2015. Combined use of principal component analysis and random forests identify population-informative single nucleotide polymorphisms: application in cattle breeds. Journal of Animal Breeding and Genetics 132 (5), 346–356.

Blanco, G., Borrell, Y.J., et al., 2006. A new set of highly polymorphic microsatellites for the white and black anglerfish (Lophiidae). Molecular Ecology Notes 6 (3), 767–769.

Borrell, Y.J., Piñera, J.A., et al., 2012. Mitochondrial DNA and microsatellite genetic differentiation in the European anchovy *Engraulis encrasicolus* L. ICES Journal of Marine Science: Journal du Conseil 69 (8), 1357–1371.

Bottero, M.T., Civera, T., et al., 2003. A multiplex polymerase chain reaction for the identification of cow's, goat's and sheep's milk in dairy products. International Dairy Journal 13 (4), 277–282.

Bowers, J.E., Bandman, E.B., et al., 1993. DNA fingerprint characterization of some wine grape cultivars. American Journal of Enology and Viticulture 44 (3), 266–274.

Bradman, H., Grewe, P., et al., 2011. Characterisation of 22 polymorphic microsatellite loci for the broadbill swordfish, *Xiphias gladius*. Conservation Genetics Resources 3 (2), 263–266.

Brumfield, R.T., Beerli, P., et al., 2003. The utility of single nucleotide polymorphisms in inferences of population history. Trends in Ecology & Evolution 18 (5), 249–256.

Bustin, S.A., 2004. AZ of Quantitative PCR. International University Line La Jolla.

Cagney, C., Crowley, H., et al., 2004. Prevalence and numbers of *Escherichia coli* O157: H7 in minced beef and beef burgers from butcher shops and supermarkets in the Republic of Ireland. Food Microbiology 21 (2), 203–212.

Cai, Y., Li, X., et al., 2014. Quantitative analysis of pork and chicken products by droplet digital PCR. BioMed Research International.

Cammá, C., Di Domenico, M., et al., 2012. Development and validation of fast real-time PCR assays for species identification in raw and cooked meat mixtures. Food Control 23 (2), 400–404.

Caramante, M., Corrado, G., et al., 2011. Simple sequence repeats are able to trace tomato cultivars in tomato food chains. Food Control 22 (3–4), 549–554.

Castigliego, L., Armani, A., et al., 2015. Two alternative multiplex PCRs for the identification of the seven species of anglerfish (*Lophius* spp.) using an end-point or a melting curve analysis real-time protocol. Food Chemistry 166, 1–9.

Collee, J.G., Bradley, R., et al., 2006. Variant CJD (vCJD) and bovine spongiform encephalopathy (BSE): 10 and 20 years on: Part 2. Folia Neuropathologica 44 (2), 102.

Comesaña, A.S., Abella, P., et al., 2003. Molecular identification of five commercial flatfish species by PCR-RFLP analysis of a 12S rRNA gene fragment. Journal of the Science of Food and Agriculture 83 (8), 752–759.

Consolandi, C., Palmieri, L., et al., 2008. A procedure for olive oil traceability and authenticity: DNA extraction, multiplex PCR and LDR-universal array analysis. European Food Research and Technology 227 (5), 1429–1438.

Conyers, C.M., Allnutt, T.R., et al., 2011. Development of a microsatellite-based method for the differentiation of European wild boar (*Sus scrofa scrofa*) from domestic pig breeds (*Sus scrofa domestica*) in food. Journal of Agricultural and Food Chemistry 60 (13), 3341–3347.

Costa, J., Mafra, I., et al., 2010. Monitoring genetically modified soybean along the industrial soybean oil extraction and refining processes by polymerase chain reaction techniques. Food Research International 43 (1), 301–306.

Costa, J., Mafra, I., et al., 2012. Advances in vegetable oil authentication by DNA-based markers. Trends in Food Science & Technology 26 (1), 43–55.

Cheng, Q., Zhu, Y., et al., 2014a. High polymorphism and moderate differentiation of chub mackerel, *Scomber japonicus* (Perciformes: Scombridae), along the coast of China revealed by fifteen novel microsatellite markers. Conservation Genetics 15 (5), 1021–1035.

Cheng, X., He, W., et al., 2014b. Multiplex real-time PCR for the identification and quantification of DNA from duck, pig and chicken in Chinese blood curds. Food Research International 60, 30–37.

Cheong, H.S., Kim, L.H., et al., 2013. Development of discrimination SNP markers for Hanwoo (Korean native cattle). Meat Science 94 (3), 355–359.

Chiu, T.S., Lee, Y.J., et al., 2002. Polymorphic microsatellite markers for stock identification in Japanese anchovy (*Engraulis japonica*). Molecular Ecology Notes 2 (1), 49–50.

Choi, J., Chung, W., et al., 2015. Whole-genome resequencing analyses of five pig breeds, including Korean wild and native, and three European origin breeds. DNA Research 22 (4), 259–267. http://dx.doi.org/10.1093/dnares/dsv011.

Chuang, P., Chen, M., et al., 2012. Identification of tuna species by a real-time polymerase chain reaction technique. Food Chemistry 133 (3), 1055–1061.

Dahinden, I., von Büren, M., et al., 2001. A quantitative competitive PCR system to detect contamination of wheat, barley or rye in gluten-free food for coeliac patients. European Food Research and Technology 212 (2), 228–233.

Dai, Z., Qiao, J., et al., 2015. Species authentication of common meat based on PCR analysis of the mitochondrial COI gene. Applied Biochemistry and Biotechnology 176 (6), 1770–1780.

Dalmasso, A., Civera, T., et al., 2011. Simultaneous detection of cow and buffalo milk in mozzarella cheese by real-time PCR assay. Food Chemistry 124 (1), 362–366.

Dalmasso, A., Fontanella, E., et al., 2007. Identification of four tuna species by means of real-time PCR and melting curve analysis. Veterinary Research Communications 31, 355–357.

Dalvit, C., De Marchi, M., et al., 2008. Genetic traceability of meat using microsatellite markers. Food Research International 41 (3), 301–307.

Danancher, D., Garcia-Vazquez, E., 2011. Genetic population structure in flatfishes and potential impact of aquaculture and stock enhancement on wild populations in Europe. Reviews in Fish Biology and Fisheries 21 (3), 441–462.

Debnath, J., Martin, A., et al., 2009. A polymerase chain reaction directed to detect wheat glutenin: implications for gluten-free labelling. Food Research International 42 (7), 782–787.

Delgado, J.V., Martínez, A.M., et al., 2012. Genetic characterization of Latin-American Creole cattle using microsatellite markers. Animal Genetics 43 (1), 2–10.

Di Finizio, A., Guerriero, G., et al., 2007. Identification of gadoid species (Pisces, Gadidae) by sequencing and PCR-RFLP analysis of mitochondrial 12S and 16S rRNA gene fragments. European Food Research and Technology 225 (3–4), 337–344.

Dimauro, C., Cellesi, M., et al., 2013. Use of the canonical discriminant analysis to select SNP markers for bovine breed assignment and traceability purposes. Animal Genetics 44 (4), 377–382.

Drummond, M.G., Brasil, B., et al., 2013. A versatile real-time PCR method to quantify bovine contamination in buffalo products. Food Control 29 (1), 131–137.

Drywa, A., Pocwierz-Kotus, A., et al., 2014. Identification of multiple diagnostic SNP loci for differentiation of three salmonid species using SNP-arrays. Marine Genomics 15, 5–6.

Edea, Z., Dadi, H., et al., 2013. Genetic diversity, population structure and relationships in indigenous cattle populations of Ethiopia and Korean Hanwoo breeds using SNP markers. Frontiers in Genetics 4.

Ema, P., Manjeli, Y., et al., 2014. Genetic diversity of four Cameroonian indigenous cattle using microsatellite markers. Journal of Livestock Science 5, 9–17.

Ensing, D., Crozier, W.W., et al., 2013. An analysis of genetic stock identification on a small geographical scale using microsatellite markers, and its application in the management of a mixed-stock fishery for Atlantic salmon *Salmo salar* in Ireland. Journal of Fish Biology 82 (6), 2080–2094.

Espiñeira, M., Alonso, M., et al., 2016. Advances in authenticity testing for fish speciation. In: Downey, G., Sykes, R. (Eds.), Advances in Food Authenticity Testing. Woodhead Publishing (in press).

Espiñeira, M., Gonzalez-Lavín, N., et al., 2009a. Development of a method for the identification of scombroid and common substitute species in seafood products by FINS. Food Chemistry 117 (4), 698–704.

Espiñeira, M., González-Lavín, N., et al., 2008a. Authentication of anglerfish species (*Lophius* spp.) by means of polymerase chain reaction-restriction fragment length polymorphism (PCR-RFLP) and forensically informative nucleotide sequencing (FINS) methodologies. Journal of Agricultural and Food Chemistry 56 (22), 10594–10599.

Espiñeira, M., González-Lavín, N., et al., 2008b. Development of a method for the genetic identification of flatfish species on the basis of mitochondrial DNA sequences. Journal of Agricultural and Food Chemistry 56 (19), 8954–8961.

Espiñeira, M., Vieites, J.M., et al., 2009b. Development of a genetic method for the identification of salmon, trout, and bream in seafood products by means of PCR-RFLP and FINS methodologies. European Food Research and Technology 229 (5), 785–793.

EU, 2011. Regulation (EU) No 1169/2011 of the European Parliament and of the Council of 25 October 2011 on the Provision of Food Information to Consumers, Amending Regulations (EC) No 1924/2006 and (EC) No 1925/2006 of the European Parliament and of the Council, and Repealing Commission Directive 87/250/EEC, Council Directive 90/496/EEC, Commission Directive 1999/10/EC, Directive 2000/13/EC of the European Parliament and of the Council, Commission Directives 2002/67/EC and 2008/5/EC and Commission Regulation (EC) No 608/2004 No 1169/2011.

Fajardo, V., González, I., et al., 2013. In: Sforza, S. (Ed.), DNA-Based Methods for Authentication of Meat and Meat Products. Food Authentication Using Bioorganic Molecules. DEStech Publications, Inc.

Fajardo, V., González, I., et al., 2010. A review of current PCR-based methodologies for the authentication of meats from game animal species. Trends in Food Science & Technology 21 (8), 408–421.

Farag, M.R., Imam, T.S., et al., 2015. Identification of some domestic animal species (camel, buffalo and sheep) by PCR-RFLP analysis of the mitochondrial cytochrome b gene. Advances in Animal and Veterinary Sciences 3 (2), 136–142.

Faria, M.A., Magalhaes, R., et al., 2000. *Vitis vinifera* must varietal authentication using microsatellite DNA analysis (SSR). Journal of Agricultural and Food Chemistry 48 (4), 1096–1100.

Faria, M.A., Nunes, E., et al., 2008. Relative quantification of *Vitis vinifera* L. varieties in musts by microsatellite DNA analysis. European Food Research and Technology 227 (3), 845–850.

Farrag, S.A., Tanatarov, A.B., et al., 2010. Using of DNA fingerprinting in poultry research. International Journal of Poultry Science 9 (5), 406–416.

Filippini, G., Cetica, V., et al., 2006. Beef traceability using molecular methodologies. Veterinary Research Communications 30, 375–377.

Floren, C., Wiedemann, I., et al., 2015. Species identification and quantification in meat and meat products using droplet digital PCR (ddPCR). Food Chemistry 173, 1054–1058.

Fontanesi, L., Scotti, E., et al., 2008. Differences of the porcine amelogenin X and Y chromosome genes (AMELX and AMELY) and their application for sex determination in pigs. Molecular Reproduction and Development 75 (11), 1662–1668.

Fuchs, M., Cichna-Markl, M., et al., 2012. Development and validation of a novel real-time PCR method for the detection of celery (*Apium graveolens*) in food. Food Chemistry 130 (1), 189–195.

García-Beneytez, E., Moreno-Arribas, M.V., et al., 2002. Application of a DNA analysis method for the cultivar identification of grape musts and experimental and commercial wines of *Vitis vinifera* L. using microsatellite markers. Journal of Agricultural and Food Chemistry 50 (21), 6090–6096.

Garoia, F., Guarniero, I., et al., 2003. Polymorphic dinucleotide microsatellites for the Mediterranean angler species (Lophiidae). Molecular Ecology Notes 3 (2), 294–296.

Giménez, M.J., Pistón, F., et al., 2010. Application of real-time PCR on the development of molecular markers and to evaluate critical aspects for olive oil authentication. Food Chemistry 118 (2), 482–487.

Girish, P.S., Anjaneyulu, A.S.R., et al., 2009. Poultry meat speciation by sequence analysis of mitochondrial 12S rRNA gene. Indian Journal of Animal Sciences 79 (2), 217–220.

Giugliano, D., Esposito, K., 2005. Mediterranean diet and cardiovascular health. Annals of the New York Academy of Sciences 1056 (1), 253–260.

Gokulakrishnan, P., Kumar, R.R., et al., 2013. Determination of sex origin of meat from cattle, sheep and goat using PCR based assay. Small Ruminant Research 113 (1), 30–33.

Gonçalves, J., Pereira, F., et al., 2012. New method for the simultaneous identification of cow, sheep, goat, and water buffalo in dairy products by analysis of short species-specific mitochondrial DNA targets. Journal of Agricultural and Food Chemistry 60 (42), 10480–10485.

González, E.G., 2003. Microsatélites: sus aplicaciones en la conservación de la biodiversidad. Graellsia 59 (2–3), 377–388.

Groenen, M.A.M., Megens, H., et al., 2011. The development and characterization of a 60K SNP chip for chicken. BMC Genomics 12 (1), 274.

Gryson, N., Ronsse, F., et al., 2002. Detection of DNA during the refining of soybean oil. Journal of the American Oil Chemists' Society 79 (2), 171–174.

Guerreiro, J.S., Fernandes, P., et al., 2012. Identification of the species of origin of milk in cheeses by multivariate statistical analysis of polymerase chain reaction electrophoretic patterns. International Dairy Journal 25 (1), 42–45.

Haider, N., Nabulsi, I., et al., 2012. Identification of meat species by PCR-RFLP of the mitochondrial COI gene. Meat Science 90 (2), 490–493.

Han, J., Wu, Y., et al., 2012. PCR and DHPLC methods used to detect juice ingredient from 7 fruits. Food Control 25 (2), 696–703.

Han, S.H., Park, S.M., et al., 2013. PCR-rflp for the identification of mammalian livestock animal species. Journal of Embryo Transfer 28 (4), 355–360.

Haunshi, S., Basumatary, R., et al., 2009. Identification of chicken, duck, pigeon and pig meat by species-specific markers of mitochondrial origin. Meat Science 83 (3), 454–459.

He, J., Xu, W., et al., 2013. Development and optimization of an efficient method to detect the authenticity of edible oils. Food Control 31 (1), 71–79.

Hernández, A., Aranda, E., et al., 2012. Efficiency of DNA typing methods for detection of smoked paprika "pimenton de la vera" adulteration used in the elaboration of dry-cured Iberian pork sausages. Journal of Agricultural and Food Chemistry 58 (22), 11688–11694.

Hernández, M., Esteve, T., et al., 2005. Real-time polymerase chain reaction based assays for quantitative detection of barley, rice, sunflower, and wheat. Journal of Agricultural and Food Chemistry 53 (18), 7003–7009.

Herrero, B., Lago, F.C., et al., 2011a. Authentication of swordfish (*Xiphias gladius*) by RT- PCR and FINS methodologies. European Food Research and Technology 233 (2), 195–202.

Herrero, B., Lago, F.C., et al., 2011b. Development of a rapid and simple molecular identification methodology for true sardines (*Sardina pilchardus*) and false sardines (*Sardinella aurita*) based on the real-time PCR technique. European Food Research and Technology 233 (5), 851–857.

Herrero, B., Lago, F.C., et al., 2012. Real-time PCR method applied to seafood products for authentication of European sole (*Solea solea*) and differentiation of common substitute species. Food Additives & Contaminants: Part A 29 (1), 12–18.

Herrero, B., Madrinán, M., et al., 2010. Authentication of Atlantic cod (*Gadus morhua*) using real time PCR. Journal of Agricultural and Food Chemistry 58 (8), 4794–4799.

Herrero, B., Royo, L.J., et al., 2013. Authentication of male beef by multiplex fast real-time PCR. Food Additives & Contaminants: Part a 30 (2), 218–225.

Herrero, B., Vieites, J.M., et al., 2011c. Authentication of Atlantic salmon (*Salmo salar*) using real-time PCR. Food Chemistry 127 (3), 1268–1272.

Herrero, B., Vieites, J.M., et al., 2011d. Duplex real-time PCR for authentication of anglerfish species. European Food Research and Technology 233 (5), 817–823.

Hersleth, M., Naes, T., et al., 2012. Lamb meat-importance of origin and grazing system for Italian and Norwegian consumers. Meat Science 90 (4), 899–907.

Hird, H.J., Chisholm, J., et al., 2012. Development of real-time PCR assays for the detection of Atlantic cod (*Gadus morhua*), Atlantic salmon (*Salmo salar*) and European plaice (*Pleuronectes platessa*) in complex food samples. European Food Research and Technology 234 (1), 127–136.

Hischenhuber, C., Crevel, R., et al., 2006. Review article: safe amounts of gluten for patients with wheat allergy or coeliac disease. Alimentary Pharmacology & Therapeutics 23 (5), 559–575.

Hsieh, H., Chai, T., et al., 2007. Using the PCR-RFLP method to identify the species of different processed products of billfish meats. Food Control 18 (4), 369–374.

Ibáñez, J., de Andrés, M.T., et al., 2003. Genetic study of key Spanish grapevine varieties using microsatellite analysis. American Journal of Enology and Viticulture 54 (1), 22–30.

Infante, C., Blanco, E., et al., 2007. Phylogenetic differentiation between Atlantic *Scomber colias* and Pacific *Scomber japonicus* based on nuclear DNA sequences. Genetica 130 (1), 1–8.

Jerome, M., Lemaire, C., et al., 2003. Direct sequencing method for species identification of canned sardine and sardine-type products. Journal of Agricultural and Food Chemistry 51 (25), 7326–7332.

Jerome, M., Martinsohn, J.T., et al., 2008. Toward fish and seafood traceability: anchovy species determination in fish products by molecular markers and support through a public domain database. Journal of Agricultural and Food Chemistry 56 (10), 3460–3469.

Kalogianni, D.P., Bazakos, C., et al., 2015. Olive oil DNA fingerprinting by multiplex SNP genotyping on fluorescent microspheres. Journal of Agricultural and Food Chemistry 63 (12), 3121–3128.

Karaiskou, N., Triantafyllidis, A., et al., 2007. Horse mackerel egg identification using DNA methodology. Marine Ecology 28 (4), 429–434.

Karlsson, A.O., Holmlund, G., 2007. Identification of mammal species using species-specific DNA pyrosequencing. Forensic Science International 173 (1), 16–20.

Kasapidis, P., Magoulas, A., 2008. Development and application of microsatellite markers to address the population structure of the horse mackerel *Trachurus trachurus*. Fisheries Research 89 (2), 132–135.

Kasapidis, P., Silva, A., et al., 2011. Evidence for microsatellite hitchhiking selection in European sardine (*Sardina pilchardus*) and implications in inferring stock structure. Scientia Marina 76 (1), 123–132.

Kathirvelpandian, A., Gopalakrishnan, A., et al., 2014. Microsatellite markers to determine population genetic structure in the Golden anchovy, *Coilia dussumieri*. Biochemical Genetics 52 (5–6), 296–309.

Kesmen, Z., Celebi, Y., et al., 2013. Detection of seagull meat in meat mixtures using real-time PCR analysis. Food Control 34 (1), 47–49.

Kijewska, A., Wiecaszek, B., et al., 2011. Analysis of population and taxonomical structure of Atlantic cod, *Gadus morhua* (Actinopterygii: Gadiformes: Gadidae) from the Baltic Sea with use of microsatellite DNA. Acta Ichthyologica et Piscatoria 41 (4), 307–314.

Kim, S., Li, X., et al., 2010. Development of SNP markers for domestic pork traceability. Journal of Animal Science and Technology 52 (2), 91–96.

Köppel, E., Stadler, M., et al., 1998. Detection of wheat contamination in oats by polymerase chain reaction (PCR) and enzyme-linked immunosorbent assay (ELISA). Zeitschrift für Lebensmitteluntersuchung und -Forschung A 206 (6), 399–403.

Köppel, R., Daniels, M., et al., 2013. Multiplex real-time PCR for the detection and quantification of DNA from duck, goose, chicken, turkey and pork. European Food Research and Technology 236 (6), 1093–1098.

Köppel, R., Dvorak, V., et al., 2010. Two tetraplex real-time PCR for the detection and quantification of DNA from eight allergens in food. European Food Research and Technology 230 (3), 367–374.

Kumar, A., Kumar, R.R., et al., 2012. Species specific polymerase chain reaction (PCR) assay for identification of pig (*Sus domesticus*) meat. African Journal of Biotechnology 11 (89), 15590–15595.

Kumar, D., Singh, S.P., et al., 2014. Authentication of beef, carabeef, chevon, mutton and pork by a PCR-RFLP assay of mitochondrial *cytb* gene. Journal of Food Science and Technology 51 (11), 3458–3463.

Kumar, S., Kahlon, T., et al., 2011. A rapid screening for adulterants in olive oil using DNA barcodes. Food Chemistry 127 (3), 1335–1341.

Lago, F.C., Alonso, M., et al., 2014. Fish and seafood authenticity – species identification. In: Boziaris, I.S. (Ed.), Seafood Processing. John Wiley & Sons, Ltd, pp. 419–452.

Lago, F.C., Herrero, B., et al., 2013a. In: Sforza, S. (Ed.), Authentication and Traceability of Fish and Seafood Species by Means of Molecular Tools. Food Authentication Using Bioorganic Molecules. DEStech Publications, Inc.

Lago, F.C., Herrero, B., et al., 2011a. Authentication of species in meat products by genetic techniques. European Food Research and Technology 232 (3), 509–515.

Lago, F.C., Herrero, B., et al., 2011b. Genetic identification of horse mackerel and related species in seafood products by means of forensically informative nucleotide sequencing methodology. Journal of Agricultural and Food Chemistry 59 (6), 2223–2228.

Lago, F.C., Herrero, B., et al., 2011c. FINS methodology to identification of sardines and related species in canned products and detection of mixture by means of SNP analysis systems. European Food Research and Technology 232 (6), 1077–1086.

Lago, F.C., Vieites, J.M., et al., 2013b. Authentication of gadoids from highly processed products susceptible to include species mixtures by means of DNA sequencing methods. European Food Research and Technology 236 (1), 171–180.

Lanzilao, I., Burgalassi, F., et al., 2005. Polymerase chain reaction-restriction fragment length polymorphism analysis of mitochondrial *cytb* gene from species of dairy interest. Journal of AOAC International 88 (1), 128–135.

Larson, W.A., Seeb, J.E., et al., 2014. Single-nucleotide polymorphisms (SNPs) identified through genotyping-by-sequencing improve genetic stock identification of Chinook salmon (*Oncorhynchus tshawytscha*) from western Alaska. Canadian Journal of Fisheries and Aquatic Sciences 71 (5), 698–708.

Li, X., Li, J., et al., 2013. Novel real-time PCR method based on growth hormone gene for identification of Salmonidae ingredient in food. Journal of Agricultural and Food Chemistry 61 (21), 5170–5177.

Lin, W., Hwang, D., 2008. Application of species-specific PCR for the identification of dried bonito product (Katsuobushi). Food Chemistry 106 (1), 390–396.

Liu, S., Xu, K., et al., 2015. Identification of five highly priced tuna species by quantitative real-time polymerase chain reaction. Mitochondrial DNA (Preprint) 1–10.

Lockley, A.K., Bardsley, R.G., 2000. DNA-based methods for food authentication. Trends in Food Science & Technology 11 (2), 67–77.

López-Calleja, I., González, I., et al., 2007. Application of a polymerase chain reaction to detect adulteration of ovine cheeses with caprine milk. European Food Research and Technology 225 (3–4), 345–349.

López-Calleja, I., Gonzalez, I., et al., 2005. Application of polymerase chain reaction to detect adulteration of sheep's milk with goat's milk. Journal of Dairy Science 88 (9), 3115–3120.

Madesis, P., Ganopoulos, I., et al., 2014. Advances of DNA-based methods for tracing the botanical origin of food products. Food Research International 60, 163–172.

Mafra, I., Ferreira, I.M., et al., 2008. Food authentication by PCR-based methods. European Food Research and Technology 227 (3), 649–665.

Mafra, I., Roxo, A., et al., 2007. A duplex polymerase chain reaction for the quantitative detection of cow's milk in goat's milk cheese. International Dairy Journal 17 (9), 1132–1138.

Mane, B.G., Mendiratta, S.K., et al., 2013. Sequence analysis of mitochondrial 16S rRNA gene to identify meat species. Journal of Applied Animal Research 41 (1), 77–81.

Maretto, F., Reffo, E., et al., 2010. Finding 16S rRNA gene-based SNPs for the genetic traceability of commercial species belonging to Gadiformes. Italian Journal of Animal Science 6 (1s), 161–163.

Mateus, J.C., Russo-Almeida, P.A., 2015. Traceability of 9 Portuguese cattle breeds with PDO products in the market using microsatellites. Food Control 47, 487–492.

McConnell, S.K., O'Reilly, P., et al., 1995. Polymorphic microsatellite loci from Atlantic salmon (*Salmo salar*): genetic differentiation of North American and European populations. Canadian Journal of Fisheries and Aquatic Sciences 52 (9), 1863–1872.

Molina-Luzán, M.J., Lopez, J.R., et al., 2012. Validation and comparison of microsatellite markers derived from Senegalese sole (*Solea senegalensis*, Kaup) genomic and expressed sequence tags libraries. Molecular Ecology Resources 12 (5), 956–966.

Montemurro, C., Miazzi, M.M., et al., 2015. Traceability of PDO olive oil "Terra di Bari" using high resolution melting. Journal of Chemistry 205, 1–7.

Mujico, J.R., Lombardía, M., et al., 2011. A highly sensitive real-time PCR system for quantification of wheat contamination in gluten-free food for celiac patients. Food Chemistry 128 (3), 795–801.

Mullen, M.P., McClure, M.C., et al., 2013. Development of a custom SNP chip for dairy and beef cattle breeding, parentage and research. Interbull Bulletin 47.

Mustalahti, K., Lohiniemi, S., et al., 2002. Gluten-free diet and quality of life in patients with screen-detected celiac disease. Effective Clinical Practice 5 (3), 105–113.

Muths, D., Grewe, P., et al., 2009. Genetic population structure of the Swordfish (*Xiphias gladius*) in the southwest Indian Ocean: sex-biased differentiation, congruency between markers and its incidence in a way of stock assessment. Fisheries Research 97 (3), 263–269.

Muzzalupo, I., Pisani, F., et al., 2015. Direct DNA amplification from virgin olive oil for traceability and authenticity. European Food Research and Technology 1–5.

Negrini, R., Nicoloso, L., et al., 2008. Traceability of four European protected geographic indication (PGI) beef products using single nucleotide polymorphisms (SNP) and Bayesian statistics. Meat Science 80 (4), 1212–1217.

Nicoloso, L., Crepaldi, P., et al., 2013. Recent advance in DNA-based traceability and authentication of livestock meat PDO and PGI products. Recent Patents on Food, Nutrition & Agriculture 5 (1), 9–18.

Nishimura, S., Watanabe, T., et al., 2013. Application of highly differentiated SNPs between Japanese black and Holstein to a breed assignment test between Japanese black and F1 (Japanese black × Holstein) and Holstein. Animal Science Journal 84 (1), 1–7.

Nollet, L.M.L., Toldrá, F., 2009. Handbook of Dairy Foods Analysis. CRC Press.

Oh, J., Song, K., et al., 2014. Genetic traceability of black pig meats using microsatellite markers. Asian-Australasian Journal of Animal Sciences 27 (7), 926.

Oka, T., Tsudzuki, M., 2014. Genetic differentiation in the Oh-Shamo (Japanese large game) breed of chickens assessed by microsatellite DNA polymorphisms. International Journal of Poultry Science 13 (6), 319.

Okuma, T.A., Hellberg, R.S., 2015. Identification of meat species in pet foods using a real-time polymerase chain reaction (PCR) assay. Food Control 50, 9–17.

Olexová, L., Dovicovicová, L., et al., 2006. Detection of gluten-containing cereals in flours and "gluten-free" bakery products by polymerase chain reaction. Food Control 17 (3), 234–237.

Orrú, L., Catillo, G., et al., 2009. Characterization of a SNPs panel for meat traceability in six cattle breeds. Food Control 20 (9), 856–860.

Orrú, L., Napolitano, F., et al., 2006. Meat molecular traceability: How to choose the best set of microsatellites? Meat Science 72 (2), 312–317.

Ortola-Vidal, A., Schnerr, H., et al., 2007. Quantitative identification of plant genera in food products using PCR and Pyrosequencing® technology. Food Control 18 (8), 921–927.

Pafundo, S., Busconi, M., et al., 2010. Storage-time effects on olive oil DNA assessed by amplified fragments length polymorphisms. Food Chemistry 123 (3), 787–793.

Pafundo, S., Gullí, M., et al., 2011. Comparison of DNA extraction methods and development of duplex PCR and real-time PCR to detect tomato, carrot, and celery in food. Journal of Agricultural and Food Chemistry 59 (19), 10414–10424.

Palmieri, L., Bozza, E., et al., 2009. Soft fruit traceability in food matrices using real-time PCR. Nutrients 1 (2), 316.

Pappalardo, A.M., Ferrito, V., 2015. A COIBar-RFLP strategy for the rapid detection of *Engraulis encrasicolus* in processed anchovy products. Food Control 57, 385–392.

Pariset, L., Mariotti, M., et al., 2011. Genetic diversity of sheep breeds from Albania, Greece, and Italy assessed by mitochondrial DNA and nuclear polymorphisms (SNPs). The Scientific World Journal 11, 1641–1659.

Pasqualone, A., 2011. Authentication of durum wheat-based foods: classical vs. innovative methods. In: Beatriz, M., Oliveira, P.P., Mafra, I., Amaral, J.S. (Eds.), Current Topics on Food Authentication. Transworld Research Network, pp. 23–39.

Pasqualone, A., Alba, V., et al., 2010. Durum wheat cultivar traceability in PDO Altamura bread by analysis of DNA microsatellites. European Food Research and Technology 230 (5), 723–729.

Pereira, L., Guedes-Pinto, H., et al., 2011. An enhanced method for *Vitis vinifera* L. DNA extraction from wines. American Journal of Enology and Viticulture 62 (4), 547–552.

Pérez, M., Presa, P., 2008. Validation of a tRNA-Glu-cytochrome b key for the molecular identification of 12 hake species (*Merluccius* spp.) and Atlantic Cod (*Gadus morhua*) using PCR-RFLPs, FINS, and BLAST. Journal of Agricultural and Food Chemistry 56 (22), 10865–10871.

Peter, C., Bruford, M., et al., 2007. Genetic diversity and subdivision of 57 European and Middle-Eastern sheep breeds. Animal Genetics 38 (1), 37–44.

Piknová, L., Brezná, B., et al., 2008. Detection of gluten-containing cereals in food by 5-nuclease real-time polymerase chain reaction. Journal of Food and Nutrition Research (Slovak Republic) 47 (3), 114–119.

Pocwierz-Kotus, A., Kijewska, A., et al., 2015. Genetic differentiation of brackish water populations of cod *Gadus morhua* in the southern Baltic, inferred from genotyping using SNP-arrays. Marine Genomics 19, 17–22.

Pollevick, G.D., Giambiagi, S., et al., 1992. Sex determination of bovine embryos by restriction fragment polymorphisms of PCR amplified ZFX/ZFY loci. Nature Biotechnology 10 (7), 805–807.

Prado, M., Boix, A., et al., 2013. Development of a real-time PCR method for the simultaneous detection of mackerel and horse mackerel. Food Control 34 (1), 19–23.

Price, M.A., 1995. Development of carcass grading and classification systems. In: Morgan Jones, S.D. (Ed.), Quality and Grading of Carcasses of Meat Animals, 27. CRC Press, New York, pp. 173–199.

Prins, T.W., van Dijk, J.P., et al., 2010. Towards a multiplex cereal traceability tool using padlock probe ligation on genomic DNA. Food Chemistry 118 (4), 966–973.

Raieta, K., Muccillo, L., et al., 2015. A novel reliable method of DNA extraction from olive oil suitable for molecular traceability. Food Chemistry 172, 596–602.

Ramos-Gómez, S., Busto, M.D., et al., 2014. Development of a method to recovery and amplification DNA by real-time PCR from commercial vegetable oils. Food Chemistry 158, 374–383.

Ramos, A.M., Megens, H.J., et al., 2011. Identification of high utility SNPs for population assignment and traceability purposes in the pig using high-throughput sequencing. Animal Genetics 42 (6), 613–620.

Rasmussen, R.S., Morrissey, M.T., 2011. Advances in DNA-based techniques for the detection of seafood species substitution on the commercial market. Journal of the Association for Laboratory Automation 16 (4), 308–321.

Rea, S., Storani, G., et al., 2009. Species identification in anchovy pastes from the market by PCR-RFLP technique. Food Control 20 (5), 515–520.

Realini, C.E., Furnols, M.F., et al., 2014. Spanish, French and British consumers' acceptability of Uruguayan beef, and consumers' beef choice associated with country of origin, finishing diet and meat price. Meat Science 95 (1), 14–21.

Recupero, M., Garino, C., et al., 2012. A method to check and discover Adulteration of Nebbiolo-based monovarietal musts: detection of Barbera and Dolcetto cv via SSR analysis coupled with Lab-On-Chip® microcapillary electrophoresis. Food Analytical Methods 6 (3), 952–962.

Rodríguez-Plaza, P., González, R., et al., 2006. Combining microsatellite markers and capillary gel electrophoresis with laser-induced fluorescence to identify the grape (*Vitis vinifera*) variety of musts. European Food Research and Technology 223 (5), 625–631.

Rodríguez-Ramírez, R., Arana, A., et al., 2011. Molecular traceability of beef from synthetic Mexican bovine breeds. Genetics and Molecular Research 10 (4), 2358–2365.

Rojas, M., González, I., et al., 2012. Authentication of meat and commercial meat products from common pigeon (*Columba livia*) woodpigeon (*Columba palumbus*) and stock pigeon (*Columba oenas*) using a TaqMan® real-time PCR assay. Food Control 23 (2), 369–376.

Sánchez, A., Quinteiro, J., et al., 2009. Identification of European hake species (*Merluccius merluccius*) using real-time PCR. Journal of Agricultural and Food Chemistry 57 (9), 3397–3403.

Sandberg, M., Lundberg, L., et al., 2003. Real time PCR for the detection and discrimination of cereal contamination in gluten free foods. European Food Research and Technology 217 (4), 344–349.

Santaclara, F.J., Cabado, A.G., et al., 2006. Development of a method for genetic identification of four species of anchovies: *E. encrasicolus, E. anchoita, E. ringens and E. japonicus.* European Food Research and Technology 223 (5), 609–614.

Sardaro, M.L.S., Marmiroli, M., et al., 2013. Genetic characterization of Italian tomato varieties and their traceability in tomato food products. Food Science & Nutrition 1 (1), 54–62.

Savazzini, F., Martinelli, L., 2006. DNA analysis in wines: development of methods for enhanced extraction and real-time polymerase chain reaction quantification. Analytica Chimica Acta 563 (1–2), 274–282.

Scali, M., Vignani, R., et al., 2012. Genetic differentiation between Cinta Senese and commercial pig breeds using microsatellite. Electronic Journal of Biotechnology 15 (2), 1.

Scott, M., Knight, A., 2009. Quantitative PCR analysis for fruit juice authentication using PCR and laboratory-on-a-chip capillary electrophoresis according to the Hardy-Weinberg law. Journal of Agricultural and Food Chemistry 57 (11), 4545–4551.

Sefc, K.M., Lefort, F., et al., 2001. Microsatellite markers for grapevine: a state of the art. In: Molecular Biology & Biotechnology of the Grapevine. Springer, pp. 433–463.

Serradilla, M.J., Hernández, A., et al., 2013. Authentication of 'Cereza del Jerte' cherry cultivars using real time PCR. Food Control 30 (2), 679–685.

Shackell, G.H., 2008. Traceability in the meat industry – the farm to plate continuum. International Journal of Food Science & Technology 43 (12), 2134–2142.

Sharma, R., Kishore, A., et al., 2015. Genetic diversity and relationship of Indian cattle inferred from microsatellite and mitochondrial DNA markers. BMC Genetics 16 (1), 73.

Siret, R., Boursiquot, J.M., et al., 2000. Toward the authentication of varietal wines by the analysis of grape (*Vitis vinifera* L.) residual DNA in must and wine using microsatellite markers. Journal of Agricultural and Food Chemistry 48 (10), 5035–5040.

Siret, R., Gigaud, O., et al., 2002. Analysis of grape *Vitis vinifera* L. DNA in must mixtures and experimental mixed wines using microsatellite markers. Journal of Agricultural and Food Chemistry 50 (13), 3822–3827.

Sonnante, G., Montemurro, C., et al., 2009. DNA microsatellite region for a reliable quantification of soft wheat adulteration in durum wheat-based foodstuffs by real-time PCR. Journal of Agricultural and Food Chemistry 57 (21), 10199–10204.

Spaniolas, S., Bazakos, C., et al., 2008. Use of lambda DNA as a marker to assess DNA stability in olive oil during storage. European Food Research and Technology 227 (1), 175–179.

Srivastava, G.K., Rajput, N., et al., 2015. Single nucleotide markers of D-loop for identification of Indian wild pig (*Sus scrofa cristatus*). Veterinary World 8 (4), 532–536.

Stroganov, A.N., Buryakova, M.E., et al., 2010. Variability of DNA microsatellite loci in populations of Pacific cod *Gadus macrocephalus* Tilesius (Gadidae). Moscow University Biological Sciences Bulletin 65 (2), 74–77.

Suh, S., Kim, Y., et al., 2014. Assessment of genetic diversity, relationships and structure among Korean native cattle breeds using microsatellite markers. Asian-Australasian Journal of Animal Sciences 27 (11), 1548.

Sun, Y.N., Qin, Y., et al., 2014. Development of polymorphic microsatellite markers and the population genetic structure of the half-fin anchovy, *Setipinna taty*. Genetics and Molecular Research: GMR 13 (3), 6293.

Tatham, A.S., Shewry, P.R., 2008. Allergens to wheat and related cereals. Clinical & Experimental Allergy 38 (11), 1712–1726.

Tedeschi, T., Calabretta, A., et al., 2011. A PNA microarray for tomato genotyping. Molecular BioSystems 7 (6), 1902–1907.

Teletchea, F., 2009. Molecular identification methods of fish species: reassessment and possible applications. Reviews in Fish Biology and Fisheries 19 (3), 265–293.

Terzi, V., Morcia, C., et al., 2005. DNA-based methods for identification and quantification of small grain cereal mixtures and fingerprinting of varieties. Journal of Cereal Science 41 (3), 213–220.

Thomas, M.R., Scott, N.S., 1993. Microsatellite repeats in grapevine reveal DNA polymorphisms when analysed as sequence-tagged sites (STSs). Theoretical and Applied Genetics 86 (8), 985–990.

Tlustos, C., 2009. The dioxin contamination incident in Ireland 2008. Organohalogen Compound 71, 1172–1176.

Tobe, S.S., Linacre, A.M.T., 2008. A multiplex assay to identify 18 European mammal species from mixtures using the mitochondrial cytochrome b gene. Electrophoresis 29 (2), 340–347.

Turan, B.C., Gurlek, M., et al., 2009. Genetic differentiation of Mediterranean horse mackerel (*Trachurus mediterraneus*) populations as revealed by mtDNA PCR-RFLP analysis. Journal of Applied Ichthyology 25 (2), 142–147.

Twito, T., Weigend, S., et al., 2007. Biodiversity of 20 chicken breeds assessed by SNPs located in gene regions. Cytogenetic and Genome Research 117 (1–4), 319–326.

Ulberth, F., Lees, M., 2003. Milk and dairy products. Food Authenticity and Traceability 357–377.

Vietina, M., Agrimonti, C., et al., 2011. Applicability of SSR markers to the traceability of monovarietal olive oils. Journal of the Science of Food and Agriculture 91 (8), 1381–1391.

Wang, C., Xu, L.L., et al., 2015. Selected representative microsatellite loci for genetic monitoring and population structure analysis of miniature swine. Genetics and Molecular Research: GMR 14 (2), 3910.

Wilkinson, S., Archibald, A.L., et al., 2012. Development of a genetic tool for product regulation in the diverse British pig breed market. BMC Genomics 13 (1), 580.

Wu, Y., Zhang, H., et al., 2011. PCR-CE-SSCP applied to detect cheap oil blended in olive oil. European Food Research and Technology 233 (2), 313–324.

Yang, S., Li, X., et al., 2014. A genome-wide scan for signatures of selection in Chinese indigenous and commercial pig breeds. BMC Genetics 15 (1), 7.

You, J., Huang, L., et al., 2014. Species-specific multiplex real-time PCR assay for identification of deer and common domestic animals. Food Science and Biotechnology 23 (1), 133–139.

Zeleny, R., Schimmel, H., 2002. Sexing of beef – a survey of possible methods. Meat Science 60 (1), 69–75.

Zeltner, D., Glomb, M.A., et al., 2009. Real-time PCR systems for the detection of the gluten-containing cereals wheat, spelt, kamut, rye, barley and oat. European Food Research and Technology 228 (3), 321–330.

Zhang, H., Wu, Y., et al., 2012. PCR-CE-SSCP used to authenticate edible oils. Food Control 27 (2), 322–329.

Zhang, L., Wu, G., et al., 2009. The gene MT3-B can differentiate palm oil from other oil samples. Journal of Agricultural and Food Chemistry 57 (16), 7227–7232.

Zhivotovsky, L.A., Shaikhaev, E.G., et al., 2013. Identification of salmonid fish using microsatellite markers with identical PCR-primers. Russian Journal of Marine Biology 39 (6), 447–454.

Near Infrared Spectroscopy and Food Authenticity

7

D. Cozzolino
Central Queensland University, Rockhampton, QLD, Australia

1. Introduction

Developments in both hardware (eg, sensors, computers, instruments) and software (eg, new algorithms) have resulted in the growth of analytical and instrumental methods based in vibrational spectroscopy techniques such as Near Infrared (NIR) and Mid Infrared (MIR). These methods have been extensively used to qualitatively (eg, traceability and authenticity) and quantitatively (eg, measure composition) analyze several food matrices (Arvantoyannis et al., 2005; Karoui et al., 2010; Weeranantanaphan et al., 2011; Cozzolino, 2009, 2012; Downey, 2013).

Food authenticity is a well-established field of research, which involves input from different analytical techniques such as Gas Chromatography (GC), Mass Spectrometry (MS), NMR, DNA fingerprinting, and vibrational spectroscopy (NIR, MIR, Raman) (Cordella et al., 2002; Downey, 2013). However, there still remains a division over which analytical strategy to use in order to develop methods to target specific authenticity issues, identify adulterant in food products, or develop nontargeted methods with the aim of detecting deviations from the norm (Downey, 2013).

This chapter will provide a general introduction to NIR spectroscopy and chemometrics. In addition, examples on the applications and uses of NIR spectroscopy targeting issues related with authenticity and traceability in several food products such as fish, meat, honey, and wine will be discussed.

2. Traceability and Authenticity

The increasing market and consumer desire for quality of food have created a need for efficient and accurate analytical methods for the authentication and traceability of food. Several analytical methods have been or are currently in use to authenticate or trace food such as UV spectroscopy, High Performance Liquid Chromatography (HPLC), GC, Liquid Chromatography (LC), and MS (Cordella et al., 2002; Downey, 2013). Although these methods provide a high level of information about the chemistry of the compound measured, the sample requires different steps of preprocessing (eg, extraction, purification) before and during the analysis. Both NIR and MIR spectroscopy, with its intrinsic benefits such as being noninvasive and rapid and requiring almost no necessary sample preparation, have been able to determine physical and chemical parameters in different foods simultaneously.

Advances in Food Traceability Techniques and Technologies. http://dx.doi.org/10.1016/B978-0-08-100310-7.00007-7

The authentication of foods and agricultural products is of primary importance to keep consumers demands and to maintain the sustainable nature of the food industry. From a legislative point of view, quality standards were established by the industry and government in order to set the requirements for quality labels that specify the chemical composition of raw materials and foods. From an economic point of view, product authentication is essential in order to avoid unfair competition that can eventually create a destabilized market and disrupt regional or national economies (Arvantoyannis et al., 2005; Karoui et al., 2010; Weeranantanaphan et al., 2011; Cozzolino, 2009, 2012).

The authenticity of foods has two different aspects: (1) authenticity with respect to production (eg, geographical origin, organic vs nonorganic) and (2) authenticity with respect to the description (eg, adulteration issues) (Arvantoyannis et al., 2005; Weeranantanaphan et al., 2011). For example, geographical and botanical origins of food, such as "natural," "organic," "raw," "unheated," and "fresh versus frozen" have been analyzed using different methods and techniques (eg, volatile compounds, minerals, carbohydrates, and protein identification) (Arvantoyannis et al., 2005; Karoui et al., 2010; Weeranantanaphan et al., 2011; Cozzolino, 2009, 2012).

Food adulteration has been practiced since ancient times; however, it has become more sophisticated. The practice of adulteration commonly arises for two main reasons: firstly it can be profitable, and secondly, adulterants can be easily mixed and are subsequently difficult to detect (Karoui et al., 2010; Weeranantanaphan et al., 2011; Cozzolino, 2009, 2012). Foods or ingredients most likely to be targets for adulteration include those which are of high value or are subject to the vagaries of weather during their growth or harvesting (Karoui et al., 2010; Weeranantanaphan et al., 2011; Cozzolino, 2009, 2012). Adulteration can also take many forms, including the addition of sugars, acids, volatile oils, overdilution of concentrate, addition of juices of other fruits, use of concentrate in a "fresh" product, or the use of low quality product recovered from what are normally waste products of manufacture as premium products (Karoui et al., 2010; Weeranantanaphan et al., 2011; Cozzolino, 2009, 2012).

3. The Tools (Hardware and Software)

3.1 Near Infrared Spectroscopy

Infrared (IR) spectroscopy is originated from the absorption measurements of different IR frequencies by a sample positioned in the path of an IR beam (eg, NIR and MIR beams). When the frequency of a specific vibration is equal to the frequency of the IR radiation directed at the molecule, this molecule absorbs the radiation (Abbas et al., 2012; Woodcock et al., 2008; Karoui et al., 2010; Rodriguez-Saona and Alledendorf, 2011; Cozzolino, 2011; Cozzolino and Murray, 2012; Downey, 2013).

Absorption spectroscopy is based on the Beer–Lambert law stating that for a homogeneous and nonscattering liquid sample, the concentration of an absorber is proportional to the sample absorbance. Near IR is characterized by the overlapping of many different overtone and combination vibrations (vibrational modes of C—H, N—H, and O—H chemical bonds) resulting in broad bands that result in NIR spectra data with low structural selectivity (compared with MIR spectra where fundamentals are more resolved, allowing the structure of a sample to be better elucidated). Overtones correspond to energy transitions that are higher than those for fundamentals (MIR), and in particular, the frequencies of the first and second overtones correspond to about two or three times that of the fundamentals observed in the MIR region. Combination bands result from transitions involving two or more different vibrational modes of one functional group occurring simultaneously: the frequency of a combination band is the sum or the multiples of the relevant frequencies. The absorption intensity decreases when the overtone level increases. However, the higher energy of NIR radiation and the implication of combination vibrations enable NIR spectroscopy to provide more complex structural information than in the MIR region. NIR spectroscopy is widely used to determine organic matter constituents in several types of samples (eg, liquids, solids, gas), as seen in Fig. 7.1. It is based on the absorption of electromagnetic radiation by a sample at wavelengths in the range between 800 and 2500 nm (Abbas et al., 2012; Woodcock et al., 2008; Karoui et al., 2010; Rodriguez-Saona and Alledendorf, 2011).

Different instruments have been used or are under development in order to target traceability and authenticity issues in several food matrices. NIR spectrophotometers are conveniently classified according to the type of monochromators. For example, in a filter instrument, the monochromator is a wheel holding a number of absorption or interference filters, while its spectral resolution is limited. In a scanning monochromator instrument a grating or a prism is used to separate the individual frequencies of the radiation either entering or leaving the sample. The wavelength divider rotates, allowing radiation of the individual wavelengths to subsequently reach the detector. The development of Micro Electro Mechanical Systems (MEMS) combine mechanical parts, sensors, actuators, and electronics on a common substrate through the use of microfabrication technology. Other developments in instrumentation include focal plane array cameras in combination with liquid crystal tunable filter, acousto-optic

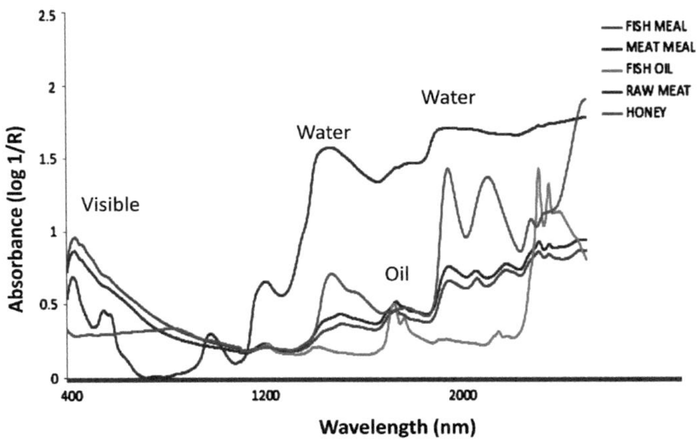

Figure 7.1 Figure shows typical NIR spectra of foods and agricultural products. The visible, water, and oil regions are also indicated.

tunable filters, the use of LEDs, or other monochromator principles which allow for much faster acquisition spectra. Multispectral and hyperspectral imaging systems are being incorporated to the myriad of instruments available. Multispectral (a few wavelengths) or hyperspectral (a continuous range of wavelengths) imaging systems produce a spectral data cube—a spectrum at every 2-D spatial position. A sequence of spatial images was acquired by means of an NIR camera and a set of band-pass filters. However, while the acquisition can be fast, the disadvantage of this approach is that only a limited number of wavelengths can be analyzed, and that they need to be known a priori. This technology represents a paradigm shift for industrial spectroscopy and enables a variety of new industrial applications (McClure, 2004; Cozzolino et al., 2003, 2009; Roggo et al., 2007; Nicolai et al., 2007; Huang et al., 2008; Manley, 2014). Table 7.1 summarizes and compares the advantages and disadvantages of NIR spectroscopy with other common methods and techniques used in authenticity.

3.2 Multivariate Data Methods

The combination of chemometrics or multivariate analysis with analytical instruments have the ability to determine more than one component at a time and can act as a support to establish links to other sample characteristics (McClure, 2003; Karoui et al., 2010; Weeranantanaphan et al., 2011; Cozzolino, 2009, 2012). Chemometrics covers quite a broad range of methods such as exploratory data analysis, Pattern Recognition (PR), and statistical Experimental Design (DoE) (Arvantoyannis et al., 1999; Naes et al., 2002; Downey, 1996, 1998).

The most commonly used techniques applied in the field of food authenticity are Principal Component Analysis (PCA), Partial Least Squares (PLS), and Principal Component Regression (PCR). Chemometrics, unlike classical statistics, considers multiple variables simultaneously and takes collinearity (the variation in one variable, or a group of variables, in terms of covariation with other variables) into account. Calibration development can mathematically describe the covariation (degree of association) between variables or find a mathematical function (regression model), by which

Table 7.1 Advantages and Disadvantages of NIR Spectroscopy Compared With Other Common Methods and Techniques Used in Authenticity

Advantages	Drawbacks
Nondestructive	Low sensitivity compared with other techniques such as NMR, MS
High reproducibility	Low selectivity compared with other methods
At, on, inline analysis	
Little or no sample preparation required	
Analysis of liquids, pastes, solids	

Adapted from Khakimov, B., Gürdeniz, G., Engelsen, S.B., 2015. Trends in the application of chemometrics to foodomics studies. Acta Alimentaria 44, 4–31.

the values of the dependent variables are calculated from values of the measured (independent) variables (Tzouros and Arvantoyannis, 2001; Naes et al., 2002; Reid et al., 2006; Oliveri et al., 2011; Blanco and Villaroya, 2002; Brereton, 2003, 2007).

PCA is used as a tool for screening, extracting, and compressing multivariate data. PCA employs a mathematical procedure that transforms a set of possibly correlated response variables into a new set of noncorrelated variables, called principal components (Fig. 7.2). PCA can be performed on either a data matrix or a correlation matrix depending on the type of variables being measured (Tzouros and Arvantoyannis, 2001; Naes et al., 2002; Reid et al., 2006; Oliveri et al., 2011; Blanco and Villaroya, 2002; Brereton, 2003, 2007). PCA produces linear combinations of variables that are useful descriptors or even predictors of some particular structure in the data matrix. However, according to some authors, PCA might not be the most adequate method to use (Downey, 2013).

Discriminant Analysis (DA), Linear Discriminant Analysis (LDA), and Partial Least Squares Discriminant Analysis (PLS-DA) can be considered qualitative calibration methods, and they are the most used methods in authenticity. Instead of calibrating for a continuous variable, one calibrates for group membership (categories). The resulting models are evaluated in terms of their predictive ability to predict the new and unknown samples (Standard Error of Prediction, SEP). Discrimination models are usually developed using PLS regression techniques, as described elsewhere. LDA is a supervised classification technique. The criterion of LDA for the selection of latent variables is maximum differentiation between the categories and minimal variance within categories. This method produces a number of orthogonal linear discriminant functions, equal to the number of categories minus one, that allow the samples to be classified in one or another category (Tzouros and Arvantoyannis, 2001; Naes et al., 2002; Reid et al., 2006; Oliveri et al., 2011).

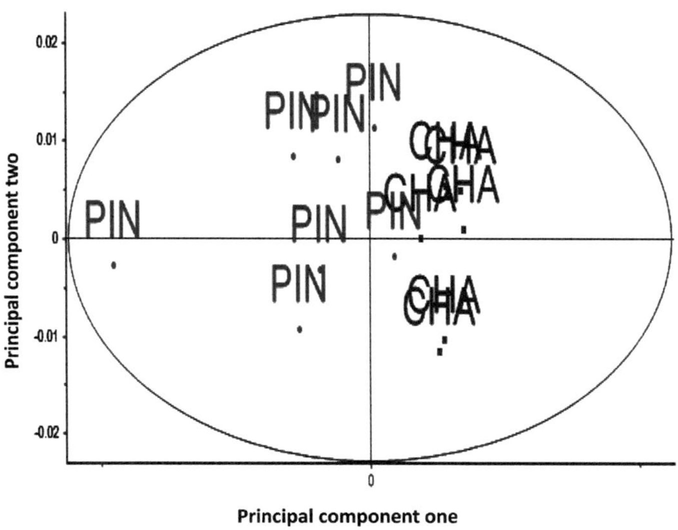

Figure 7.2 A typical example on the use of principal component score plots to visualize the patterns related with different wine styles. *PIN*, Pinot; *CHA*, Chardonnay.

The application of Artificial Neural Networks (ANN) is a technique for data and knowledge processing that is characterized by its analogy with a biological neuron (Otto, 1999; Naes et al., 2002). Unlike linear regression, PCR, and PLS, ANN can deal with nonlinear relationships between variables. Both PLS and PCA are the two most powerful tools for data analysis which require instrumental methods to extract the information about quality attributes hidden in the data ("model calibration") (Tzouros and Arvantoyannis, 2001; Naes et al., 2002; Reid et al., 2006; Oliveri et al., 2011).

4. Examples and Applications

Near infrared spectroscopy has been used as tool to authenticate, trace, or discriminate several foods and agricultural products. This section illustrates, with examples, different uses and applications of NIR spectroscopy to authenticate or trace fish, meat, wine, and honey products.

4.1 Fish and Fish By-Products

4.1.1 Geographical Origin

The proximate chemical composition of tilapia fillets sourced from four distinctive geographical regions of China (Guangdong, Hainan, Guangxi, and Fujian Province) was assessed using NIR spectroscopy. The method proposed by Liu et al. (2006) provided an example of fish traceability. The tilapia origin traceability model was developed using NIR spectroscopy coupled with Soft Independent Modeling of Class Analogy (SIMCA). The authors reported that when classifying tilapia by means of SIMCA, more than 80% of samples from the Guangdong, Hainan, and Fujian regions and 75% of fillets from the Fujian region were correctly and exclusively assigned to the corresponding clusters. Overall, only 1% to 2% of the fish samples were classified incorrectly. The results derived from this study indicated that NIR spectroscopy coupled with pattern recognition methods such as SIMCA were a feasible way to determine the origin and traceability of export tilapia fillets (Liu et al., 2006).

4.1.2 Fresh Versus Frozen

The potential of Visible (VIS) and NIR (400–1000 nm) hyperspectral imaging as a rapid and noninvasive method was investigated to differentiate freshness in prawn samples. Unfrozen and frozen groups, as well as two different levels of freshness, were used for classification. Spectral data were obtained from whole prawn samples and preprocessed using derivatives (first and second). The Successive Projections Algorithm (SPA) was used to select the individual wavelengths for classification. Several other algorithms were also used, such as Least Squares-Support Vector Machine (LS-SVM), Adaptive Boosting (AdaBoost), and Back-Propagation Neural Networks (BP-NN). The results from this study demonstrated that SPA-LS-SVM achieved satisfactory average correct classification rates (98.33% and 95%) for the prediction of samples in the unfrozen and frozen groups, respectively. The overall results showed that the hyperspectral imaging technique is promising for the rapid and noninvasive freshness classification of prawns (Dai et al., 2015).

The substitution of fresh fish with frozen-thawed fish is a typical fraud that not only damages consumers from an economical point of view, but also causes safety issues. It is in this context that fish authentication is important for correct product labeling, as promoted by recent regulatory actions. The results of the authentication of fresh West African goatfish (*Pseudupeneus prayensis*) fillets using a portable VIS-NIR instrument, a compact digital camera, and a texture analyzer were reported by Ottavian et al. (2014). The classification performance of the different analytical technologies and how the fusion of different technologies can be effectively used to improve the classification accuracy was reported by the authors. The combination of VIS-NIR and the digital camera was considered as very promising for cost-effective online/at-line applications, as both sensors are rapid and noninvasive and do not require sample preparation (Ottavian et al., 2014).

The rapid and nondestructive measurement of freshness is essential for the control of fish and the quality and safety of its products. The prediction of freshness in silver chub samples were developed using fourier transform near infrared (FT-NIR) spectroscopy using several algorithms: PLS, interval-PLS, synergy interval-PLS, and SVM. The results reported by these authors indicated that FT-NIR spectroscopy together with SVM could be a reliable method for the detection of fish freshness (Ding et al., 2014).

4.1.3 Authentication

The potential use of NIR for the authentication of raw and cooked freeze-dried rainbow trout (*Oncorhynchus mykiss*) fillets was evaluated using different classification techniques. Latent variable models applied on the spectral data were developed and used to estimate proximate composition, fatty acid profile, fillet yield, and cooking loss, and to classify the available dataset by the rearing farm and genetic strain of each sample. The results from this study showed that NIR spectra can be used both to accurately estimate several chemical properties and to classify samples by rearing farm and by processing method (raw vs cooked) (Zotte et al., 2014).

The use of NIR spectroscopy was investigated for the analysis of whole fish and fish fillets from different species. O'Brien et al. (2013) have evaluated a miniaturized, handheld instrument to discriminate between high-quality and expensive from lower-quality, less expensive fillets. According to these authors, such issues related with authenticity can occur in practice due to the mislabeling of products in the harvesting or processing plant or to the fraudulent substitution of high-quality and low-quality species at the restaurant level. To test the possibility of distinguishing superior from lower-quality fish species, NIR spectra were measured in diffuse reflection from the skin and meat of the investigated fish. Subsequently, the spectra were evaluated using PCA and further classified using SIMCA. The authors reported good classification results in order to distinguish between skin and meat and to discriminate between different fish species, such as salmon trout versus samlet, cod versus winter cod, and between mullet versus red mullet, as well as to differentiate between the different fish origins (France, Senegal, Norway, Iceland, and Italy) (O'Brien et al., 2013).

4.2 Meat and Meat By-Products

More than ever, consumers today demand clear and reliable information about the food they consume. This issue has a great impact on the economy since the consumer's choice is greatly influenced by the food composition detailed in the label. In the case of processed meat products, this is going to be especially important because a simple visual inspection would not allow us to discriminate between the different components so easily (as in the case of fresh meat). In order to assure fair trade, food safety, and freedom of choice, honest and accurate food labels are a requisite that must be assured by legal authorities. In order to achieve this, robust and reliable analysis methodologies must be implemented in control laboratories. Many consumers place particular emphasis on noncompositional aspects of meat as food that are not related to composition (eg, protein, fat, color), and they include geographical origin, production methods, rearing, or feeding systems (Weeranantanaphan et al., 2011; Sentandreu and Sentandreu, 2014; Kamruzzaman et al., 2015).

4.2.1 Raw Versus Processed Meats

A nondestructive method based on VIS-NIR spectroscopy to rapidly and accurately discriminate between raw meat and meat injected with water was evaluated. Water-injected meat samples were built by injecting water into nondestructed meat samples consisting of pigskin, fat layers, and muscle layers. The VIS-NIR spectra of raw meat and meat samples injected with water were collected and analyzed. Effective spectral NIR bands were extracted from the preprocessed spectral data and used to classify samples according to raw meat or water-injected meat, as well as different levels of water-injected meat. These results showed that the overall recognition accuracy to distinguish between raw and water-injected meat was higher than 90%, using SVM. For water-injected meat samples with different water injection rates, the recognition accuracy showed a positive correlation with the water injection rate difference. The results also demonstrated that the proposed method can be effectively used for discriminating water-injected meat from raw meat (Hag et al., 2015).

4.2.2 Fresh Versus Frozen

Fresh and frozen-thawed pork meats were classified using VIS-NIR hyperspectral imaging using eight wavelengths from the VIS range as follows: 624, 673, 460, 588, 583, 448, 552, and 609 nm. The first three Principal Components (PCs) obtained using PCA accounted for over 99.98% of the variance. Classification models were developed using PLS-DA with the reflected spectra at full wavelengths and those at the optimal wavelengths, the extracted textures based on the PC images, and the fused variables combining spectra at the optimal wavelengths and textures. The results showed that the best classification rate (97.73%) was achieved, confirming the high potential of textures for fresh and frozen-thawed meat discrimination (Ma et al., 2015).

4.2.3 Adulteration

Multispectral imaging supported by multivariate data analysis was used to detect minced beef fraudulently substituted with pork and vice versa. Multispectral images in 18 different wavelengths of 220 meat samples in total from four independent experiments (55 samples per experiment) were acquired for this study. The appropriate amount of beef and pork-minced meat was mixed in order to achieve nine different proportions of adulteration and two categories of pure pork and beef. After an image processing step, data from the first three experiments were analyzed using PLS-DA and LDA. The results reported by Ropodi et al. (2015) showed very good discrimination between pure and adulterated samples, for PLS-DA and LDA, yielding 98.48% overall correct classification. Additionally, 98.48% and 96.97% of the samples were classified within a ± 10% category of adulteration for LDA and PLS-DA, respectively. Lastly, the models were further validated using the data of the fourth experiment for independent testing, where all pure and adulterated samples were classified correctly in the case of PLS-DA, while LDA was proved to be less accurate (Ropodi et al., 2015).

Donkey meat samples from different parts of a donkey body (neck, rump, and tendon), beef, pork, and mutton samples were used to develop classification models based in NIR spectroscopy. The accuracies of classification models were developed using Mahalanobis distances, SIMCA, and LS-SVM, respectively, in combination with pretreatment such as derivatives (first and second), Multiplicative Scatter Correction (MSC), and Standard Normal Variate (SNV). The optimal models for intact samples obtained by Mahalanobis distances correctly classified 100% of the samples in the calibration set and 98.96% in the prediction set. The authors of this study also examined different minced samples with different accuracies using different algorithms where donkey meat samples were correctly classified with 100% rate in either calibration or prediction. Overall, these results showed that it is feasible that NIR spectroscopy combined with chemometrics discriminate donkey meat from other sources of meat (Niu et al., 2014).

The potential of VIS-NIR spectroscopy to discriminate between dark cuts from normal cuts of beef samples was examined using commercial meat samples. One hundred and twenty beef carcass sides were selected from a slaughter plant by experienced graders according to their carcass grade, such as 60 A grade carcasses (normal) and 60 B4 grade carcasses (dark cuts) were classified using PLS-DA. Two different instruments were evaluated delivering similar classification rates (90–95%). According to Prieto et al. (2014), further work remains to be carried out to develop robust VIS-NIR models to be implemented online in the abattoir, where portable equipment applied directly on the carcass could objectively assist in dark-cutting carcass segregation.

4.3 Meals and Feeds

4.3.1 Identification of Animal Species

Since the bovine spongiform encephalopathy (BSE) crisis, the use of animal proteins in animal feed has been prohibited by the European Union (EU) (Cozzolino and Murray, 2004, 2012). NIR microscopy was used to identify different classes of animal by-products. Samples of fish meals and meals of land animal origin were ground, sedimented, and analyzed using an Auto Image Microscope connected to an FT-NIR. Sediment fraction particles were spread on a Spectralon plate, presented to the NIR microscope, and scanned in the region between 1112 and 2500 nm. The Support Vector

Machine (SVM) algorithm was used to construct models to identify class origin. Models correctly classified 100% of the samples in the calibration set and between 95% and 95.5% in the validation set. The results demonstrated the potential of FT-NIR microscopy as a rapid method for distinguishing between fish and land animal particles (De la Haba et al., 2007).

The use of NIR spectroscopy combined with chemometrics was used to identify and authenticate fishmeal batches made with different fish species. Samples from a commercial fishmeal factory were scanned in the NIR region (1100–2500 nm) in reflectance. Several methods such as PCA, DPLS, and LDA based on PCA scores were used to identify the origin of fishmeal produced using different fish species. The DPLS modes correctly classified 80% and 82% of the fishmeal samples while LDA models correctly classified >80% of fishmeal samples according to fish species (Cozzolino et al., 2002, 2005; Cozzolino and Murray, 2012).

Current and future legislation regarding the use of processed animal proteins in animal nutrition requires the availability of robust analytical methods that allow for the proper implementation of corresponding legal restrictions. The use of NIR microscopy (NIRM) is a method that allows for the differentiation between meat and bone meal and fishmeal, and it is assumed that the different content and composition of the fat is one of the factors responsible for the observed differences. The use of NIRM has been evaluated in order to check for the influence of intentionally introduced reduction of the fat content on the capability of the NIRM method to correctly classify defatted samples. This has practical implications, since processed animal proteins may be defatted by solvents under real-world conditions. The results reported by Tena et al. (2014) confirmed that the scope of the NIRM method could be successfully extended to samples that have been previously extracted with non-polar solvents. Only after the use of stricter techniques, such as extraction with chlorinated solvents or hydrolysis, the NIRM method produced a higher portion of wrong classifications. However, since these extraction techniques are not often used under real-world conditions, the impact upon the use of the NIRM method in the feed sector for the specific application of the differentiation between meat and bone meal and fishmeal is minor (Tena et al., 2014).

4.4 Wine

4.4.1 Identification of Grape Variety

Verification of authenticity of wine and other alcoholic beverages by means of NIR spectroscopy has become a field of increase expansion in both research and in the wine industry. Examples on the use of VIS-NIR spectroscopy to discriminate between wines made from several grape varieties (eg, Riesling, unwooded Chardonnay, Shiraz, Sauvignon Blanc, Tempranillo) or produced from different geographical origins (Australia, New Zealand, Spain, Italy) where different classification accuracies were reported by several authors (Fig. 7.3). Different multivariate data methods were used including PCA, LDA, PLS-DA, and SIMCA (Cozzolino et al., 2003, Liu et al., 2008; Arvantoyannis, 2010; Riovanto et al., 2011; Casale et al., 2010).

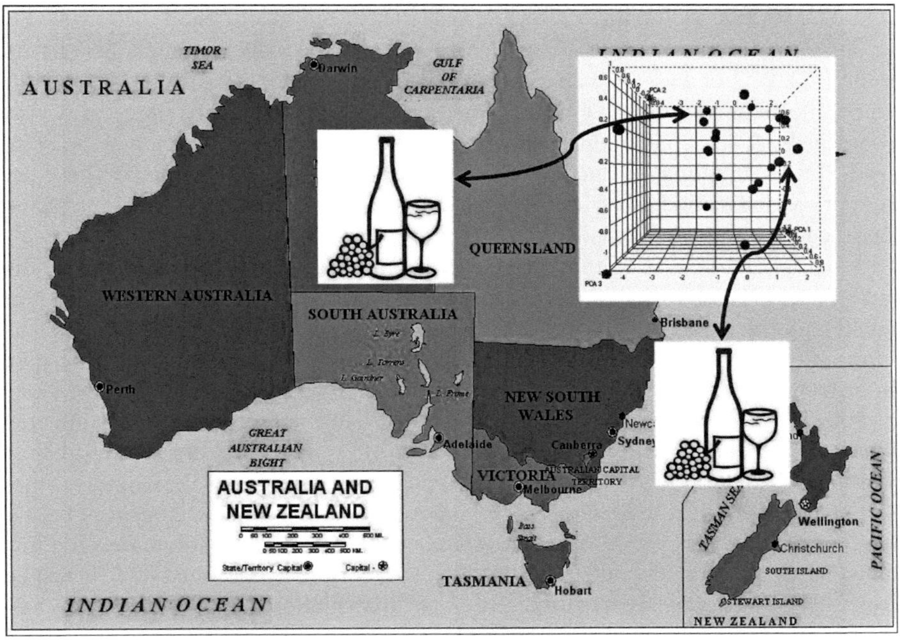

Figure 7.3 Schematic representation of the use of NIR spectroscopy to target wine authenticity. The use of a principal component score based on the NIR spectra of wines allows the identification of the origin of wines from either Australia or New Zealand.

4.4.2 Denomination, Designation, and Geographical Origin

Chemometrics combined with ultraviolet (UV) and VIS-NIR spectroscopy were evaluated to classify wines from different Spanish-controlled Designation of Origin (DO), namely, Rías Baixas (North Spain). Wine samples made solely from the Albariño cultivar from subzones Condado, Salnés, Rosal, and Ribeira do Ulla were analyzed in transmittance mode using the UV, VIS, and NIR ranges alone or in combination. PCA was used to explore and determine patterns in the data where SIMCA, LDA, and SVM classification methods were used to classify the samples. According to the authors the different combinations of spectral ranges and classification methods yield different rates of classification depending on the region. For example, wines originating from the Condado region were completely classified using SIMCA and combinations of the UV/VIS/NIR, NIR, or VIS/NIR spectral ranges, while wines from Salnés and Ribeira do Ulla were completely classified using SIMCA and the UV/VIS spectral range. These results showed that combinations of UV/VIS/NIR spectra and chemometric methods can be used as a rapid technique to classify subzones into a DO such as Rías Baixas. This technique may also assist authentication because the models also discriminated wines from outside Rías Baixas (Martelo Vidal et al., 2013).

The use of NIR spectroscopy was evaluated to distinguish the wines produced in two close provinces of the Abruzzo region of Italy. A collection of 32 wines was considered, 18 of which were produced in the province of Chieti, while the other 14 were from the province of Teramo. A conventional dual-beam spectrophotometer was used for absorption measurements in the 1300–1900 nm spectroscopic range. Score maps in the PC1-PC2 or PC2-PC3 spaces were obtained, which successfully grouped the wine samples in two distinct clusters, corresponding to Chieti and Teramo provinces, respectively. A modeling of dual-band spectroscopy was also proposed, making use of two LEDs for illumination and a PIN detector instead of the spectrometer. These data were processed using Linear Discriminant Analysis, which demonstrated satisfactory classification results (Mignani et al., 2014).

Geographical origins of wine grapes are significant factors affecting wine quality and wine prices, and it is important to discriminate different wine original regions quickly and accurately. The determinations of wine regions based on Bayesian information fusion that fused NIR transmission spectra information and MIR spectra information of wines were reported. In this method, Xiang et al. (2014) aimed to improve the determination results by expanding the sources of analysis information (combination of NIR and MIR). The four different regions are Huailai, Yantai, Gansu, and Changli, which are all typical geographical originals for Chinese wines. NIR and MIR discriminant models for wine regions were established using PLS-DA based on NIR spectra and MIR spectra separately. The discriminant models reported yield in average accuracy rates of calibration sets between 78.1% and 82.7%, respectively. These results suggested that Bayesian information fusion of NW and MW spectra is feasible for fast identification of wine original regions (Xiang et al., 2014).

4.5 Honey

4.5.1 Floral Origin

Similar to wine, the use of vibrational spectroscopy to authenticate honey has been expanded (Ruoof et al., 2006; Woodcock et al., 2007, 2009; Cozzolino et al., 2011). The feasibility of NIR spectroscopy and multivariate analysis as tools to classify Chinese honey samples according to their different floral origins was explored. Five kinds of honey, namely, acacia, linden, rape, vitex, and jujube, were analyzed using an NIR spectrophotometer with a fiber optic probe. Classification models based on the NIR spectra were developed using Mahalanobis-Distance Discriminant Analysis (MD-DA) and a Back Propagation Artificial Neural Network (BP-ANN). By the MD-DA model, total correct classification rates of 87.4% and 85.3% were observed for the calibration and validation samples, respectively, while the ANN model resulted in total correct classification rates of 90.9% and 89.3% for the calibration and validation sets, respectively. By ANN, the respective correct classification rates of linden, acacia, vitex, rape, and jujube were 97.1%, 94.3%, 80.0%, 97.1%, and 85.7% in calibration, and 100%, 93.3%, 80.0%, 100%, and 73.3% in validation. The results indicated that NIR combined with a classification technique could be a suitable technology for the classification of Chinese honeys from different botanical origins (Chen et al., 2012).

4.5.2 Adulteration With Sugars

The implementation of VIS-NIR spectroscopy for the detection of glucose concentration in a mixture of Saudi and imported honey samples were adulterated by glucose syrup using five concentrations: 0, 5, 12, 19, and 33 g/100 g, respectively. The results obtained by these authors suggested that NIR spectroscopy is a powerful technique for the quantification of glucose adulteration in Saudi honey (Mouazen and Wallan, 2014).

The combination of NIR spectroscopy and chemometrics has been evaluated to detect adulteration in honey samples. Several algorithms and preprocessing methods were used such as wavelet transformation (WT), PCA, LS-SVM, SVM, BP-ANN, LDA, and *K* nearest neighbors. These authors reported that WT was the most effective method for variable selection, and the best classification models were achieved using LS-SVM (accuracy of 95.1%) (Zhu et al., 2010).

4.5.3 Geographical Origin

The information contained in the NIR spectra of honey samples sourced from Protected Geographical Indication (PGI), namely, "Mel de Galicia," was processed by means of different chemometric techniques to develop an authentication system for this type of sample. According to these, authors classification based on SIMCA achieved the best classification model with 93.3% of sensitivity and 100% of specificity. The main conclusion from these authors was that the combination of NIR information data with SIMCA allowed the development of a single and fast method to differentiate between genuine PGI-Galician honey samples and other commercial honey samples from other origins (Herrero Latorre et al., 2013).

4.5.4 Brand Identification

The discrimination of four commercial brands of Portuguese honey samples according to their botanical origin using a combination of sensors such as fusion of impedance Electronic Tongue (E-Tongue) and optical spectroscopy (UV–VIS–NIR) with PCA and Cluster Analysis (CA) was explored. A new technique for variable selection through one-dimensional clustering, which proved very useful for data fusion, was also evaluated. The results were validated against the standard sample identification method using melissopalynology analysis. Both the electronic and optical spectra were fitted to analytical models, and the model coefficients were used as new variables for PCA and CA. According to Ulloa et al. (2013), this approach has improved honey classification by the E-Tongue but not by the optical methods. The use of Multiway PCA (MPCA) was the best option for data fusion, yielding 100% classification success. The authors have demonstrated that the use sensor fusion (electronic and optical spectroscopy) data and variable selection methods significantly improved the classification of the honey samples through multivariate statistical analysis (Ulloa et al., 2013).

5. Concluding Remarks

This chapter has shown how NIR spectroscopy has been developed as one of the most versatile methods to target issues associated with authenticity, discrimination, or traceability in several foods. The main advantages of NIR spectroscopy over traditional chemical and chromatographic methods (eg, HPLC, GC, GC–MS) are the speed, minimal sample preparation, and ease to use in an industrial setting or routine operations. However, adapting and applying this method to efficiently and consistently monitor authenticity, we need to increase our understanding about the chemical and biochemical basis associated with origin/authenticity/traceability derived from the NIR spectra, in order to maintain a sustainable food production and to guarantee to the consumers the origin of the foods. However, some of the disadvantages and limitations of NIR spectroscopy are related with the lack of sensitivity and selectivity compared to other tools such as MIR and Raman spectroscopy. NIR absorption bands are generally broad and overlapping therefore to extract useful information from NIR spectral data, and multivariate data analysis tools such as PCA and PLS are necessary. In addition, the use of the multivariate data analysis makes the method validation very challenging.

New approaches, such as the use of new algorithms, data preprocessing, the combination of different sensors (sensor fusion), and the developments in hyperspectral spectroscopy proved to be an alternative to the "classical" use of NIR spectroscopy in order to improve the classification accuracy of the models.

In the near future, we can imagine that portable and easy to use NIR instruments (eg, instruments and apps in mobile phones) will allow consumers to verify the authenticity of their foods as well as to trace the origin of the product at the supermarket, acting as the first line of defense against food fraud.

References

Abbas, O., Dardenne, P., Baeten, V., 2012. Near-infrared, mid-infrared, and Raman spectroscopy. In: Chemical Analysis of Food: Techniques and Applications. Elsevier Science, Pico Y. Burlington, pp. 59–91.

Arvantoyannis, I.S., Chalhoub, C., Gotsiou, P., Lydakis-Simantiris, N., Kefalas, P., 2005. Novel quality control methods in conjunction with chemometrics (multivariate analysis) for detecting honey authenticity. Critical Reviews in Food Science and Nutrition 45, 193–203.

Arvantoyannis, I., Katsot, M.N., Psarra, P., Soufleros, E., Kallinthraka, S., 1999. Application of quality control methods for assessing wine authenticity: use of multivariate analysis (chemometrics). Trends in Food Science & Technology 10, 321–329.

Arvantoyannis, I., 2010. Wine Authenticity, traceability and safety monitoring. In: Reynolds, A.G. (Ed.), Managing Wine Quality. CRC Pres, Boca Raton, USA.

Blanco, M., Villaroya, I., 2002. NIR spectroscopy: a rapid-response analytical tool. Trends in Analytical Chemistry 21, 40–50.

Brereton, R.G., 2003. Chemometrics: Data Analysis for the Laboratory and Chemical Plant. John Wiley & Sons, Ltd, Chichester, UK.

Brereton, R.G., 2007. Applied Chemometrics for Scientist. John Wiley & Sons, Ltd, Chichester, UK.

Casale, M., Oliveri, P., Armanino, C., 2010. NIR and UV-vis spectroscopy, artificial nose and tongue: comparison of four fingerprinting techniques for the characterisation of Italian red wines. Analytica Chimica Acta 668, 143–147.

Cordella, Ch, Moussa, I., Martel, A.-C., Sbirrazzuoli, N., Lizzani-Cuvelier, L., 2002. Recent developments in food characterisation and adulteration detection: technique-oriented perspectives. Journal of Agricultural and Food Chemistry 50, 1751–1764.

Cozzolino, D., 2009. Near infrared spectroscopy in natural products analysis. Planta Medica 75, 746–757.

Cozzolino, D., 2011. Infrared methods for high throughput screening of metabolites: food and medical applications. Combinatorial Chemistry & High Throughput Screening 14, 125–131.

Cozzolino, D., 2012. Recent trends on the use of infrared spectroscopy to trace and authenticate natural and agricultural food products. Applied Spectroscopy Reviews 47, 518–530.

Cozzolino, D., Murray, I., 2012. A review on the application of infrared technologies to determine and monitor composition and other quality characteristics in raw fish, fish products, and seafood. Applied Spectroscopy Reviews 47, 207–218.

Cozzolino, D., Corbella, E., Smyth, H.E., 2011. Quality control of honey using infrared spectroscopy: a review. Applied Spectroscopy Reviews 46, 523–538.

Cozzolino, D., Murray, I., 2004. Analysis of animal by-products. In: Roberts, C.A., Workman, J., Reeves III, J.B. (Eds.), Near Infrared Spectroscopy in Agriculture. American Society of Agronomy, Crop Science Society of America, Soil Science Society of America, USA, pp. 647–662.

Cozzolino, D., Chree, A., Murray, I., Scaife, J.R., 2002. The assessment of the chemical composition of fishmeal by near infrared reflectance spectroscopy. Aquaculture Nutrition 8, 149–155.

Cozzolino, D., Chree, A., Scaife, J.R., Murray, I., 2005. Usefulness of near infrared reflectance (NIR) spectroscopy and chemometrics to discriminate fishmeal batches made with different fish species. Journal of Agricultural and Food Chemistry 53, 4459–4463.

Cozzolino, D., Smyth, H.E., Gishen, M., 2003. Feasibility study on the use of visible and near-infrared spectroscopy together with chemometrics to discriminate between commercial white wines of different varietal origins. Journal of Agricultural and Food Chemistry 51, 7703.

Chen, L., Wang, J., Ye, Z., Zhao, J., Xue, X., Vander Heyden, Y., Sun, Q., 2012. Classification of Chinese honeys according to their floral origin by near infrared spectroscopy. Food Chemistry 135 (2), 338–342.

Dai, Q., Cheng, J.-H., Sun, D.-W., Pu, H., Zeng, X.-A., Xiong, Z., 2015. Potential of visible/near-infrared hyperspectral imaging for rapid detection of freshness in unfrozen and frozen prawns. Journal of Food Engineering 149, 97–104.

De la Haba, M.J., Pierna, J.A.F., Fumiere, O., Garrido-Varo, A., Guerrero, J.E., Perez-Marin, D.C., Dardenne, P., Baeten, V., 2007. Discrimination of fish bones from other animal bones in the sedimented fraction of compound feeds by near infrared microscopy. Journal of Near Infrared Spectroscopy 15, 81–88.

Ding, R., Huang, X., Han, F., Dai, H., Teye, E., Xu, F., 2014. Rapid and nondestructive evaluation of fish freshness by near infrared reflectance spectroscopy combined with chemometrics analysis. Analytical Methods 6, 9675–9683.

Downey, G., 1996. Authentication of food and food ingredients by near infrared spectroscopy. Journal of Near Infrared Spectroscopy 4, 47–61.

Downey, G., 1998. Food and food ingredient authentication by mid-infrared spectroscopy and chemometrics. Trends in Analytical Chemistry 17, 418–424.

Downey, G., 2013. Vibrational spectroscopy in studies of food origin. In: Brereton, R.G. (Ed.), New Analytical Approaches for Verifying the Origin of Food. Woodhead, Publisher.

Hao, D.-M., Zhou, Y.-N., Wang, Y., Zhang, S., Yang, Y.-M., Lin, L., Li, G., Wang, X.-L., 2015. Recognition of water-injected meat based on visible/near-infrared spectrum and sparse representation. Spectroscopy and Spectral Analysis 35, 93–98.

Herrero Latorre, C., Pena Crecente, R.M., Garcia Martin, S., 2013. A fast chemometric procedure based on NIR data for authentication of honey with protected geographical indication. Food Chemistry 141, 3559–3565.

Huang, H., Yu, H., Xu, H., Ying, Y., 2008. Near infrared spectroscopy for on/in-line monitoring of quality in foods and beverages: A review. Journal of Food Engineering 87, 303–313.

Kamruzzaman, M., Makino, Y., Oshita, S., 2015. Non-invasive analytical technology for the detection of contamination, adulteration, and authenticity of meat, poultry, and fish: a review. Analytica Chimica Acta 853, 19–29.

Karoui, R., Downey, G., Blecker, C., 2010. Mid-infrared spectroscopy coupled with chemometrics: a tool for the analysis of intact food systems and the exploration of their molecular structure–quality relationships – a review. Chemical Reviews 110, 6144–6168.

Khakimov, B., Gürdeniz, G., Engelsen, S.B., 2015. Trends in the application of chemometrics to foodomics studies. Acta Alimentaria 44, 4–31.

Liu, L., Cozzolino, D., Cynkar, W.U., Gishen, M., Colby, C.B., 2006. Geographic classification of Spanish and Australian Tempranillo red wines by visible and near infrared spectroscopy combined with multivariate analysis. Journal of Agricultural and Food Chemistry 54, 6754.

Liu, L., Cozzolino, D., Cynkar, W.U., Dambergs, R.G., Janik, L., O'Neill, B.K., Colby, C.B., Gishen, M., 2008. Preliminary study on the application of visible–near infrared spectroscopy and chemometrics to classify Riesling wines from different countries. Food Chemistry 106, 781.

Ma, J., Pu, H., Sun, D.-W., Gao, W., Qu, J.-H., Ma, K.-Y., 2015. Application of Vis-NIR hyperspectral imaging in classification between fresh and frozen-thawed pork Longissimus Dorsi muscles. International Journal of Refrigeration-Revue Internationale du Froid 50, 10–18.

Manley, M., 2014. Near-infrared spectroscopy and hyperspectral imaging: non-destructive analysis of biological materials. Chemical Society Reviews 43, 8600.

Martelo-Vidal, M.J., Domínguez-Agis, F., Vázquez, M., 2013. Ultraviolet/visible/near-infrared spectral analysis and chemometric tools for the discrimination of wines between subzones inside a controlled designation of origin: a case study of Rías Baixas. Australian Journal of Grape and Wine Research 19, 62–67.

McClure, W.F., 2003. Review: 204 years of near infrared technology: 1800–2003. Journal of Near Infrared Spectroscopy 11, 487–518.

Mignani, A.G., Ciaccheri, L., Gordillo, B., Mencaglia, A.A., Gonzalez-Miret, M.L., Heredia, F.J., Cichelli, A., Vo-Dinh, T., Lieberman, R.A., Gauglitz, G.G., 2014. Near-infrared spectroscopy and pattern-recognition processing for classifying wines of two Italian provinces. In: Advanced Environmental, Chemical, and Biological Sensing Technologies XI. Proceedings of SPIE, 9106.

Mouazen, A.M., Al-Walaan, N., 2014. Glucose adulteration in Saudi honey with visible and near infrared spectroscopy. International Journal of Food Properties 17, 2263–2274.

Naes, T., Isaksson, T., Fearn, T., Davies, T., 2002. A User-Friendly Guide to Multivariate Calibration and Classification. NIR Publications, Chichester, UK.

Nicolai, B.M., Beullens, K., Bobelyn, E., Peirs, A., Saeys, W., Theron, K.I., Lammertyn, J., 2007. Non-destructive measurement of fruit and vegetable quality by means of NIR spectroscopy: a review. Postharvest Biology and Technology 46, 99–118.

Niu, X.-Y., Shao, L.-M., Dong, F., Zhao, Z.-L., Zhu, Y., 2014. Discrimination of donkey meat by NIR and chemometrics. Spectroscopy and Spectral Analysis 34, 2737–2742.

O'Brien, N., Hulse, C.A., Pfeifer, F., Siesler, H.W., 2013. Technical note near infrared spectroscopic authentication of seafood. Journal of Near Infrared Spectroscopy 21, 299–305.

Oliveri, P., Di Egidio, V., Woodcock, T., Downey, G., 2011. Application of class-modelling techniques to near infrared data for food authentication purposes. Food Chemistry 125, 1450–1456.

Ottavian, M., Fasolato, L., Serva, L., Facco, P., Barolo, M., 2014. Data fusion for food authentication: fresh/Frozen-thawed discrimination in west African goatfish (*Pseudupeneus prayensis*) fillets. Food and Bioprocess Technology 7, 1025–1036.

Otto, M., 1999. Chemometrics: Statistics and Computer Application in Analytical Chemistry, second ed. Wiley-VCH, Federal Republic of Germany, p. 343.

Prieto, N., Lopez-Campos, O., Zijlstra, R.T., Uttaro, B., Aalhus, J.L., 2014. Discrimination of beef dark cutters using visible and near infrared reflectance spectroscopy. Canadian Journal of Animal Science 94, 445–454.

Reid, L.M., O'Donnell, C.P., Downey, G., 2006. Recent technological advances for the determination of food authenticity. Trends in Food Science & Technology 17, 344–353.

Riovanto, R., Cynkar, W.U., Berzaghi, P., Cozzolino, D., 2011. Discrimination between shiraz wines from different Australian regions: the role of spectroscopy and chemometrics. Journal of Agricultural and Food Chemistry 59, 10356–10359.

Rodriguez-Saona, L.E., Allendorf, M.E., 2011. Annual Review of Food Science and Technology 2, 467.

Roggo, Y., Chalus, P., Maurer, L., Lema-Martinez, C., Edmond, A., Jent, N., 2007. A review of near infrared spectroscopy and chemometrics in pharmaceutical technologies. Journal of Pharmaceutical and Biomedical Analysis 44, 683–690.

Ropodi, A.I., Pavlidis, D.E., Mohareb, F., Panagou, E.Z., Nychas, G.-J.E., 2015. Multispectral image analysis approach to detect adulteration of beef and pork in raw meats. Food Research International 67, 12–18.

Ruoff, K., Luginbuhl, W., Bogdanov, S., 2006. Authentication of the botanical origin of honey by near-infrared spectroscopy. Journal of Agricultural and Food Chemistry 54, 6867–6872.

Sentandreu, A.M., Sentandreu, E., 2014. Authenticity of meat products: tools against fraud. Food Research International 60, 19–29.

Tena, N., Fernandez Pierna, J.A., Boix, A., Baeten, V., von Holst, C., 2014. Differentiation of meat and bone meal from fishmeal by near-infrared spectroscopy: extension of scope to defatted samples. Food Control 43, 155–162.

Tzouros, N.E., Arvantoyannis, I.S., 2001. Agricultural produces: synopsis of employed quality control methods for the authentication of foods and application of chemometrics for the classification of foods according to their variety or geographical origin. Critical Reviews in Food Science and Nutrition 41, 287–319.

Ulloa, P.A., Guerra, R., Cavaco, A.M., Rosa da Costa, A., Figueira, A.C., Brigas, A.F., 2013. Determination of the botanical origin of honey by sensor fusion of impedance e-tongue and optical spectroscopy. Computers and Electronics in Agriculture 94, 1–11.

Woodcock, T., Downey, G., O'Donnell, C.P., 2008. Better quality food and beverages: the role of near infrared spectroscopy. Journal of Near Infrared Spectroscopy 16, 1–29.

Woodcock, T., Downey, G., O'Donnell, C.P., 2009. Near infrared spectral fingerprinting for confirmation of claimed PDO provenance of honey. Food Chemistry 114, 742–746.

Woodcock, T., Downey, G., Kelly, J.D., 2007. Geographical classification of honey samples by near-infrared spectroscopy: a feasibility study. Journal of Agricultural and Food Chemistry 55, 9128–9134.

Weeranantanaphan, J., Downey, G., Allen, P., Da-Wen, S., 2011. A review of near infrared spectroscopy on muscle food analysis: 2005–2010. Journal of Near Infrared Spectroscopy 19, 61–104.

Xiang, L.-L., Li, M.-H., Li, J.-M., Li, J.-H., Zhang, L.-D., Zhao, L.-L., 2014. Determination of wine original regions using information fusion of NIR and MIR spectroscopy. Spectroscopy and Spectral Analysis 34, 2662–2666.

Zhu, X., Li, S., Shan, Y., Zhang, Z., Li, G., Su, D., Liu, F., 2010. Detection of adulterants such as sweeteners materials in honey using near-infrared spectroscopy and chemometrics. Journal of Food Engineering 101, 92–97.

Zotte, A.D., Ottavian, M., Concollato, A., Lorenzo, S., Martelli, R., Parisi, G., 2014. Authentication of raw and cooked freeze-dried rainbow trout (*Oncorhynchus mykiss*) by means of near infrared spectroscopy and data fusion. Food Research International 60, 180–188.

The Use of ICP-MS in Food Traceability

8

M. Aceto
Università degli Studi del Piemonte Orientale, Viale Teresa Michel, Alessandria, Italy

1. A Brief Introduction to ICP-MS Technique

The coupling of mass spectrometry to plasma methods of atomization/ionization of samples provides one of the most sensitive techniques for the determination of trace and ultra-trace elements: Inductively Coupled Plasma-Mass Spectrometry or ICP-MS. This technique has been used since the 1980s in several fields of application, and it has been at least since the 1990s that it has shown its potentiality in the authentication of foodstuffs (McCurdy et al., 1992).

The ICP-MS technique exploits particular features: (1) high rate of conversion of analytes from solution into elemental ions, yielded by the plasma temperature (6.000–10.000K); (2) high selectivity of the mass spectrometer filter; and (3) high sensitivity of the detection system. The fundamental parts of an ICP-MS instrument are the following:

- *Introduction system*: a device carrying samples to the nebulizer, usually in liquid form. A peristaltic pump is the most common device, providing a constant uptake flow rate. An automatized system, such as an *autosampler*, can be present in order to optimize analysis time and consumption of reagents.
- *Nebulization system*: its task is transforming the sample from a liquid solution to a spray composed of light drops by means of an incident argon flow. The lightest are the drops, where the most analytes are carried to the plasma.
- *ICP compartment*: the argon plasma is generated on top of a torch, and it is fed by interaction with the electromagnetic field generated by a radiofrequency source. When the nebulized sample enters in the plasma, it is instantaneously desolvated and atomized; then atoms are turned into ions depending on the ionization potential of the respective elements.
- *Interface*: the vapor containing ions and atoms is driven by argon through an interface made of two subsequent cones, the *skimmer* and the *sampler* cone, which allow the focalization of ions into a small volume.
- *Mass spectrometer*: a quadrupole or a hexapole allowing the separation of ions according to their m/z ratios.
- *Detection system*: an electron multiplier that transform ion signals into electric pulses.

Fig. 8.1 shows a generic scheme of an ICP-MS system.

ICP-MS instruments with low-resolution quadrupole mass spectrometers allow determination of trace and ultra-trace elements with good accuracy and sensitivity. Using high-resolution MS systems, higher accuracy, precision, and sensitivity are guaranteed and, in addition, it becomes possible to determine isotope ratios of heavy elements (strontium, lead, etc.) which are considered among the best chemical markers

Advances in Food Traceability Techniques and Technologies. http://dx.doi.org/10.1016/B978-0-08-100310-7.00008-9

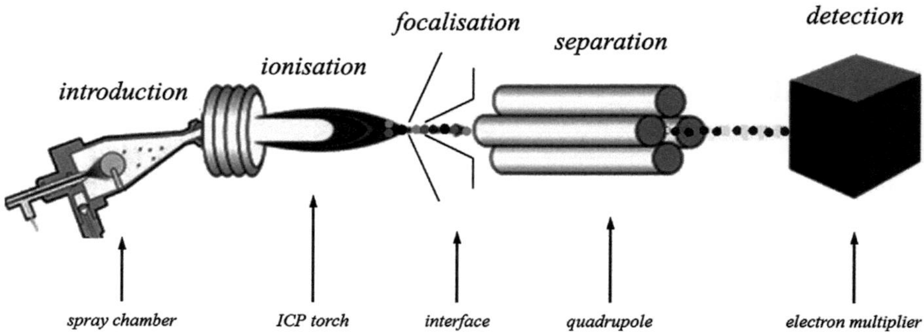

Figure 8.1 Scheme of an ICP-MS system.

for food authentication and traceability; with these features, ICP-MS becomes a member of the group of techniques allowing the determination of isotope ratios which is collectively termed Isotope Ratio Mass Spectrometry, or IRMS, which in turn is a major part of Stable Isotope Analysis, or SIA. In this chapter, only isotopic analysis of heavy elements will be dealt with, since analysis of light elements is not performed by means of ICP-MS instrumentations.

The scientific literature on ICP-MS is at present huge, both in terms of books and journal articles. In-depth analysis on the theoretical basis of the technique and on its applications can be found, among others, in Montaser (1998) and in Thomas (2013).

The application of ICP-MS to food authentication and traceability has rapidly increased through the years (Gonzálvez et al., 2009), as a consequence of three main factors: (1) a greater comprehension of the relationship among soil and plants, from the point of view of minerals; (2) a greater comprehension of the processes involved in food production chains; and (3) a strong increase in the analytical performances of ICP-MS instrumentations. A great number of studies have been published, and several methods have been proposed, based mainly on the determination of trace element concentrations or of isotope ratios of heavy elements. The need of determining these parameters can be optimally addressed by the ICP-MS technique, thanks to its sensitivity (down to ppq, or part per quadrillion or pg/L, on the instruments with the best performances), its mass resolution (R = 10.000, defined as 10% peak valley, on high-resolution instruments), and its ability of determining many elements in a relatively short time.

2. Classification Studies on Food

Before describing the contribution of ICP-MS to food analysis, it is important to point out the difference between *traceability* and *authentication*: they are not synonyms, but rather complementary concepts. In the following text the sum of traceability and authentication will be expressed with the term *classification*.

Traceability is a procedure that aims at finding a link among a foodstuff and the soil where its raw matters come from (Aceto et al., 2009). It is a highly demanding process: samples of different natures, generated inside the same production chain, must be compared. In order to verify the traceability of foodstuffs, inorganic parameters are mostly useful, with particular reference to the following:

- *Correlation of single elements among soil and foodstuffs*: this can be verified with several statistical tests (eg, analysis of variance or ANOVA, Pearson correlation analysis, etc.); useful variables at the purpose can be the total concentration of an element or one determinate fraction (eg, exchangeable, available).
- *Comparison of the distribution of certain elements among the different stages of a production chain*: rather than the concentration of a single element, the concentrations of a determinate group of elements together with their relative ratios, ie, the *distribution* of these elements, could be used. A production chain can be traced when the distribution of certain elements maintains its trend along the chain stages.
- *Comparison of isotope ratios among the different stages of a production chain*: isotope ratios of heavy elements (lead, strontium) derive their value from the soil and can remain unaltered along the production chain; a similar behavior is attributed to boron. Light elements (hydrogen, oxygen, nitrogen, etc.), on the contrary, are subjected to several phenomena that can strongly influence the final values. A production chain can be traced when a determinate isotope ratio maintains its value all along the stages of the chain. Traditionally, isotope ratios are expressed as δ values, ie, the ratio of isotopes in a sample to the ratio in a standard.

Authentication is a procedure that aims at highlighting differences among foodstuffs of the same type which differ on a geographical (eg, red wines from different regions), varietal (eg, red wines from different varieties), or technological (eg, red wines obtained with different methods) basis. Authentication is a less demanding process because it is not limited to inorganic parameters: organic compounds or spectral features can be used as well. Thanks to their known role in geochemical processes, though, trace elements are particularly suitable as markers for authentication. It is particularly advantageous in determining the highest possible number of elements (*multielement profiling* or *fingerprinting*) so as to have the highest possible number of variables for discrimination.

ICP-MS methods are particularly suitable for the classification of foodstuffs, considering that trace element concentrations and isotope ratios of heavy elements are among the most useful parameters, as proved by several publications issued on the subject. Reviews on this subject can be found in Gonzálvez and de la Guardia (2013a) with concern to multielement profiling, in Rodrigues et al. (2013) and Zhao et al. (2014) with concern to isotopic analysis, and in Drivelos and Georgiou (2012), Ariyama (2014), and Georgiou and Danezis (2015) with concern to both elemental and isotopic analysis.

Following the determination of chemical markers with ICP-MS methods, a large amount of data is usually available. Consider the quantification of the concentrations of 30 elements in 100 food samples. In order to visualize the information contained in the resulting 3.000 data, not less than 435 biplots could be produced, which an unpractical situation. This is because the determination stage is usually followed by a mathematical stage in which *chemometrics* are applied. In particular, *pattern recognition* chemometric methods are used. Techniques such as Principal Components Analysis (PCA), Cluster Analysis (CA), Linear Discriminant Analysis (LDA), or Soft

Independent Modeling of Class Analogy (SIMCA) are routinely applied to food classification. Reviews of chemometric methods in food classification have been issued by Bertacchini et al. (2013) and by Gonzálvez and de la Guardia (2013b).

3. Where Do Elements Come From?

Nearly all of it starts from the soil. It is possible to demonstrate that most foodstuffs derive their elemental content firstly from soil, whether of vegetal origin (fruits and vegetables, oils, etc.) or animal origin (meat, milk, and dairy products). Whether we are interested in trace element concentrations or in isotope ratios, the role of each chemical element in the interaction among soil and plants is the key for understanding whether a certain element could act as a marker for classification. In this context, elements can be divided into three groups (Merian et al., 2004):

- *Macroessentials* or *nutrients*: these elements are indispensable for the life of plants and therefore they are actively assumed from soil. Alkaline, alkaline earths, and some transition metals belong to this group.
- *Microessentials*: these elements have a minor role, if any, in the physiology of plants, so they are usually assumed in a passive way. Rare earth elements (REE) belong to this group.
- *Toxic*: these elements are dangerous for plants, which are able to avoid picking them up from soil. Heavy metals such as cadmium, mercury, thallium, or lead belong to this group.

Successively to the interaction soil plants, which we can imagine as the *natural* source of elements in foodstuffs, other factors or secondary sources give their contributions: farming practices, animal metabolism, anthropic sources, etc (Aceto, 2003). *Farming practices* are important in determining the total availability of metal ions for uptake by plants. As an example, the use of fertilizers rich in certain metals can be a relevant artificial source of those metals; a typical case is the so-called *Bordeaux mixture*, a mixture of $CuSO_4$ and $Ca(OH)_2$ used as fungicide in enological practices, which eventually acts as an extra source of copper and calcium ions. The *metabolism of animals* is a factor of relevance for specific foodstuffs only, of course: meat, dairy products, honey, and other natural products of animal origin. The transformations occurring inside animals' organisms are complex and in some cases unexplained, but in several cases, it is possible to trace the path of elements of relevance in the diet of animals. Finally, there are the contributions from *anthropic sources*, which can be split into two directions. *Intentional* sources are those due to the production chains, in which specific compounds must be added along the chain in order to perform specific tasks; typical examples are the addition of salt and spices to meat in the making of salami or the addition of tartrates to wine for leveling off the amount of metal ions. Unintentional sources are those deriving from environmental pollution: acid rains, pollution of aquifers, fraudulent disposal of wastes nearby agricultural crops, etc. All these phenomena can add extra amounts of metal ions, with particular concern to transition and heavy metals, to food/raw matters.

In traceability schemes, the ideal markers are those elements which (1) act as microessentials for plants and (2) whose concentrations are not influenced by secondary sources. The concentrations of such elements in foodstuffs or in their raw matters can proportionally reflect the concentrations in soil, due to the fact that they are assumed without active intervention of plants. In this context, the concept of *distribution*, previously expressed, is important. Obviously, the concentrations of elements are lower in foodstuffs or raw matters than in soil, but what is most informative is the distribution of the elements of interest. When all the elements considered are assumed proportionally, we have no *fractionation* and therefore there is similarity of distribution among soil and food/raw matters (Fig. 8.2A). When a

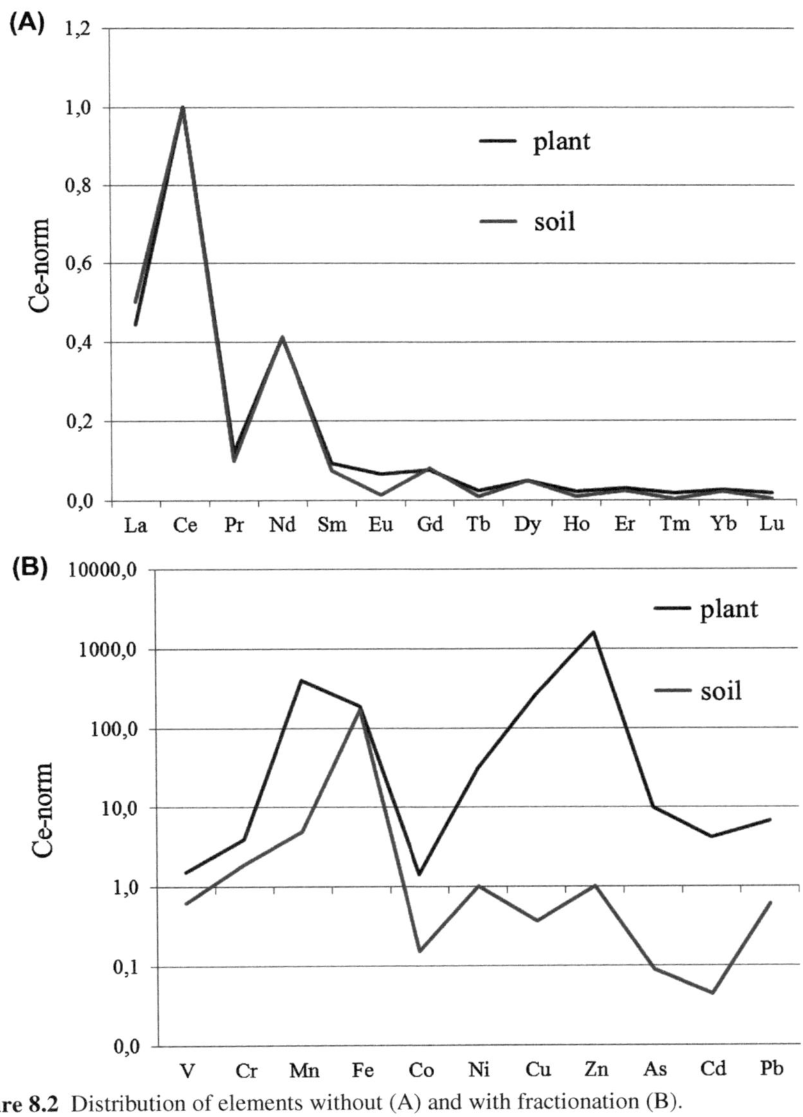

Figure 8.2 Distribution of elements without (A) and with fractionation (B).

certain element is assumed in a preferential way because it is a nutrient, or it is excluded from absorption because it is toxic, we have fractionation from the original distribution in soil (Fig. 8.2B). To evaluate the extent of fractionation, data are usually normalized; geochemists make reference to particular substrates, such as chondrites, a meteoric rock (Coryell et al., 1963), but other methods are available. Oddone et al. (2009) suggested normalization to a rare earth element, in order to better appreciate the saw tooth pattern caused by the Oddo–Harkins rule and its possible modifications in the stages of a production chain.

The Rare Earth Elements (REE) are considered to be interesting as traceability markers (Aceto et al., 2009). Geochemists already use them as geological markers in the identification of the origin of rocks, but REE can display similar features in biological systems as they do not seem to have an active role in plant physiology. All these features derive, of course, from their similar electronic configuration from which a very similar chemical behavior descends. There are many scientific indications (Brown et al., 1990; Tyler, 2004; Liang et al., 2008) on the fact that plants tend to absorb REE ions from soil with little or no fractionation of the original distribution, and this behavior could be transmitted to foodstuffs with simple production chains (see later in this chapter).

Similar considerations are valid with concern to isotope ratios: heavy elements such as strontium and lead provide efficient and reliable traceability markers (respectively $^{87}Sr/^{86}Sr$ and any combination of ^{204}Pb, ^{206}Pb, ^{207}Pb, and ^{208}Pb nuclides) because the ratios of their isotopes are hardly, if ever, fractionated in the terrestrial ecosystem, having a marginal role in the physiology of plants (Graustein, 1988). These ratios are consolidated in soil after conversion from rocks and are transmitted nearly unaltered through plants and from here to food or raw matters, provided no external sources are present; in this context, Sr isotope ratio seems to be more powerful than Pb ratios, because it is less affected by anthropic sources, including environmental pollution; indeed, the geological character of a production area is inherited in this parameter. This has been well demonstrated in a fundamental study by Voerkelius et al. (2010), in which authors showed the potentiality of building an $^{87}Sr/^{86}Sr$ values database in surface waters in order to predict the geographic origin of some types of food. A different behavior is the one of light elements: hydrogen, carbon, oxygen, nitrogen, and sulfur isotopic ratios are subjected to wide variations because they are strongly influenced by chemical, physical, and, above all, biological phenomena; for this reason, they cannot be considered reliable markers for traceability.

Different, less strict requisites can be evaluated for authentication. A certain element, or its isotopic ratio, could act as authentication marker even if its concentration had many sources, provided that the contributions external to the natural one are systematic and representative of geographic, varietal, or technological features. In this case, it is difficult establishing what the best markers, ie, the most suitable elements, are: each classification scheme accounts for itself. Elements suitable for classification of one determinate food could result as useless for classification of another food. Major, minor, trace, or ultra-trace elements could be efficient markers at the same time, but since major/minor elements can be active components in the technological processes involved in production chains, a simple rule could be the following: in every element group (ie, alkaline, alkaline earths, or transition metals) the most abundant components are generally less reliable than the scarcest ones as authentication markers. So lithium and rubidium are far more useful than sodium and potassium, as are beryllium, strontium, and barium with respect to calcium and magnesium, and the elements of the second and third transition metals series (REE among them) with respect to those of the first series. To resume, major/minor elements seem to work better in authentication schemes based on varietal or technological discrimination, while trace and ultra-trace elements are mostly suitable in order to discriminate on a geographic basis, which is consistent with their role as geochemical markers as well (White, 2013).

As it will become apparent in the following, in the scientific literature few classification studies are *true* traceability studies in the sense soon described, ie, a comparison of distributions along the production chain. Relatively more traceability studies exploit a comparison of isotope ratios of heavy elements. Indeed, most of the studies labeled as "traceability" are more precisely *authentication* studies, ie, discrimination of food productions on a geographic, varietal, or technological basis.

3.1 The Influence of the Production Chain

In order to develop schemes for food classification, it is mandatory to evaluate how the secondary sources act, which eventually means evaluating how the production chain of a certain food is made. Particularly relevant is the knowledge of all the sources along the chain that can potentially contribute to element concentrations or isotope ratios in the products located at the end of the chain. With concern to trace element concentrations, three situations can be singled out:

- *Very short or nonexistent chains*: foodstuffs that are obtained without any chemical transformations from raw matters, such as fruit and vegetable products. These are ideal subjects for classification studies.
- *Short or medium chains*: foodstuffs that are obtained with few passages, each of which is known and under control: vegetable oils, wines and spirits, milk, etc. While authentication schemes can usually be developed on every instance, traceability schemes are fairly possible provided a deep knowledge of the entire chain is guaranteed.
- *Long and complex chains*: foodstuffs produced with many passages, mostly involving chemical transformations: meat, salami, dairy products, etc. In these cases, traceability schemes cannot be developed because too many sources overlap with the initial contribution from soil. On the contrary, it is possible (but not guaranteed) that authentication schemes be possible.

Indeed, the knowledge of the overall production chain is less relevant in authentication studies: a successful scheme can be developed even in cases where a complex chain is present, with a lot of secondary sources of elements. In several cases, inorganic chemical markers useful for classification can be singled out, provided a reliable enough calibration had been obtained.

4. The Most Important Food Production Chains: Description and Case Studies

In the following paragraphs the most important food sectors will be analyzed in order to verify which of them are eligible for classification, putting particular focus on the production chains and on the difference between authentication and traceability. Several case studies will be presented.

4.1 Fruit and Vegetable Foodstuffs

Under the caption *fruit and vegetable foodstuffs*, it is possible to include fresh fruits and vegetables, spices, flavors, herbs, tea, and coffee, which result from mostly mechanical transformations of raw matter, but also derivatives, ie, foodstuffs resulting from mechanical and chemical transformations of vegetable food, such as vegetable oils, fruit juices, etc.

4.1.1 Fresh Fruits and Vegetables

The production chains of fresh fruit and vegetable foodstuffs are most probably the simplest in the whole agro-industrial sector. This is because these products do not generally undergo chemical changes in the passage from soil to the marketed product, so that determinate elemental distributions and isotopic signatures of heavy elements can be considered as fingerprint of crops. Factors such as fertilization, climate, agricultural practices, pollution, and botanical varieties must be taken into account, but some studies (Ariyama et al., 2006, 2007) suggest that these variables are less relevant in influencing the element composition of plants than the geological features of soil.

According to this assumption, it is usually easy to find chemical markers useful for both traceability and authentication. The role of $\delta^{87}Sr$ as marker has been cited by Zhao et al. (2014) in their review on the application of IRMS to the classification of vegetal foodstuffs, and it can be exemplified by the work of Trincherini et al. (2014) on the geographical classification of tomato samples. Several studies have been proposed involving the determination of trace elements of isotope ratios as chemical markers of fruit and vegetable foodstuffs. Most of these studies were obviously devoted to the classification of high-quality products, such as typical productions labeled with Protected Designation of Origin (PDO, the equivalent of Italian DOP or *Denominazione di Origine Protetta*, or French AOC, *Appellation d'Origine Contrôlée*), Protected Geographical Indication (PGI), or Traditional Specialty Guaranteed (TSG), can be strongly supported when it becomes possible to recognize imitations and frauds by means of classification methods. This is typically the case of studies such as those of Oddone et al. (2009) on PGI "Nocciola Piemonte" hazelnuts from Piedmont (Italy), Furia et al. (2011) on PGI Tropea Red Onion from Calabria (Italy), Benabdelkamel et al. (2012) on PGI "Clementine di Calabria" tangerines, and Drivelos et al. (2014) on PDO "Fava Santorinis" peas from Greece: in all these cases, authors developed classification methods based on multielement profiling and were able to distinguish between typical productions and their fraudulent imitations. In most cases the importance of the contribution of REE in the classification has been highlighted: their distribution can be used in order to verify the link between soil and vegetable food, as shown by Bettinelli et al. (2005) in their study on a tomato production chain.

Spices are considered as being more prized than fresh fruits and vegetables, since their production involves a strict selection of raw matter and a low yield in the final product. Particular cases could reach extra high prices: saffron from *Crocus sativus*, which is perhaps the most expensive food in the world, can cost as high as 15.000–20.000 €/Kg. If we also consider that the production chains of such foods is usually limited to mechanical transformations (collection, filtering, grinding, etc.), they can be considered as the ideal candidates for classification. Multielement profiling was applied by D'Archivio et al. (2014), who analyzed Italian saffron samples from Abruzzi, Sardinia, and Umbria. A good classification on a geographical basis was obtained using B, Na, Rb, and Sr as variables. Other spices subjected to classification studies have been vanilla (Hondrogiannis et al., 2013), for which a good geographic classification was obtained with multielement profiling, and Hungarian PDO *Szegedi Füszerpaprika* (Brunner et al., 2010), for which a combined use of $\delta^{87}Sr$ and multielement profiling allowed distinguishing true PDO paprika samples from imitations and hot pepper. Song et al. (2014) found a good correlation between water, soil exchangeable fraction, and the plant.

Another category of vegetable foodstuffs of high commercial relevance is the one composed by tea and coffee. Their production chains are simple as the final products are obtained by a collection of leaves (tea) and seeds (coffee) that are respectively dried and milled. Several studies were issued on the possibility of classifying samples of these products on a geographic basis using inorganic parameters. The classification of tea samples from different continents or, in some cases, from different countries, was obtained with multielement profiling (Moreda-Piñeiro et al., 2003) or combined to isotope analysis of light elements (Pilgrim et al., 2010); Lagad et al. (2013) also evidenced a good correlation between $\delta^{87}Sr$ values of soil and the corresponding tea plants. The possibility of classifying tea samples according to cultivar (Chen et al., 2009) or to type (black, green, oolong, etc.) has been reported (Han et al., 2014). In the classification of coffee samples, it was shown that the geographic provenance, linked to soil typical composition, could be more characteristic in determining the elemental composition than agronomic parameters, such as the harvest year and the degree of ripeness, or technological parameters, such as the application (or not) of roasting (Valentin and Watling, 2013). Similar results were obtained using strontium isotope ratios combined with oxygen (Rodrigues et al., 2011) or boron (Liu et al., 2014).

4.1.2 Derivative Products

Starting from fresh fruits or vegetables, a large number of products can be obtained. In this context, different situations can be singled out. Products may result from mechanical operations on fresh matter, eg, squeezing, pressing, or milling, with relatively simple production chains; such are the cases of vegetable oils, pure tomato juices, or pure fruit juices. In other cases, more complex production chains are used that involve chemical transformations and the addition of various substances, ie, fruits juices obtained from concentrate, tomato pastas, etc. It is apparent that in the first case the possibility of both traceability and authentication can be pursued, while the second case is much more complicated for what concerns traceability.

Let us start from vegetable oils. These are usually produced by the simple grinding of fruits or seeds, and the juice obtained should be very similar to raw matter for what concerns the elemental composition. Contamination could occur from use of fertilizers and pesticides or with contact from the metallic surfaces of the processing equipment or storage material, but considering the high hydrophobicity of oils, a very low solubilization of metal ions can be expected. The last factor, ie, high hydrophobicity, is also a limiting factor from the analytical point of view, since the concentration of most elements (eg, REE) is in the ng/L range or less. Among vegetable oils, extra-virgin olive oil (EVO) can be considered as one of the most appreciated foodstuffs thanks to its dietary values; therefore, it is also one of the most counterfeit, both with concern to its geographic provenance and to adulteration with lesser value oils. It is not surprising that a highly relevant part of scientific research on food had been devoted to EVO classification. Camin et al. in a reference study (2010), clearly demonstrated the potential of SIA of light elements in the geographic classification of EVOs, but they also showed the complementary contribution of trace elements in the classification of oil samples from eight European countries. Trace elements were particularly useful in distinguishing samples according to soil type: shale/clay, limestone, or acid magmatic. The role of soil was confirmed in the studies by Benincasa et al. (2007) on EVOs from central and southern Italy and by Beltrán et al. (2015) on virgin olive oils from four areas of southwestern Spain: the geographic classification was not influenced by the different olive cultivars involved. On the other hand, Llorent-Martínez et al. (2011) classified different types of oil produced in Spain: virgin olive, olive, pomace-olive, corn, sunflower, and soybean oils. Cr, Cu, Fe, and Mn were the most useful variables in the classification. With concern to SIA of heavy elements, Medini et al. (2015) investigated the role of $\delta^{87}Sr$ as a marker of geographic classification for EVOs, discussing in depth the analytical aspects bound to the very low amounts of Sr in oil.

A typical vegetable with many derivative products is tomato. The preparation of such products involves the addition of several materials, such as salt, olive oil, flavors, etc., which can alter the elemental composition. Lo Feudo et al. (2010) analyzed tomato samples from four Italian regions and triple concentrate tomato pastes from Italy, California, China, and Greece using only multielement profiling with ICP-MS. The two groups of samples were analyzed separately; in both instances, they obtained a good geographic classification. A more thorough study was carried out by Bontempo et al. (2011), who analyzed several samples of tomato produced in three Italian regions, together with samples of juice, *passata*, and paste derivatives, using IRMS for light elements and ICP-MS for multielement profiling. Apart from the usual high contribution to geographic classification provided by isotopic ratio of light elements, authors evidenced the strong complementary contribution of trace elements and, in particular, of REE; it resulted that the geographic classification could be obtained for all samples, irrespective of the type of product (raw tomato, juice, *passata*, or paste) and harvest period of tomatoes. Similar results have been obtained by Fragni et al. (2015) in their study on *passata*, double, and triple concentrate samples from Italy, China, and Spain, using only multielement profiling with ICP-MS.

More difficult is the classification of highly processed foodstuffs such as fruit juices. Industrial products result from the addition of several compounds such as preservatives, flavor enhancers, vitamins, mineral salts, etc. In some cases, juice is reconstructed from a concentrate by the addition of external water, which can add trace elements. For this reason, classification can be better yielded with a combined approach. Rummel et al. (2010) used SIA of light elements and δ^{87}Sr to obtain a geographic classification of orange juice samples from different regions in North and South America, Africa, and Europe.

4.2 Cereals

Cereals have been and are a highly relevant part of the diet of most cultures all over the world. Intermediates obtained from milling cereals, ie, flours, are the bases for foodstuffs consumed at every latitude, such as bread, pasta, and bakery products. The production chains involved in the industrial processing of cereals come through mostly mechanical procedures (collection, milling, refining, etc.), so that raw cereals and flours could be easily traced. When it comes to bread and other bakery products, several materials could be added to flour (yeasts, fats, eggs, carbohydrates), including some with a relevant inorganic part (salts). All these additions are negative for the traceability process.

Despite the potential ease in tracing raw cereals and flours, these products have not attracted much attention from the researchers because, from the merceological point of view, their classification is not a relevant issue. Ariyama et al. (2011) showed the potential of δ^{87}Sr versus total [Sr] and of lead isotope ratios as markers of geographic provenance for different cereals. Podio et al. (2013) exploited the elemental and isotopic (δ^{87}Sr) signatures as markers to distinguish wheat cultivated in three regions of Argentina; they also demonstrated a good correlation between soil, water, and wheat for some elements and isotopes. Relatively more attention has been devoted to rice, because some varieties (eg, Basmati, Italian rice) are recognized as high quality and therefore subjected to fraud. Some studies demonstrated the possibility of distinguishing the geographic provenance of rice by means of determination of boron, strontium, or lead isotope ratios (Oda et al., 2001) in addition to trace element concentrations (Ariyama et al., 2012); in particular the more effective markers in the discrimination were found to be δ^{87}Sr, total [Sr], and [Rb], while Pb isotope ratios were less useful due to the influence of anthropogenic sources of pollution. Shen et al. (2013) demonstrated the correlation of the available fractions of some trace elements (Be, Ca, Cd, Cu, K, Mg, Mn, Na, and Ni) among soil and rice, thus allowing a good discrimination of rice samples coming from four provinces of China. In a similar study by Zhao et al. (2012) on wheat samples collected in two different harvest seasons from four regions in China, authors found that the elements Ca, Fe, Mo, Na, and Zn were good markers of geographical origin, but while Ca, Fe, and Mo were mostly correlated to soil (therefore valid as traceability markers), Na and Zn were possibly influenced by climate and agricultural practices. In addition, Ba, Cu, Mg, and Mn were found to be correlated to wheat variety.

Reviews of the classification of rice (Vlachos and Arvanitoyannis, 2008) and maize (Arvanitoyannis and Vlachos, 2009) have been published.

4.3 Animal Products

This category comprises foodstuffs obtained by living animals and their derivatives. The most relevant are eggs, honey, and milk (the last one will be actually the subject of paragraph 4.5). The production chain of these products is mainly dominated by animal metabolism; therefore, their elemental composition is mostly influenced by the diet of the producing animals and, because nearly all the species involved are herbivores, by the features of crops on which they are living. A different matter is of course the destiny of derivatives (ie, sweets, egg pastas, etc.), for which a higher amount of sources must be considered.

The commercial value of eggs does not justify classification studies, if not in the very particular case of *caviar*, ie, roe obtained from certain fish species. Caviar is one of the most expensive foodstuffs in the world, with concern mainly to *caviar Volga* produced from sturgeons of the Caspian Sea; consequently, it is greatly subjected to counterfeit. The production chain of caviar is complicated by the addition of salt, used for both taste and preservation, and by packaging processes. Rodushkin et al. (2007) in a rare classification study, classified samples of vendace and whitefish caviar according to geographic origin (Sweden, Finland, and the United States) and to water type (brackish vs. freshwater source). Authors determined with ICP-MS more than 70 elements at concentrations from ppm to ppt; in order to verify the effect of the various processes involved in the manufacturing of caviar, unprocessed vendace roe and salt samples were included in the study. It was found that As, Ba, Br, I, Li, Mo, Se, and Sr were useful in the geographic classification, while Fe, Al, Ti, and V could be strongly influenced by processing and packaging. The best markers of geographical provenance were element ratios between alkaline earth elements (eg, Sr/Ca), as a consequence of the fact that in otoliths, these ratios can reflect their relative proportion in the ambient waters, as suggested by Bath et al. (2000). Finally, Sr isotope ratio allowed a good classification between caviar from brackish and freshwater sources. In another study (DePeters et al., 2013) on the possibility of distinguishing caviar obtained from farm-raised versus wild white sturgeons living in California, authors found that most of the trace elements determined did not significantly contribute to the classification, with the only exception being arsenic that was found to be absent in eggs from farm-raised sturgeons.

Honey, on the contrary, is the subject of several classification studies on a geographical or botanical basis. Determination of parameters with ICP-MS could be mostly useful in geographical classification, as the element composition of honey should be connected with the soil on which is active melliferous flora, ie, the whole of plants producing compounds that can be collected by insects and turned into honey. The production chain of honey, however, can be complicated by several factors: both vegetal (melliferous plants) and animal (the bees) metabolisms are involved; environmental pollution or use of fertilizers can increase the content of transition and heavy metals (Batista et al., 2012); additional substances such as antibiotics, even if not allowed in certain official regulations, eg, the European Union, could be used by keepers in order to protect bees from diseases and to get queen bees to lay more eggs; finally, it can be hard establishing an accurate reference to the territory on which bees stand on, in particular for the *1000 flower type* (multifloral) honey produced from a variety of flowers. For honey produced according to high purity standards, however, multielement profiling or isotope ratio analysis could provide good markers for classifying the geographic provenance. In the already cited study by Voerkelius et al. (2010), authors demonstrated the good correlation of δ^{87}Sr value between honey samples and the corresponding local surface water samples, establishing a geographic classification. With concern to multielement profiling, Baroni et al. (2015) analyzed honey samples from three areas in Argentina and, despite the samples being mostly multifloral, it was found that they could be correctly classified on a geographic basis with trace elements and δ^{87}Sr; in addition, there was a good correlation among honey and soil (the variables mainly contributing to the correlation were Co, Ni, Sr, and K/Rb in soil and Mn and Rb in honey) and among honey and local water (Ca, B, and Ca/Sr contents in water as well as Rb and Mg contents in honey). Similar results were obtained by Di Bella et al. (2015) in the classification of Italian honeys produced in Calabria and Sicily: it was possible to distinguish honey samples produced in the two regions according to multielement profiling. The classification of honey samples according to the botanical species by means of inorganic parameters was reported to be harder than according to the geographic provenance; nevertheless, some works (Chudzinska and Baralkiewicz, 2011; Chen et al., 2014) seem to suggest this possibility by proper application of multivariate chemometric analysis.

The classification of honey has been reviewed by Camiña et al. (2012). A more specific review, with concern to the role of metals in honey and their determination, was published by Pohl et al. (2009).

4.4 Wine and Alcoholic Drinks

Wine is, with no doubt, the foodstuff which most of the classification studies are devoted to, and surely it has been one of the first subjects of the application of sophisticated analytical techniques. With reference to the subject of the present chapter, the works by McCurdy et al. (1992), and by Stroh et al. (1994) on the determination of trace elements in wine are among the first suggesting their potential as classification markers after analysis by means of ICP-MS.

The classification of wine samples using inorganic parameters can be obtained by the proper selection of variables. These can be different whether the objective is tracing a wine sample, ie, finding its link with soil, or distinguishing samples on a geographical basis (varietal and technological classification can be obtained more profitably using organic variables). Before showing examples of traceability and authentication of wines, it is mandatory to describe how the production chain of wine influences its final elemental composition, therefore addressing the selection of variables.

Many processes, either natural or artificial, are involved along the chain that starts from the collection of grapes to bottled wine. As with other plants, the grapevine absorbs mineral substances from the soil according to its physiological needs, which implies fractionation of the distribution: essential elements (ie, Ca, Na, K, etc.) are favored, toxic elements (ie, Cd, Hg, Pb, etc.) are depleted, and nonessential elements (ie, Ba, Rb, Sr, etc.) can be absorbed due to their chemical affinity to essential elements. The elemental composition of grapes can partially reflect the geochemical features of the corresponding soil: Protano and Rossi (2014), in their study on Sangiovese grapes cultivated in Tuscany, identified Ba, Rb, and Sr as representative of the geochemistry of different vineyard soils. As to the conversion of grapes into must, some studies (Oddone et al., 2007; Aceto et al., 2013) demonstrated that the distribution of REE in soil is well maintained in the corresponding must (Fig. 8.3).

Thereafter, the role of technology must be taken into account, evaluating all the passages involved in the winemaking technology. This can comprehend several transformations, according to the different methods of winemaking (ie, red, white, or rosé vinification). The passages that can mostly influence the final composition of wine are the following:

- *The use of substances prior to harvest*: fertilizers, pesticides, Bordeaux mixture, sulfur, etc., which can be sources of transition metals.
- *The addition of substances to must or wine*: this long list may include phosphates, tartrates, SO_2, etc., which provide many elements, mostly alkaline and alkaline-earth elements.

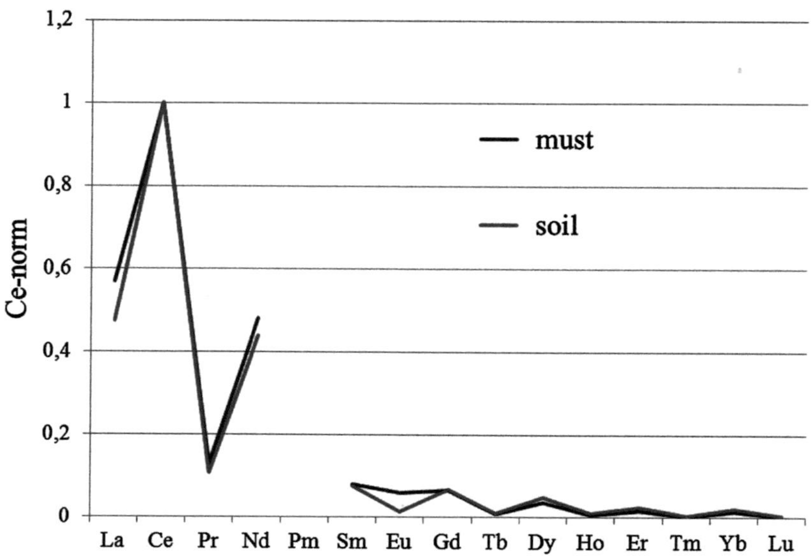

Figure 8.3 Distribution of REE in soil and in wine must.

- *Contact with substrates displaying various tasks*: bentonites, diatomaceous earths, filters, etc., which are used to level off the amount of metals and are known to cause fractionation of REE (Jakubowski et al., 1999; Nicolini et al., 2004; Catarino et al., 2007; Aceto et al., 2013), while other elements seem to remain unaffected (Castiñeira Gómez Mdel et al., 2004).
- *Contact with tanks*: the storage of wine before bottling is usually performed in steel or concrete tanks; because wine is a slightly acidic solution, the release of metal ions from tank surfaces is to be expected. Contributions from packaging must also be expected (Hopfer et al., 2013).

The ideal variables for wine classification are the elements completely unaffected by the winemaking process, which therefore reflect the geochemical peculiarity of soil. It is not easy to individuate such variables. In many cases, multielement profiling for geographic classification has been used as a sort of *untargeted* approach, determining the highest possible number of elements irrespectively of their role in the chain. Limiting the examples to the very last years, the following can be cited: Di Paola-Naranjo et al. (2011) on Argentinean red wines; Zou et al. (2012) on Chinese wines; Martin et al. (2012) on Australian wines; Geana et al. (2013) on Romanian wines; Kruzlicova et al. (2013) on Croatian wines; Šelih et al. (2014) on Slovenian wines; Sen and Tokatli (2014) on Turkish wines; Coetzee et al. (2014) on wines from South Africa; and Azcarate et al. (2015) on Argentinean white wines. Other references can be found in Versari et al. (2014). In each case, authors highlighted what the best variables were in order to have a good classification, but it is apparent that there is no general consensus. At any rate, a comparison of these works evidences that the most recalled variables are alkaline and alkaline-earth minor elements, ie, Li, Rb, Sr, Ba, and of course REE.

To resume, trace elements seem to be good markers for wine authentication on a geographical basis, while it is not clear at present whether they could also work for traceability.

With concern to isotope analysis, it is at least since the 1990s (Horn et al., 1993) that the role of strontium as a robust classification marker for wine provenance has been consolidated. The premises for this result are that (1) the original $\delta^{87}Sr$ signature of soil varies from one crop to another, according to the age of the corresponding bedrock, (2) this signature is not significantly influenced by the production chain of wine, and (3) high-precision analytical methods for determining $\delta^{87}Sr$ are now available. The last point is not trivial as the $^{87}Sr/^{86}Sr$ ratio becomes useful as a marker when it is possible, highlighting differences in the fourth decimal place. The scientific literature on the use of $\delta^{87}Sr$ as a marker for wine is vast; the most recent studies (Marchionni et al., 2013; Martins et al., 2014; Mercurio et al., 2014; Petrini et al., 2015; Durante et al., 2015; Marchionni et al., 2016) evidence the strong relationship between $\delta^{87}Sr$ in grapes and in wine, indicating that the vinification process does not alter the isotopic signature, as already suggested in the first works by Almeida and Vasconcelos (2001, 2004); with concern to the link between grapes and soil (and therefore between wine and soil), they suggest that the isotopic signature of grapes is strongly related to the *bioavailable* fraction of the soil rather than to its bulk.

The potentialities for the classification of wine provided by strontium isotopes seem at present to overshadow the use of other elements. While in the past, several studies proved the interest for the isotopic signature of lead (Almeida and Vasconcelos, 1999; Larcher et al., 2003), in recent years the number of publications on this subject has quickly decreased. The same holds true for the use of boron isotopes, to which a very minor part of research on food authentication has been devoted (Coetzee and Vanhaecke, 2005; Rosner et al., 2011; Coetzee et al., 2011). The role of SIA in the field of enology and viticulture has been reviewed by Santesteban et al. (2015).

With concern to other alcoholic drinks, classification studies on distilled spirits were reviewed in Szymczycha-Madeja et al. (2015), while Mahmood et al. (2012) classified samples of commercial beers.

4.5 Milk Products

The production chain of milk is relatively easy to manage. From grass to milk passing through the stomach of the producing animal, inorganic compounds (be them nutrients or contaminants) undergo a fast transfer. Therefore, milk composition should reflect well the environmental conditions to which the animal is exposed, including soil, water, and diet. It must be considered, though, that the commercial value of milk hardly justifies a classification study. A particular case is the one of buffalo milk, which is used to produce *Mozzarella di Bufala Campana* DOP cheese, a well-known product of the Italian dairy industry. Benincasa et al. (2008) analyzed cow and buffalo milk samples produced in the same farm under identical environmental, feeding, and animal husbandry conditions; multielement profiling from major to trace elements and subsequent chemometric treatment of data with LDA discriminated the two animal species, providing a good way of identifying the fraudulent labeling of milk and associated by-products such as *Mozzarella* cheese.

Definitely more complicated is the production chain of dairy products. The composition of such products starts from milk but further on is strongly influenced by factors involving chemical transformations: processing, aging, activity and quality of microbial flora, addition of substances with various roles, etc. In addition, most dairy products are indeed *alive products* in the sense that they host microbial species breeding on a rich substrate; their metabolism generates continuously evolving chemical compounds. According to these features, it is hard to find the link with soil; a rare case is the one presented by Suhaj and Koreňovská (2008), who analyzed *Bryndza* sheep cheese produced in nine regions of Slovakia, finding good correlations among trace element concentrations in pasture soils, grasses, milk and cheese products, and by-products; apparently, that particular cheese-making process did not significantly affect the link between the soil and the final product. Similar results were obtained by Stevenson et al. (2015), who found an excellent agreement in the values of $\delta^{87}Sr$ in soils, milk, and cheese samples from the same site, confirming the role of strontium isotopes as soil tracers.

Authentication schemes can be developed more easily on the basis of inorganic parameters. Rossmann et al. (2000) suggested the use of isotope ratio analysis, including $\delta^{87}Sr$, in order to determine the geographic provenance of butter. Pillonel et al. (2003) were able to distinguish Emmental cheeses produced in six European regions (Allgau in Germany; Bretagne and Savoie in France, Finland, and Switzerland; and

Vorarlberg in Austria) with multielement profiling coupled with stable isotope ratios ($\delta^{87}Sr$) and determination of radioactive elements (^{90}Sr, ^{234}U, ^{238}U). Classification succeeded using all the variables determined, including isotope ratios of light elements, but it was only partially efficient when a single class of parameters was used alone. Moreno-Rojas et al. (2012) correctly classified a large number of PDO cheeses produced in Northern Spain using Ca, P, and Na as variables.

4.6 Meats and Fishes

The sector of meat and fishes is possibly the most complex from the point of view of the production chain. The livestock system is composed by many variables (environment, race, diet, drinking water features, conjugal conditions, etc.) which can act as potential secondary sources of elements. Diet is one of the most relevant factors, and it is frequently out of control in analytical terms. Despite the fact that meat is almost entirely obtained from herbivore species, and therefore the primary source should be soil (actually grass, which is the same), livestock can be fed with different diets or in different farms, with potentially different methods. One typical comparison is extensive systems of production (ie, *organic farming*) versus intensive systems: in the first case the diet should be dominated by local grass consumption, with few supplements from sources external to farms, while in the second case feeding can come from several sources and therefore reflect a mixture of provenances. Additionally, the mineral elements assumed by animals tend to fix onto muscle fibers in the long-term, contrary to milk, which quickly reflects the minerals assumed by animals from the diet. Finally, another secondary source could be pollution from industrial, mining, or accidental anthropic activity (disasters such as Chernobyl). Some studies have exploited polluting elements as markers of the geographic origin of meat (Bruce et al., 2003).

The type of livestock feeding (extensive vs. intensive) can be identified with SIA of light elements, while geographic classification of meat samples can be obtained by accurate selection of inorganic variables to complement light elements. As an example, Heaton et al. (2008) correctly classified beef meat samples from Europe, the United States, South America, Australia, and New Zealand using both IRMS and ICP-MS: Fe, Rb, Se, and Sr were found to be complementary to light element isotope ratios as variables. The relevance of Rb and Se as geochemical markers has already been highlighted by Franke et al. (2005). Zhao et al. (2013) obtained similar results in a study on beef meat samples from China. In an in-depth study by Rummel et al. (2012), the potential of $\delta^{87}Sr$ as a geographic marker for beef meat has been shown, but it was found that the range of variability was too large to establish a reliable protocol, possibly due to the fact that the phenomena of mineral assumption from the diet are not entirely under control. Relatively more promising results were shown by Baroni et al. (2011) in their study on beef meat samples from three areas in Argentina: authors were able to distinguish meat samples from the different areas using elemental and isotopic parameters and also to find clear coincidences between the meat elemental/isotopic composition and those from soil and water where cattle grew. Of particular relevance were the variables Rb, Ca/Sr, and $\delta^{87}Sr$, apart from $\delta^{13}C$ and $\delta^{15}N$. In a similar way,

Evans et al. (2015) analyzed meat products from farm livestock raised in Great Britain; authors found that the isotopic signature of Pb in samples could be attributed to geogenic sources only with no contribution from anthropogenic sources, and that there was a good correlation in Pb isotope ratios among meat samples and crop samples, therefore distinguishing meat samples from three different regions in Great Britain.

The situation is the worst in the case of sausages. The preparation of these products involves the addition of different substances used for both conservation and aroma, such as salt, spices, additives, and selected microorganisms. All these secondary sources can heavily influence the elemental composition of raw meat. Indeed, no classification study based on inorganic parameters is known at present.

The fish sector is a very important one, considering that it provides a large part of the human diet in many cultures. Mislabeling is a common practice in fraudulent activity, so there is a great interest in the classification of samples, concerning three main issues: identifying the correct species, the geographic provenance, and the method of production (wild vs. farmed or organic vs. intensive). The life cycle of fishery products is much more complicated than other foodstuffs. With concern to element composition, most minerals should come ideally from the water fishes live in, but several additional factors must be considered, such as geomorphology, lithology, contamination from external sources, species, size, age, life history, sex and sexual maturity, salinity, climate, and, above all, feed, which is an out-of-control parameter for caught individuals. The situation could be less problematic in aquaculture products, in which the area of interaction is more limited and the feed delivery more controlled. Despite this, fish feeds have extremely variable composition; in addition, the same fish feed, produced on a large scale, could be used for fishes cultured in different geographic areas. Classification of fishery products with trace elements has been reviewed by Li et al. (2016); from the several studies available, it appears that trace elements profiling could be a reliable method for geographic provenance, identification of fish species, or of method of production, but the application of supervised chemometric procedures (ie, LDA, SIMCA) is mandatory in order to reach good results. As to IRMS analysis, a study by Barnett-Johnson et al. (2008) showed that δ^{87}Sr values measured in otoliths (ie, fish ear bones), due to their metabolic inertness, could be used to trace the geographic provenance and the movement of salmonids in freshwater, estuarine, and marine environments. A combined approach of multielement and isotopic analysis allowed Ortea and Gallardo (2015) to classify shrimp samples according to geographic provenance, biological species, and production method (wild vs. farmed). Additional references on fish classification by means of inorganic parameters can be found in the review by Lavilla et al. (2013).

4.7 Other Foodstuffs

A highly relevant food sector, encompassing all those previously cited, is the one of *organic food*. This term defines foodstuffs produced according to strict rules with concern to biological equilibria, health of both humans and animals, and protection of the environment. Agrochemicals and synthetically produced fertilizers must be avoided all along the production chain, which is a positive feature from the classification point of view because it reduces the impact of secondary sources in the elemental composition. Due to the high commercial value of organic food, several studies have been devoted to the possibility of distinguishing organic products from conventional or industrially made products.

Most of the analytical methods for identifying true organic products involve the determination of organic molecules (eg, agrochemicals, pesticides, antibiotics, etc.). The role of inorganic parameters (trace element determination and isotope ratios) has been discussed, and it was found that the differences inferred by the agricultural practice can be partially obscured by other variables such as geographic provenance, climate, and plant species, so that it is sometimes hard to obtain a clear classification on the basis of the scheme organic versus conventional regime only. A careful selection of inorganic variables must be performed, together with the mandatory application of chemometric pattern recognition methods. A misleading factor could be the consideration that conventional agriculture is based on the use of inorganic fertilizers and pesticides which are rich in heavy metals and phosphates; therefore, the products cultivated on such crops should behave consequently, displaying a higher amount of these elements. As an example, Barbosa et al. (2014) found higher levels of most elements in conventional coffee samples from Brazil than in organic samples. Similar considerations could be drawn for animal products with concern to feeding. Notwithstanding, Borges et al. (2015b) showed that in rice samples cultivated in Brazil, only phosphorus was higher in conventional products than in organic products, while heavy metals were not significantly different between the two categories. A similar result was obtained by Borges et al. (2015a) in a study on chicken eggs: a comparison was made on the element composition in organic versus nonorganic samples, and it was found that organic eggs could be discriminated according to their higher concentrations of As, Co, Fe, Mn, Rb, Se, Tl, and V, while nonorganic eggs had higher levels only for of Cr and P, which are known to be added in the poultry feed. In a thorough study by Laursen et al. (2011), authors showed that it is possible to distinguish organic and conventional agricultural productions using multielement profiling coupled with chemometric routines; the most selective variables seem to be microessential or toxic elements such as cadmium and chlorine, which can occur in crops as a consequence of the use of synthetic fertilizers. Different conclusions were drawn by Kelly and Bateman (2010), who highlighted the role of arbuscular mycorrhizal fungi (AMF) in soils: these organisms occur at elevated levels in crops run with organic systems and seem to cause differences in the concentrations of nutrients and microessentials such as manganese, calcium, copper, and zinc. For what concerns isotope ratios, however, light elements only were found to be useful in the discrimination (Laursen et al., 2013).

Capuano et al. (2013) have reviewed the analytical methods for the classification of organic foodstuffs, while Laursen et al. (2014) have focused the attention on the role of inorganic parameters as markers.

5. Conclusions

There are good perspectives for the application of ICP-MS in the field of food classification. Its use for the determination of inorganic parameters in foodstuffs will surely increase, in particular to complement other techniques such as SIA of light elements. Instrumentations of higher sensitivity and resolution are being developed that yield better analytical results in terms of reliability of data. The great potential of strontium and trace elements as geochemical markers will be surely improved by these performances. A key role will also be played by a greater comprehension of the mechanisms involved in the relationship among soil and plants and of the processes involved in food production chains.

Acknowledgments

The author wanted to acknowledge the role of Dr. Massimo Baldizzone (ABB) and Dr. Matteo Oddone (Thermo Fisher Scientific) for their invaluable help in carrying out research on food classification and for being such great friends.

References

Aceto, M., 2003. Metals in wine. In: Preedy, V.R., Watson, R.R. (Eds.), Reviews in Food and Nutrition Toxicity. Taylor and Francis, London.

Aceto, M., Baldizzone, M., Oddone, M., 2009. Keeping the track of quality: authentication and traceability studies on wine. In: O'byrne, P. (Ed.), Red Wine and Health. Nova Science Publishers, Hauppauge (NY).

Aceto, M., Robotti, E., Oddone, M., Baldizzone, M., Bonifacino, G., Bezzo, G., di Stefano, R., Gosetti, F., Mazzucco, E., Manfredi, M., Marengo, E., 2013. A traceability study on the Moscato wine chain. Food Chemistry 138, 1914–1922.

Almeida, C.M., Vasconcelos, M.T.S.D., 1999. Determination of lead isotope ratios in port wine by inductively coupled plasma mass spectrometry after pre-treatment by UV-irradiation. Analytica Chimica Acta 396, 45–53.

Almeida, C.M., Vasconcelos, M.T.S.D., 2001. ICP-MS determination of strontium isotope ratio in wine in order to be used as a fingerprint of its regional origin. Journal of Analytical Atomic Spectrometry 16, 607–611.

Almeida, C.M., Vasconcelos, M.T.S.D., 2004. Does the winemaking process influence the wine $^{87}Sr/^{86}Sr$? A case study. Food Chemistry 85, 7–12.

Ariyama, K., Nishida, T., Noda, T., Kadokura, M., Yasui, A., 2006. Effects of fertilization, crop year, variety, and provenance factors on mineral concentrations in onions. Journal of Agricultural and Food Chemistry 54, 3341–3350.

Ariyama, K., Aoyama, Y., Mochizuki, A., Homura, Y., Kadokura, M., Yasui, A., 2007. Determination of the geographic origin of onions between three main production areas in Japan and other countries by mineral composition. Journal of Agricultural and Food Chemistry 55, 347–354.

Ariyama, K., Shinozaki, M., Kawasaki, A., Ishida, Y., 2011. Strontium and lead isotope analyses for determining the geographic origins of grains. Analytical Sciences 27, 709–713.

Ariyama, K., Shinozaki, M., Kawasaki, A., 2012. Determination of the geographic origin of rice by chemometrics with strontium and lead isotope ratios and multielement concentrations. Journal of Agricultural and Food Chemistry 60, 1628–1634.

Ariyama, K., 2014. Determination techniques of the geographical origins of agricultural products by elemental and heavy element isotopic compositions. Bunseki Kagaku 63, 205–220.

Arvanitoyannis, I.S., Vlachos, A., 2009. Maize authentication: quality control methods and multivariate analysis (chemometrics). Critical Reviews in Food Science and Nutrition 49, 501–537.

Azcarate, S.M., Martinez, L.D., Savio, M., Camina, J.M., Gil, R.A., 2015. Classification of monovarietal Argentinean white wines by their elemental profile. Food Control 57, 268–274.

Barbosa, R.M., Batista, B.L., Varrique, R.M., Coelho, V.A., Campiglia, A.D., Barbosa Jr., F., 2014. The use of advanced chemometric techniques and trace element levels for controlling the authenticity of organic coffee. Food Research International 61, 246–251.

Barnett-Johnson, R., Pearson, T.E., Ramos, F.C., Grimes, C.B., Macfarlane, R.B., 2008. Tracking natal origins of salmon using isotopes, otoliths, and landscape geology. Limnology and Oceanography 53, 1633–1642.

Baroni, M.V., Podio, N.S., Badini,´R.G., Inga, M., Ostera,´H.A., Cagnoni, M., Gallegos, E., Gautier, E.A., Peral García, P., Hoogewerff, J., Wunderlin, D.A., 2011. How much do soil and water contribute to the composition of meat? A case study: meat from three areas of Argentina. Journal of Agricultural and Food Chemistry 59, 11117–11128.

Baroni, M.V., Podio, N.S., Badini,´R.G., Inga, M., Ostera,´H.A., Cagnoni, M., Gautier, E.A., Peral García, P., Hoogewerff, J., Wunderlin, D.A., 2015. Linking soil, water, and honey composition to assess the geographical origin of Argentinean honey by multielemental and isotopic analyses. Journal of Agricultural and Food Chemistry 63, 4638–4645.

Bath, G.E., Thorrold, S.R., Jones, C.M., Campana, S.E., Mclaren, J.W., Lam, J.W., 2000. Strontium and barium uptake in aragonitic otoliths of marine fish. Geochimica et Cosmochimica Acta 64, 1705–1714.

Batista, B.L., da Silva, L.R.S., Rocha, B.A., Rodrigues, J.L., Berretta-Silva, A.A., Bonates, T.O., Gomes, V.S.D., Barbosa, R.M., Barbosa, F., 2012. Multi-element determination in Brazilian honey samples by inductively coupled plasma mass spectrometry and estimation of geographic origin with data mining techniques. Food Research International 49, 209–215.

Beltrán, M., Sánchez-Astudillo, M., Aparicio, R., García-González, D.L., 2015. Geographical traceability of virgin olive oils from south-western Spain by their multi-elemental composition. Food Chemistry 169, 350–357.

Benabdelkamel, H., di Donna, L., Mazzotti, F., Naccarato, A., Sindona, G., Tagarelli, A., Taverna, D., 2012. Authenticity of PGI "Clementine of Calabria" by multielement fingerprint. Journal of Agricultural and Food Chemistry 60, 3717–3726.

Benincasa, C., Lewis, J., Perri, E., Sindona, G., Tagarelli, A., 2007. Determination of trace element in Italian virgin olive oils and their characterization according to geographical origin by statistical analysis. Analytica Chimica Acta 585, 366–370.

Benincasa, C., Lewis, J., Sindona, G., Tagarelli, A., 2008. The use of multi element profiling to differentiate between cow and buffalo milk. Food Chemistry 110, 257–262.

Bertacchini, L., Cocchi, M., Li Vigni, M., Marchetti, A., Salvatore, E., Sighinolfi, S., Silvestri, M., Durante, C., 2013. The impact of chemometrics on food traceability. Data Handling in Science and Technology 28, 371–410.

Bettinelli, M., Spezia, S., Baffi, C., Beone, G.M., Rocchetta, R., Nassisi, A., 2005. ICP-MS determination of REEs in tomato plants and related products: a new analytical tool to verify traceability. Atomic Spectroscopy 26, 41–50.

Bontempo, L., Camin, F., Manzocco, L., Nicolini, G., Wehrens, R., Ziller, L., Larcher, R., 2011. Traceability along the production chain of Italian tomato products on the basis of stable isotopes and mineral composition. Rapid Communications in Mass Spectrometry 25, 899–909.

Borges, E.M., Volmer, D.A., Gallimberti, M., De Souza, D.F., De souza, E.L., Barbosa, F.A.F., 2015a. Evaluation of macro- and microelement levels for verifying the authenticity of organic eggs by using chemometric techniques. Analytical Methods 7, 2577–2584.

Borges, E.M., Gelinski, J.M.L.N., de Oliveira Souza, V.C., Barbosa Jr., F., Batista, B.L., 2015b. Monitoring the authenticity of organic rice via chemometric analysis of elemental data. Food Research International 77, 299–309.

Brown, P.H., Rathjen, A.H., Graham, R.D., Tribe, D.E., 1990. Rare earth elements in biological systems. In: Gschneider, K.A., Eyring, L. (Eds.), Handbook on the Physics and Chemistry of Rare Earths, vol. 13. Elsevier, Amsterdam.

Bruce, S.L., Noller, B.N., Grigg, A.H., Mullen, B.F., Mulligan, D.R., Ritchie, P.J., Currey, N., Ng, J.C., 2003. A field study conducted at Kidston gold mine to evaluate the impact of arsenic and zinc from mine tailing to grazing cattle. Toxicology Letters 137, 23–34.

Brunner, M., Katona, R., Stefanka, Z., Prohanska, T., 2010. Determination of the geographical origin of processed spice using multielement and isotopic pattern on the example of *Szegedi paprika*. European Food Research and Technology 231, 623–634.

Camin, F., Larcher, R., Nicolini, G., Bontempo, L., Bertoldi, D., Perini, M., Schlicht, C., Schellenberg, A., Thomas, F., Heinrich, K., Voerkelius, S., Horacek, M., Ueckermann, H., Froeschl, H., Wimmer, B., Heiss, G., Baxter, M., Rossmann, A., Hoogewerff, J., 2010. Isotopic and elemental data for tracing the origin of European olive oils. Journal of Agricultural and Food Chemistry 58, 570–577.

Camiña, J.M., Pellerano, R.G., Marchevsky, E.J., 2012. Geographical and botanical classification of honeys and apicultural products by chemometric methods. A review. Current Analytical Chemistry 408–425.

Capuano, E., Boerrigter-Eenling, R., Van Der Veer, G., Van Ruth, S.M., 2013. Analytical authentication of organic products: an overview of markers. Journal of the Science of Food and Agriculture 93, 12–28.

Castiñeira Gómez Mdel, M., Brandt, R., Jakubowski, N., Anderson, J.T., 2004. Changes of metal composition in German white wines through the winemaking process. A study of 63 elements by inductively coupled plasma-mass spectrometry. Journal of Agricultural and Food Chemistry 52, 2953–2961.

Catarino, S., Madeira, M., Monteiro, F., Rocha, F., Curvelo-Garcia, A.S., De Sousa, R.B., 2007. Effect of bentonite characteristics on the elemental composition of wine. Journal of Agricultural and Food Chemistry 56, 158–165.

Chen, Y.X., Yu, M.G., Xu, J., Chen, X.C., Shi, J.Y., 2009. Differentiation of eight tea (*Camellia sinensis*) cultivars in China by elemental fingerprint of their leaves. Journal of the Science of Food and Agriculture 89, 2350–2355.

Chen, H., Fan, C., Chang, Q., Pang, G., Hu, X., Lu, M., Wang, W., 2014. Chemometric determination of the botanical origin for Chinese honeys on the basis of mineral elements determined by ICP-MS. Journal of Agricultural and Food Chemistry 62, 2443–2448.

Chudzinska, M., Baralkiewicz, D., 2011. Application of ICP-MS method of determination of 15 elements in honey with chemometric approach for the verification of their authenticity. Food and Chemical Toxicology 49, 2741–2749.

Coetzee, P.P., Vanhaecke, F., 2005. Classifying wine according to geographical origin via quadrupole-based ICP-mass spectrometry measurements of boron isotope ratios. Analytical and Bioanalytical Chemistry 383, 977–984.

Coetzee, P.P., Greeff, L., Vanhaecke, F., 2011. ICP-MS measurement of B-11/B-10 isotope ratios in grapevine leaves and the investigation of possible boron isotope fractionation in grapevine plants. South African Journal of Enology and Viticulture 32, 28–34.

Coetzee, P.P., Van Jaarsveld, F.P., Vanhaecke, F., 2014. Intraregional classification of wine via ICP-MS elemental fingerprinting. Food Chemistry 164, 485–492.

Coryell, C.D., Chase, J.W., Winchester, J.W., 1963. A procedure for geochemical interpretation of terrestrial rare-earth abundance patterns. Journal of Geophysical Research 68, 559–566.

D'archivio, A.A., Giannitto, A., Incani, A., Nisi, S., 2014. Analysis of the mineral composition of Italian saffron by ICP-MS and classification of geographical origin. Food Chemistry 157, 485–489.

Depeters, E.J., Puschner, B., Taylor, S.J., Rodzen, J.A., 2013. Can fatty acid and mineral compositions of sturgeon eggs distinguish between farm-raised versus wild white (*Acipenser transmontanus*) sturgeon origins in California? Preliminary report. Forensic Science International 229, 128–132.

Di Bella, G., Lo Turco, V., Potortì, A.G., Bua, G.D., Fede, M.R., Dugo, G., 2015. Geographical discrimination of Italian honey by multi-element analysis with a chemometric approach. Journal of Food Composition and Analysis 44, 25–35.

Di Paola-Naranjo, R.D., Baroni, M.V., Podio, N.S., Rubinstein, H.R., Fabani, M.P., Badini, R.G., Inga, M., Ostera, H.A., Cagnoni, M., Gallegos, E., Gautier, E., Peral-García, P., Hoogewerff, J., Wunderlin, D.A., 2011. Fingerprints for main varieties of Argentinean wines: terroir differentiation by inorganic, organic, and stable isotopic analyses coupled to chemometrics. Journal of Agricultural and Food Chemistry 59, 7854–7865.

Drivelos, S.A., Georgiou, C.A., 2012. Multi-element and multi-isotope-ratio analysis to determine the geographical origin of foods in the European Union. Trends in Analytical Chemistry 40, 38–51.

Drivelos, S.A., Higgins, K., Kalivas, J.H., Haroutounian, S.A., Georgiou, C.A., 2014. Data fusion for food authentication. Combining rare earth elements and trace metals to discriminate "Fava Santorinis" from other yellow split peas using chemometric tools. Food Chemistry 165, 316–322.

Durante, C., Baschieri, C., Bertacchini, L., Bertelli, D., Cocchi, M., Marchetti, A., Manzini, D., Papotti, G., Sighinolfi, S., 2015. An analytical approach to Sr isotope ratio determination in Lambrusco wines for geographical traceability purposes. Food Chemistry 173, 557–563.

Evans, J.A., Pashley, V., Richards, G.J., Brereton, N., Knowles, T.G., 2015. Geogenic lead isotope signatures from meat products in Great Britain: potential for use in food authentication and supply chain traceability. Science of the Total Environment 537, 447–452.

Fragni, R., Trifirò, A., Nucci, A., 2015. Towards the development of a multi-element analysis by ICP-oa-TOF-MS for tracing the geographical origin of processed tomato products. Food Control 48, 96–101.

Franke, B.M., Gremaud, G., Hadorn, R., Kreuzer, M., 2005. Geographic origin of meat-elements of an analytical approach to its authentication. European Food Research and Technology 221, 493–503.

Furia, E., Naccarato, A., Sindona, G., Stabile, G., Tagarelli, A., 2011. Multielement fingerprinting as a tool in origin authentication of PGI food products: Tropea Red Onion. Journal of Agricultural and Food Chemistry 59, 8450–8457.

Geana, I., Iordache, A., Ionete, R., Marinescu, A., Ranca, A., Culea, M., 2013. Geographical origin identification of Romanian wines by ICP-MS elemental analysis. Food Chemistry 138, 1125–1134.

Georgiou, C.A., Danezis, G.P., 2015. Elemental and isotopic mass spectrometry. In: Picò, Y. (Ed.), Advanced Mass Spectrometry for Food Safety and Quality. Comprehensive Analytical Chemistry, vol. 68. Elsevier, Amsterdam.

Gonzálvez, A., Armenta, S., De La Guardia, M., 2009. Trace-element composition and stable-isotope ratio for discrimination of foods with protected designation of origin. Trends in Analytical Chemistry 28, 1295–1311.

Gonzálvez, A., De La Guardia, M., 2013a. Mineral profile. In: De La Guardia, M., Gonzálvez, A. (Eds.), Food Protected Designation of Origin: Methodologies and Applications. Comprehensive Analytical Chemistry, vol. 60. Elsevier, Amsterdam.

Gonzálvez, A., De La Guardia, M., 2013b. Basic chemometric tools. In: De La Guardia, M., Gonzálvez, A. (Eds.), Food Protected Designation of Origin: Methodologies and Applications. Comprehensive Analytical Chemistry, vol. 60. Elsevier, Amsterdam.

Graustein, W.C., 1988. $^{87}Sr/^{86}Sr$ ratios measure the sources and flow of strontium in terrestrial ecosystems. In: Rundel, P.W., Ehleringer, J.R., Nagy, K.A. (Eds.), Stable Isotopes in Ecological Research. Springer-Verlag, Berlin.

Han, Q., Mihara, S., Hashimoto, K., Fujino, T., 2014. Optimization of tea sample preparation methods for ICP-MS and application to verification of Chinese tea authenticity. Food Science and Technology Research 20, 1109–1119.

Heaton, K., Kelly, S.D., Hoogewerff, J., Woolfe, M., 2008. Verifying the geographical origin of beef: the application of multi-element isotope and trace element analysis. Food Chemistry 107, 506–515.

Hondrogiannis, E.M., Ehrlinger, E., Poplaski, A., Lisle, M., 2013. Use of laser ablation–inductively coupled plasma–time of flight–mass spectrometry to identify the elemental composition of vanilla and determine the geographic origin by discriminant function analysis. Journal of Agricultural and Food Chemistry 61, 11332–11337.

Hopfer, H., Nelson, J., Mitchell, A.E., Heymann, H., Ebeler, S.E., 2013. Profiling the trace metal composition of wine as a function of storage temperature and packaging type. Journal of Analytical Atomic Spectrometry 28, 1288–1291.

Horn, P., Schaaf, P., Holbach, B., Hölzl, S., Eschnauer, H., 1993. $^{87}Sr/^{86}Sr$ from rock and soil into vine and wine. Zeitschrift für Lebensmittel-Untersuchung und –Forschung 196, 407–409.

Jakubowski, N., Brandt, R., Stuerwer, D., Eschnauer, H.R., Görtges, S., 1999. Analysis of wines by ICP-MS: is the pattern of the rare earth elements a reliable fingerprint for the provenance? Fresenius' Journal of Analytical Chemistry 364, p424–428.

Kelly, S.D., Bateman, A.S., 2010. Comparison of mineral concentrations in commercially grown organic and conventional crops–tomatoes (*Lycopersicon esculentum*) and lettuces (*Lactuca sativa*). Food Chemistry 119, 738–745.

Kruzlicova, D., Fiket, Ž., Kniewald, G., 2013. Classification of Croatian wine varieties using multivariate analysis of data obtained by high resolution ICP-MS analysis. Food Research International 54, 621–626.

Lagad, R.A., Alamelu, D., Laskar, A.H., Rai, V.K., Singh, S.V., Aggarwal, S.K., 2013. Isotope signature study of the tea samples produced at four different regions in India. Analytical Methods 5, 1604–1611.

Larcher, R., Nicolini, G., Pangrazzi, P., 2003. Isotope ratios of lead in Italian wines by inductively coupled plasma mass spectrometry. Journal of Agricultural and Food Chemistry 51, 5956–5961.

Laursen, K.H., Schjoerring, J.K., Olesen, J.E., Askegaard, M., Halekoh, U., Husted, S., 2011. Multielemental fingerprinting as a tool for authentication of organic wheat, barley, faba bean, and potato. Journal of Agricultural and Food Chemistry 59, 4385–4396.

Laursen, K.H., Mihailova, A., Kelly, S.D., Epov, V.N., Bérail, S., Schjoerring, J.K., Donard, O.F.X., Larsen, E.H., Pedentchouk, N., Marca-Bell, A.D., Halekoh, U., Olesen, J.E., Husted, S., 2013. Is it really organic?–multi-isotopic analysis as a tool to discriminate between organic and conventional plants. Food Chemistry 141, 2812–2820.

Laursen, K.H., Schjoerring, J.K., Kelly, S.D., Husted, S., 2014. Authentication of organically grown plants – advantages and limitations of atomic spectroscopy for multi-element and stable isotope analysis. Trends in Analytical Chemistry 59, 73–82.

Lavilla, I., Costas-Rodriguez, M., Bendicho, C., 2013. Authentication of fishery products. In: De La Guardia, M., Gonzálvez, A. (Eds.), Food Protected Designation of Origin: Methodologies and Applications. Comprehensive Analytical Chemistry, vol. 60. Elsevier, Amsterdam.

Li, L., Boyd, C.E., Sun, Z., 2016. Authentication of fishery and aquaculture products by multi-element and stable isotope analysis. Food Chemistry 194, 1238–1244.

Lo Feudo, G., Naccarato, A., Sindona, G., Tagarelli, A., 2010. Investigating the origin of tomatoes and triple concentrated tomato pastes through multielement determination by inductively coupled plasma mass spectrometry and statistical analysis. Journal of Agricultural and Food Chemistry 58, 3801–3807.

Liang, T., Ding, S., Song, W., Chong, Z., Zhang, C., Li, H., 2008. A review of fractionations of rare earth elements in plants. Journal of Rare Earths 26, 7–15.

Liu, H., You, C., Chen, C., Liu, Y., Chung, M., 2014. Geographic determination of coffee beans using multi-element analysis and isotope ratios of boron and strontium. Food Chemistry 142, 439–445.

Llorent-Martínez, E.J., Ortega-Barrales, P., Fernández-De Córdova, M.L., Domínguez-Vidal, A., Ruiz-Medina, A., 2011. Investigation by ICP-MS of trace element levels in vegetable edible oils produced in Spain. Food Chemistry 127, 1257–1262.

Mahmood, N., Petraco, N., He, Y., 2012. Elemental fingerprint profile of beer samples constructed using 14 elements determined by inductively coupled plasma–mass spectrometry (ICP-MS): multivariation analysis and potential application to forensic sample comparison. Analytical and Bioanalytical Chemistry 402, 861–869.

Marchionni, S., Braschi, E., Tommasini, S., Bollati, A., Cifelli, F., Mulinacci, N., Mattei, M., Conticelli, S., 2013. High-precision $^{87}Sr/^{86}Sr$ analyses in wines and their use as a geological fingerprint for tracing geographic provenance. Journal of Agricultural and Food Chemistry 61, 6822–6831.

Marchionni, S., Buccianti, A., Bollati, A., Braschi, E., Cifelli, F., Molin, P., Parotto, M., Mattei, M., Tommasini, S., Conticelli, S., 2016. Conservation of $^{87}Sr/^{86}Sr$ isotopic ratios during the winemaking processes of 'Red' wines to validate their use as geographic tracer. Food Chemistry 190, 777–785.

Martin, A.E., Watling, R.J., Lee, G.S., 2012. The multi-element determination and regional discrimination of Australian wines. Food Chemistry 133, 1081–1089.

Martins, P., Madeira, M., Monteiro, F., De Sousa, R.B., Curvelo-Garcia, A.S., Catarino, S., 2014. $^{87}Sr/^{86}Sr$ ratio in vineyard soils from Portuguese denominations of origin and its potential for origin authentication. Journal International des Sciences de la Vigne et du Vin 48, 21–29.

Mccurdy, E., Potter, D., Medina, M., 1992. Trace elements in wine. Laboratory News 9, 10–11.

Medini, S., Janin, M., Verdoux, P., Técher, I., 2015. Methodological development for $^{87}Sr/^{86}Sr$ measurement in olive oil and preliminary discussion of its use for geographical traceability of PDO Nimes (France). Food Chemistry 171, 78–83.

Mercurio, M., Grilli, E., Odierna, P., Morra, V., Prohaska, T., Coppola, E., Grifa, C., Buondonno, A., Langella, A., 2014. A 'Geo-Pedo-Fingerprint' (GPF) as a tracer to detect univocal parent material-to-wine production chain in high quality vineyard districts, Campi Flegrei (Southern Italy). Geoderma 230–231, 64–78.

Merian, E., Anke, M., Ihnat, M., Stoeppler, M. (Eds.), 2004. Elements and Their Compounds in the Environment: Occurrence, Analysis and Biological Relevance, second ed. Wiley-VCH, New York.

Montaser, A. (Ed.), 1998. Inductively Coupled Plasma – Mass Spectrometry, third ed. Wiley-VCH, New York.

Moreda-Piñeiro, A., Fisher, A., Hill, S.J., 2003. The classification of tea according to region of origin using pattern recognition techniques and trace metal data. Journal of Food Composition and Analysis 16, 195–211.

Moreno-Rojas, R., Cámara-Martos, F., Sánchez-Segarra, P.J., Amaro-López, M.A., 2012. Influence of manufacturing conditions and discrimination of Northern Spanish cheeses using multi-element analysis. International Journal of Dairy Technology 65, 594–602.

Nicolini, G., Larcher, R., Pangrazzi, P., Bontempo, L., 2004. Changes in the contents of micro- and trace-elements in wine due to winemaking treatments. Vitis 43, 41–45.

Oda, H., Kawasaki, A., Hirata, T., 2001. Determination of the geographic origin of brown-rice with isotope ratios of $^{11}B/^{10}B$ and $^{87}Sr/^{86}Sr$. Analytical Science 17, i1627–i1630.

Oddone, M., Robotti, E., Marengo, M., Baldizzone, M., Aceto, M., 2007. Studio della tracciabilità sulla filiera del vino mediante determinazione dei lantanidi con ICP-MS. In: Coisson, J.D., Arlorio, M., Martelli, A. (Eds.), Proceedings of VI Italian Congress "Chimica Degli Alimenti" (Alessandria: Taro).

Oddone, M., Aceto, M., Baldizzone, M., Musso, D., Osella, D., 2009. Authentication and traceability study of hazelnuts from Piedmont, Italy. Journal of Agricultural and Food Chemistry 57, 3404–3408.

Ortea, I., Gallardo, J.M., 2015. Investigation of production method, geographical origin and species authentication in commercially relevant shrimps using stable isotope ratio and/or multi-element analyses combined with chemometrics: an exploratory analysis. Food Chemistry 170, 145–153.

Petrini, R., Sansone, L., Slejko, F.F., Buccianti, A., Marcuzzo, P., Tomasi, D., 2015. The $^{87}Sr/^{86}Sr$ strontium isotopic systematics applied to Glera vineyards: a tracer for the geographical origin of the Prosecco. Food Chemistry 170, 138–144.

Pilgrim, T.S., Watlinga, R.J., Grice, K., 2010. Application of trace element and stable isotope signatures to determine the provenance of tea (*Camellia sinensis*) samples. Food Chemistry 118, 921–926.

Pillonel, L., Badertscher, R., Froidevaux, P., Haberhauer, G., Hölzl, S., Horn, P., Jakob, A., Pfammatter, E., Piantini, U., Rossmann, A., Tabacchi, I., Bosset, J.O., 2003. Stable isotope ratios, major, trace and radioactive elements in emmental cheeses of different origins. Lebensmittel-Wissenschaft und -Technologie 36, 615–623.

Podio, N.S., Baroni, M.V., Badini, R.G., Inga, M., Ostera, H.A., Cagnoni, M., Gautier, E.A., Peral García, P., Hoogewerff, J., Wunderlin, D.A., 2013. Elemental and isotopic fingerprint of Argentinean wheat. Matching soil, water, and crop composition to differentiate provenance. Journal of Agricultural and Food Chemistry 61, 3763–3773.

Pohl, P., Sergiel, I., Stecka, H., 2009. Determination and fractionation of metals in honey. Critical Reviews in Analytical Chemistry 39, 276–288.

Protano, G., Rossi, S., 2014. Relationship between soil geochemistry and grape composition in Tuscany (Italy). Journal of Plant Nutrition and Soil Science 177, 500–508.

Rodrigues, C., Máguas, C., Prohaska, T., 2011. Strontium and oxygen isotope fingerprinting of green coffee beans and its potential to proof authenticity of coffee. European Food Research and Technology 232, 361–373.

Rodrigues, C., Maia, R., Lauteri, M., Brugnoli, E., Máguas, C., 2013. Stable isotope analysis. In: De La Guardia, M., Gonzálvez, A. (Eds.), Food Protected Designation of Origin: Methodologies and Applications. Comprehensive Analytical Chemistry, vol. 60. Elsevier, Amsterdam.

Rodushkin, I., Bergman, T., Douglas, G., Engström, E., Sörlin, D., Baxter, D.C., 2007. Authentication of Kalix (N.E. Sweden) vendace caviar using inductively coupled plasma-based analytical techniques: evaluation of different approaches. Analytica Chimica Acta 583, 310–318.

Rosner, M., Pritzkow, W., Vogl, J., Voerkelius, S., 2011. Development and validation of a method to determine the boron isotopic composition of crop plants. Analytical Chemistry 83, 2562–2568.

Rossmann, A., Haberhauer, G., Hölzl, S., Horn, P., Pichlmayer, F., Voerkelius, S., 2000. The potential of multielement stable isotope analysis for regional assignment of butter. European Food Research and Technology 211, 32–40.

Rummel, S., Hoelzl, S., Horn, P., Rossmann, A., Schlicht, C., 2010. The combination of stable isotope abundance ratios of H, C, N and S with $^{87}Sr/^{86}Sr$ for geographical origin assignment of orange juices. Food Chemistry 118, 890–900.

Rummel, S., Dekant, C.H., Hölzl, S., Kelly, S.D., Baxter, M., Marigheto, N., Quetel, C.R., Larcher, R., Nicolini, G., Fröschl, H., Ueckermann, H., Hoogewerff, J., 2012. Sr isotope measurements in beef — analytical challenge and first results. Analytical and Bioanalytical Chemistry 402, 2837–2848.

Santesteban, L.G., Miranda, C., Barbarin, I., Royo, J.B., 2015. Application of the measurement of the natural abundance of stable isotopes in viticulture: a review. Australian Journal of Grape and Wine Research 21, 157–167.

Šelih, V.S., Šala, M., Drgan, V., 2014. Multi-element analysis of wines by ICP-MS and ICP-OES and their classification according to geographical origin in Slovenia. Food Chemistry 153, 414–423.

Sen, I., Tokatli, F., 2014. Characterization and classification of Turkish wines based on elemental composition. American Journal of Enology and Viticulture 65, 134–142.

Shen, S., Xia, L., Xiong, N., Liu, Z., Sun, H., 2013. Determination of the geographic origin of rice by element fingerprints and correlation analyses with the soil of origin. Analytical Methods 5, 6177–6185.

Song, B.Y., Ryu, J.S., Shin, H.S., Lee, K.S., 2014. Determination of the source of bioavailable Sr using $^{87}Sr/^{86}Sr$ tracers: a case study of hot pepper and rice. Journal of Agricultural and Food Chemistry 62, 9232–9238.

Stevenson, R., Desrochers, S., Helie, J.F., 2015. Stable and radiogenic isotopes as indicators of agri-food provenance: insights from artisanal cheeses from Quebec, Canada. International Dairy Journal 49, 37–45.

Stroh, A., Bruckner, P., Vollkopf, U., 1994. Multielement analysis of wine using ICP-MS. Atomic Spectroscopy 15, 100–106.

Suhaj, M., Koreňovská, M., 2008. Correlation and distribution of elemental markers of origin in the production of Bryndza sheep cheese. Food Chemistry 107, 551–557.

Szymczycha-Madeja, A., Welna, M., Jamroz, P., Lesniewicz, V., Pohl, P., 2015. Advances in assessing the elemental composition of distilled spirits using atomic spectrometry. Trends in Analytical Chemistry 64, 127–135.

Thomas, R., 2013. Practical Guide to ICP-MS: A Tutorial for Beginners, third ed. CRC Press, Boca Raton, FL.

Trincherini, P.R., Baffi, C., Barbero, P., Pizzoglio, E., Spalla, S., 2014. Precise determination of strontium isotope ratios by TIMS to authenticate tomato geographical origin. Food Chemistry 145, 349–355.

Tyler, G., 2004. Rare earth elements in soil and plant systems: a review. Plant Soil 267, 191–206.

Valentin, J.L., Watling, R.J., 2013. Provenance establishment of coffee using solution ICP-MS and ICP-AES. Food Chemistry 141, 98–104.

Versari, A., Laurie, V.F., Ricci, A., Laghi, L., Parpinello, G.P., 2014. Progress in authentication, typification and traceability of grapes and wines by chemometric approaches. Food Research International 60, 2–18.

Vlachos, A., Arvanitoyannis, I.S., 2008. A review of rice authenticity/adulteration methods and results. Critical Reviews in Food Science and Nutrition 48, 553–598.

Voerkelius, S., Lorenz, G.D., Rummel, S., Quétel, C.R., Heiss, G., Baxter, M., Brach-Papa, C., Deters-Itzelsberger, P., Hoelzl, S., Hoogewerff, J., Ponzevera, E., Van Bocxstaele, M., Ueckermann, H., 2010. Strontium isotopic signatures of natural mineral waters, the reference to a simple geological map and its potential for authentication of food. Food Chemistry 118, 933–940.

White, W.M., 2013. Geochemistry. Wiley-Blackwell, New York.

Zhao, H., Guo, B., Wei, Y., Zhang, B., 2012. Effects of wheat origin, genotype, and their interaction on multielement fingerprints for geographical traceability. Journal of Agricultural and Food Chemistry 60, 10957–10962.

Zhao, Y., Zhang, B., Chen, G., Chen, A., Yang, S., Ye, Z., 2013. Tracing the geographic origin of beef in China on the basis of the combination of stable isotopes and multielement analysis. Journal of Agricultural and Food Chemistry 61, 7055–7060.

Zhao, Y., Zhang, B., Chen, G., Chen, A., Yang, S., Ye, Z., 2014. Recent developments in application of stable isotope analysis on agro-product authenticity and traceability. Food Chemistry 145, 300–305.

Zou, J.F., Peng, Z.X., Du, H.J., Duan, C.Q., Reeves, M.J., Pan, Q.H., 2012. Elemental patterns of wines, grapes, and vineyard soils from Chinese wine-producing regions and their association. American Journal of Enology and Viticulture 63, 232–240.

Future Technology in Tracing Animals on the Food Chain

9

T. Pizzuti, G. Mirabelli
University of Calabria, Rende, Italy

1. General Introduction

The food sector has been faced with a number of dangerous consumer scares. More recently, serious concerns associated with a large number of food safety and livestock epidemics, most notably Bovine Spongiform Encephalopathy (BSE), have led to the intensification of legislative requirements governing animal identification and tracing in countries worldwide.

The reasons for identifying and tracking animals and their products have evolved and include rapid response to animal health and/or food safety concerns, as well as verification of recognized premium commercial production processes as specified on qualifying product labels.

According to Bowling et al. (2008), animals must be reliably and uniquely identified. This enables tracking, which is the ability to follow animals down the food chain from producer to slaughter or to the retail outlet, and tracing, which is the ability to follow a meat product up the supply chain by means of the records which have been kept at each stage of the chain (Bass et al., 2008). Reliable tracing, which is critical in the event of an animal disease outbreak or public health issue, depends on thorough tracking.

Many countries have defined some traceability programs and have implemented their own traceability systems. Each system is different and has its own characteristics. Some traceability programs have the ability to trace animals from birth to harvest, including animal location movements in between (eg, swine traceability systems in the United Kingdom and Denmark).

According to Geiger and Hayes (2003), three different types of traceability systems can be distinguished: farm to retail traceability, batch traceability, and hypothetical traceability.

A *farm to retail* traceability program traces an animal from birth to an individual package bought at the retail level. It is very expensive because it requires extensive capital investment and data infrastructure. It is very rare in the EU, although many consumers think this is the system in place for beef and pork. The traceability systems imposed at the European level refer to the *batch traceability*. Systems based on batch traceability involve traceability from farm to carcass. In this case the history of the animal is tracked for each carcass or primary cut, while the meat is cut and processed in batches. The third type of traceability, the

Advances in Food Traceability Techniques and Technologies. http://dx.doi.org/10.1016/B978-0-08-100310-7.00009-0

hypothetical traceability, is more complex and it involves the verification of the association of a product to a specific animal through the matching of DNA swabs. This method provides an incentive to producers and processors to take care and adjust their behavior in case of fraud.

The pillars of a traceability system are founded upon the identification of individual animals or homogenous groups of animals, the ability to track their movements, proper identification of premises, and recording of this information in appropriate registers.

In general, two main components are required in the process of tracing animals: the technologies for animal identification and technologies for the storage and management of information. The working principles of these technologies are described in the following subsection, with a focus on the main issues that deal with the implementation of effective systems for animal traceability.

1.1 Individual Identification of Animals

Technology-wise, there is a range of solutions for identifying and tracking animals (Wogerwerf, 2011). According to Merchant (2002), individual animal identification can be made through

- mechanical methods;
- electronic methods; and
- biometric methods.

Typical mechanical methods include branding, tattoos, ear notching, and ear tags. Electronic identification techniques are usually based on Radio Frequency Identification (RFID). Accurate biometric identification methods include iris scanning, retinal images, and DNA analysis.

Moreover, there are some disadvantages related to the use of the different traceability methods.

The mechanical methods (eg, using ear notching, ear tags, branding, and tattoos), for example, are invasive methods and are not good enough for traceability purposes, while electronic-based methods using external RFID tags to recognize animals (Shanahan, 2009) are subject to loss, removal, or damage.

The electronics ear tags are, on balance, the best devices for individual animal identification. RFID technology enables accurate and rapid data capture for use in emergency situations. According to Schroeder (2011), electronic systems appear to be the best suited to current needs and are used in countries that have the most developed identification and traceability systems, although needs and solutions may continue to evolve. Systems based on electronic gadgets rely on RFID devices.

According to Shanahan et al. (2009), verification system for the identity of bovine animals should be included to reduce the risk of fraudulent activities. This verification system should be based on biometric identifiers that are unique to each animal and the use of RFID tags to store and record relevant data in accordance with international standards and best industry practice. The biometric methods could give high identification rates, but they are intrusive for the animals and are not cost-effective compared to other approaches (image-processing methods) (Ahrendt et al., 2011).

The following subsection provides a deep review of the modern technologies that can be involved for the electronic identification of animals. These technologies are mainly based on the use of RFID systems.

1.1.1 Alternative Electronic ID Technologies

The development in information technology has led to the introduction of RFID tags as valid electronic data carriers for use in animal identification. McMeekin et al. (2006) state that "the use of RFID tags and readers will revolutionize the way the supply chain data is captured and communicated." The International Organization for Standardization (ISO) has published standards for use of the identification of animals using electronic tags in ISO 11784 and ISO 11785. ISO11784 refers to the Code Structure, while ISO 11785 refers to the Technical Concept. For animal traceability, two types of RFID are currently in use: full duplex (FDX-B) and half duplex (HDX). Readers that conform to ISO standards should be able to read both FDX-B and HDX RFIDs.

The transmission frequency is the most important technology parameter for the correct application of RFID technology (De Cindio et al., 2011a,b). Low-frequency tags are used for the identification of animals because they are less susceptible to environmental interference from water and metal than RFID tags that operate at higher frequencies (Shanahan, 2009).

An RFID system used in animal identification is generally formed by four basic components: the transponder, the transceiver, the data accumulator, and the processing software (Grooms, 2007). The transponder is the electronic data source that stays with the animal to identify. The transponder contains a microchip in which the identification number of the animal is stored. The transceiver (or reader) is used to retrieve the information stored in the transponder. This device is essentially a radio transmitter and an antenna. The transceiver could be part of a handheld unit that is taken to the location of the animals with the RFID tags or a stationary unit that the animals pass by. The data accumulator is a device where information received by the transceiver is accumulated or stored. The data accumulator could be a computer, a Personal Digital Assistant (PDA), a scale head, or a variety of other electronic devices. The processing software transforms the accumulated data into recognizable and useful information. In case of animal identification, the useful information accumulated into the transponder is a 15-digit number that uniquely identifies the animal. In a separate process, this information could then be imported into a variety of data management programs and related to other information about the animals and their lifecycle.

The TAG or transponder is physically inserted in the animal to be identified, in different anatomical sites. Each time that a transponder is located in proximity to an

RFID terminal the signal is picked up and the animal is recognized, allowing it to trace back to all the information relating to the animal to which it refers.

RFID transponders are available as ear tags, ruminal boluses, and injectables. Ear tags and boluses are generally used for identifying livestock, and injectable transponders are used for identifying companion animals, horses, fish, and endangered species (Hogewerf, 2013). A leg tag transponder is an alternative method for identifying livestock (Carné et al., 2010).

The ruminal bolus is a device consisting of a capsule of biocompatible, nonporous material (which consists generally in derivatives and compounds of ceramic) in which a microchip is internally accommodated (Fig. 9.1), containing an electronic identifier. The microchip mainly consists in a radio frequency transponder capable of automatically communicating information to the reader. The bolus is designed to lodge in the rumen of the animal: the animal ingests the bolus, and the bolus is consequently retained in the reticulum or the second stomach of the ruminant, where it remains for the entire life of the animal.

There are different types of ruminal bolus: today, commercially ruminal boluses are ceramic, monolithic (compounds of thermoplastic material), and steel-weighted (Fallon, 2002). A cylindrical shape that facilitates the oral administration in young animals characterizes the ceramic boluses. On the other hand, the steel-weighted bolus consists of a transmitter enclosed in glass, which is then enclosed in plastic.

Different types of boluses have been used in cattle, sheep, and goats without any negative effects on animals. However, there is a unanimous agreement about the negative effects of the bolus on the food intake, the digestion, and the morphology of the ruminal wall. Nevertheless, Voulodimos et al. (2010) mentioned that foreign bodies could be introduced in the rumen without any negative effects on the ruminant's physiology.

Microchipping is the insertion of an electronic transponder, the injectable transponder, directly under the skin of the animal, where it remains a permanent identifier that can be read with an appropriate electronic reader. The microchip is usually encased in a biocompatible (nonallergenic) material that promotes the

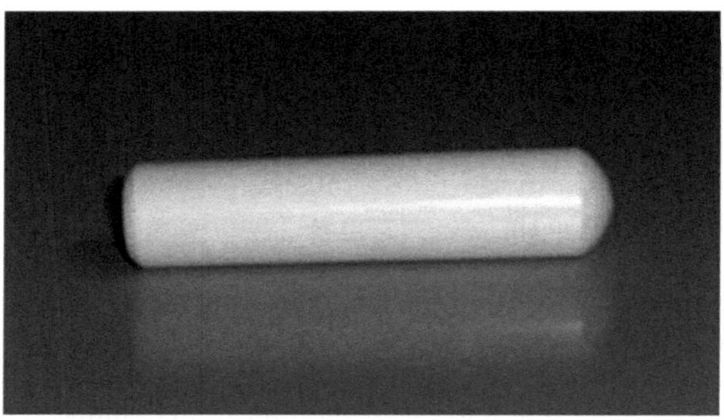

Figure 9.1 Representation of a Ceramic bolus.
From Olivieri, M., 2007. Sistemi di tracciabilità delle carni bovine mediante identificazione elettronica e genetica. Biologi Italiani, anno XXXVII n. 2.

regrowth of surrounding tissues. Microchipping is widely used for companion animals (eg, dogs) and equines; in the former the microchip is inserted under the skin, but in equines, it is inserted into the ligamentum nuchae on the side of the neck, halfway between the withers and the poll (Fig. 9.2).

In general, the preferable locations in which the injectable transponder can be implanted are the forehead, the external ear lobule, the posterior auricular base, and in the intraperitoneal cavity. Moreover, there is the risk that the microchip will migrate, making it more difficult to locate. For this reason, microchipping is not commonly used for animals that may be slaughtered for human consumption. Two main problems are directly related with the recovery of this electronic implants in abattoirs: the first problem is identified when the microchip cannot be quickly recovered, and it produces a consequence interference with the abattoir throughput; the second problem is observed when the implanted transponder is not recovered with the risk of remaining in the food products and of being consumed by humans (Sheridan, 1991).

Electronic ear tags are the most common methods used for marking cattle, sheep, and goats and are sometimes used for pigs. Ear tags are generally of plastic materials, especially in the case of electronic tags for which metal would interface with the signal from the RFID.

Electronic ear tags are generally inductively coupled transponders (Dan Grooms, 2007). This means that the energy required to run these types of transponders is supplied by electrical currents produced by electromagnetic fields created by components embedded in the transponder. The main elements of a typical inductively coupled RFID are showed in Fig. 9.3.

In particular, this type of RFID is formed by three main elements:

a. The microchip or integrated circuit that represents the data source component of the RFID. The microchip contains a unique 15-digit Animal Identification Number.

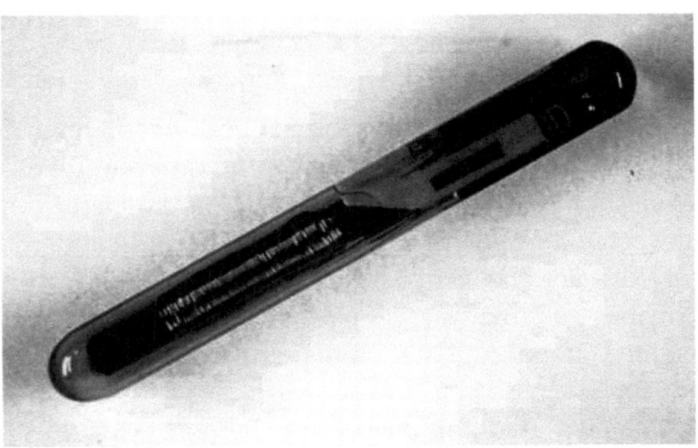

Figure 9.2 Representation of an injectable transponder.
From Ribó, O., Korn, C., Meloni, U., Cropper, M., DeWinne, P., Cuypers, M., 2001. IDEA: a large-scale project on electronic identification of livestock. Revue scientifique et technique 20 (2), 426–436.

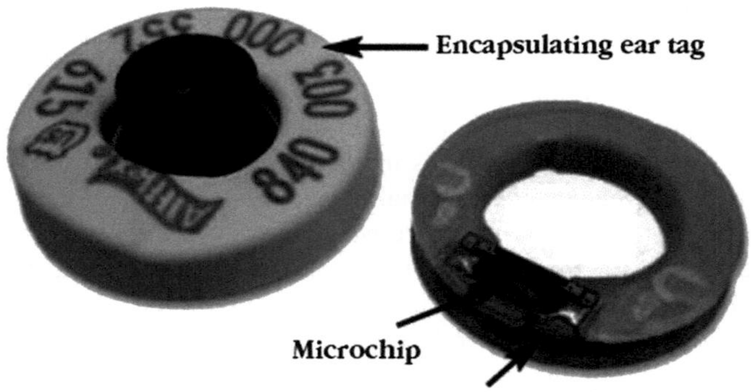

Figure 9.3 Representation of the main components of an electronic ear tag.
From Grooms, D., January 2007. Radio Frequency Identification (RFID) Technology for Cattle. Published in Extension Bulletin E-2970. Michigan State University.

b. The air or metal coil that acts as the tag's antenna. The tag transmits signals to the reader, with read distance determined by the size of the coil antenna (generally the read distance is 18 in. or less). The metal coil also creates the magnetic field that provides the electrical current to power the transponder.
c. Encapsulating material or plastic cover used to isolate the RFID system for cattle identification. The transponder, in fact, is most often embedded in a hard polymer shaped in the form of a button ear tag.

The electronic ear is a potential alternative to boluses and injectable transponders. It has been the subject of experimentation in many livestock species, with encouraging results, especially in the swine industry. Moreover, one of the main disadvantages of ear tags is that they are vulnerable to loss, removal, or damage. To this end, efforts to develop permanent electronic identification are focusing on internal forms, such as injectable transponders (Stanford et al., 2001).

1.2 Information Management

The system of identification and electronic monitoring of animals represents an advance in tracking production animals. Among the different types of identification technologies that can be used, RFID technology is increasingly becoming the gold standard-enabling technology for individual animal identification. Nevertheless, the RFID transponder does not carry any substantial information about the animal or its past and current owners and relatives. It only links the animal with the different databases that provide the storage, consistency, and usability of a large mass of data about animals and their life.

In general, a database used for traceability purposes should maintain the link between the identification code of the animal and the physical characteristics of the animal itself, along with the conditions of the infrastructure and of the environment in which the animal is grown, that is to say the productive and reproductive conditions, the breeder conditions, the transportation conditions, and the slaughtering conditions.

In particular, two types of databases are needed for supporting a traceability system (Andrés Rius, 2015). A database for animals identification is necessary in order to link the single animals to their owners. The government, private, or mixed type organizations may own this type of database. This database can be seen as a repository of all farms that are registered for the animal tracking system. It includes information about the identification of each actor involved in the system, owner information, and animal identifications associated with the owners.

A second type of database is required for maintaining all the information related to a single animal; it may be possible to maintain this type of database at the company level. It stores specific information about the animals associated to the company such as veterinarian treatments, animal movements, and animal nutrition.

According to the second type of databases, three different classes of data should be stored:

- information about animal parameters;
- information about animal movements; and
- information about animal welfare, as well as productive and reproductive data.

The first class of data can be classified referring to the actor that should record data. In particular, you can refer to the breeder or to the abattoir. Data recorded at the breeder includes the identification code of the animal along with which gender, breed, date of birth, identification code of the mother of the animal, code of the country of origin, date of application of the brand, and date of death. Data recorded at the slaughterhouses includes date of slaughter, the number of animal slaughtered, carcass weight and classification, specification if the operation has been carried for slaughtering, or killing due to a veterinary order.

In case of changes in the animal holder, information about date and identification of the breeder should be recorded in order to maintain the information about the animal transitions.

According to the third class of data, information that should be recorded includes breeding technique, method of fattening, information related to the animal feeds, methods of reproduction, information about the dates of inspection, and name of the official from the inspection authority, pathological conditions of the animal, reproductive performance, number of neonatal, and duration of the lactic period.

1.3 Open Issues

The different types of electronic identifiers have been studied and analyzed in order to evaluate their advantages and disadvantages and their application field. In particular, the final report of the IDEA program (Identification Electronic des Animaux), a four-year program of the European Commission's Joint Research Center (Ispra, Italy) points out that some identification technologies are specific for certain animals, and they can be applied only under certain conditions. For example, electronic ear tags are preferable for cattle, sheep, and goats of all ages, though ruminal boluses can be applied at a live weight of 25 kg for sheep and goats and in cattle less than 20 days old.

Analyzing the identification technologies based on RFID presented in the previous section, some considerations could be made on their application.

Referring to the ear tag, this identification technology is the easiest to apply, requires no special tricks, and is a practice commonly performed in the identification of cattle, buffalo, sheep, and goats, even in the course of ordinary traceability. In contrast, ear tags are subject to losses, and the animal holder can easily replace them.

On the other hand, the bolus cannot be altered or replaced and in practice is not subject to loss. Moreover, it can only be applied to ruminants and animals with rumen already developed, and after weaning. In adult animals, in fact, restraint for the inoculation of the bolus is difficult. In addition, good dexterity is required in order to not cause esophageal injury. One of the most important disadvantages of the boluses it that they must be recovered at the slaughterhouse, with considerable loss of time, and they create considerable problems related to the disposal of the capsule.

The injectable transponder, or subcutaneous tag, summarizes the advantages of ear tags and boluses, because it can be easily applied and it cannot be adulterated or lost; in addition, it can also be used in monogastric species. In addition, in the case of BSE, if the region of inoculum is localized in areas such as the head region or anatomical parts are unmarketable (ears, the distal part of the limbs), it is not necessary to recover during slaughter and there are no problems of disposal.

Nevertheless, several studies have shown migration of tags under the skin, with great difficulty in recovering slaughter; therefore in the case of animals for slaughter, both the EU and the Italian Ministry of Health advise against the adoption of this technology to prevent the discovery of the tag in the final product.

The identification by electronic devices is feasible even in the later stages of the animal for slaughter. Downstream of the slaughter, in order to maintain carcasses' traceability, the electronic identifiers used for animal identification must be transferred to the electronic identifier of carcasses, using special electronic tags or "smart labels," whose release is promoted by modern electronic systems. These smart labels contain, in addition to standard bar codes used in logistics, the signs of demographic leaders. According to Olivieri (2007) the labeling of various anatomical cuts can use this kind of technology at a reduced cost.

In summary, although RFID technologies are used for the identification, traceability, and monitoring of animals, there are some issues that must be resolved. One of the main problems is related to the lack of a global standard for information recording, covering all types of animals and establishing a common ground on which information regarding animal traceability can be collected (De Cindio et al., 2011a,b). In addition, another issue investigated from different researchers is the identification of the optimal part of the animal body where the injectable capsule should be implanted. As mentioned before, different opinions have been expressed, and results of the researches carried out highlight that there is no definite or unique answer. A complete guideline for the selection of the best RFID system to use and the location in which the transponder should be implanted is needed.

In conclusion, there is a danger of loss of product traceability, both during antemortem and postmortem. The ordinary and electronic systems, as defined by current standards, may be subject to error, resulting in the loss of the link between the identification number of the animal and the animal itself. The critical points can be analyzed from the different point of view of the supply chain. In particular, during the phases of breeding and transportation, ear tags can be accidentally detached, even if this condition is not detected in case of bolus or subcutaneous microchip, with a consequently loss in the identity of the animal. Also at the slaughterhouse it is possible to assist in the loss of the connection between the half carcass and the animal of origin or to the incorrect relation between the identification of the new slaughter animal and the identification number of the animal of origin.

Analogous errors can be made during the phase of working, packaging, and sales, when you assist to an improper relationship between the new identification code of the processed/packaged product and the original serial number of the animal.

Therefore the traditional traceability systems based on paper records and/or electronic equipment should be supported by other tools able to ensure the undeniable origin of the meat. This condition is required in order to have the absolute control of the supply chain and to quickly identify any anomalies. The limitation described can be overcome by use of advanced technologies for animal identification, such as Retinal Imaging, DNA identification analysis, and Alternative Electronic ID technology based on RFID technology. A detailed description of these technologies is provided in the next section.

2. Advances in Animal Tracing

2.1 Advances in Technology

The electronic auto-identification technologies are generally used to uniquely identify animals. Moreover, auto-identification technologies can be manipulated in case of fraud or loss. In order to overcome the limitations of the identification technologies discussed earlier, various biometric identifiers have been suggested in literature. Nevertheless, biometric identifiers are mainly used as an instrument for the verification of the animal identity. Biometric identifiers include retinal scans, muzzle patterns in cattle (or nose prints), facial recognition, immunological labeling, and DNA analysis. They are considered more reliable than electronic identification because they cannot be falsified. The transponder of an RFID, in fact, can be easily removed and inserted into the tag of another animal (Gonzales-Barron and Ward, 2005).

As stated by Whittier et al. (2003) retinal imaging is a form of identification that can be used both in animals and humans. The image of the blood vessel pattern on the retina is analogous to a fingerprint, and it does not change throughout the animal's lifetime, but remains the same from birth. In addition, using a computer algorithm, the retinal image can be converted into a unique ID, as suggested by the company Optibrand (2005). Each image could be acquired in two minutes, although the animal had to be restrained in a crush. As stated by Allen et al. (2008), this system has the potential to improve the performance of ear tag-based identification systems. Retinal imaging may be a real-time alternative to currently available biomarkers requiring the collection of samples for laboratory analysis; moreover, some precautions have to be taken into account as a function of the period of life of the animal (Rojas-Olivares et al., 2011).

The muzzle pattern or nose print of breeds has been studied since 1921. Cattle and dairy animals possess distinct impression patterns in their muzzle. The muzzles of bovine have some linear features with a specific pattern. The science of such features with a specific pattern on the skin is known as dermatoglyphics. The study of the features that characterize the skin of muzzle is known as muzzle dermatoglyphics. Muzzle dermatoglyphics could play a major role in the identification of bovine animals. Recent reports indicate that it may serve as an efficient tool for identification and correct breed differentiation. Baranov et al. (1993) has proven that muzzle prints from various breeds are mostly different, similar to the human's fingerprint.

The digital format of the muzzle print consists of beads and ridges, as shown in Fig. 9.4. The bead is an irregular region that looks like an island, and the ridge is an elongated region that looks like a river with an irregular width. There are two ways to capture a noise print into digital format: printing the muzzle data on paper and taking a photo for the muzzle. Fig. 9.5 shows the muzzle pattern obtained by printing a muzzle image with ink and paper.

Tarvat et al. (2014) have demonstrated that digital imaging of muzzle patterns could achieve a recognition rate of almost 99%.

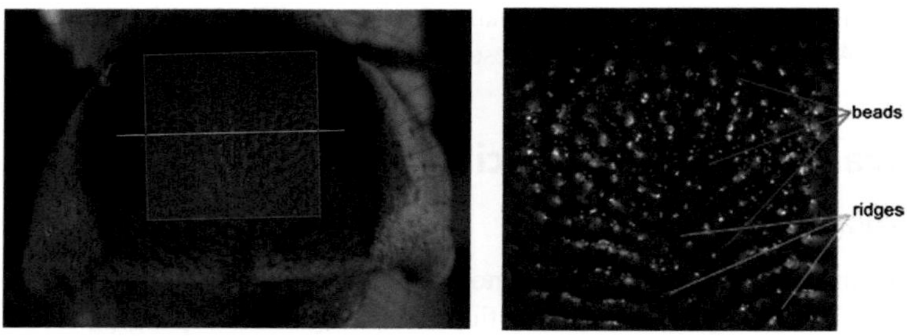

Figure 9.4 Beads and ridges in a muzzle photo.
From Noviyanto, A., Arymurthy, A.M., 2012. Automatic cattle identification based on muzzle photo using speed-up robust features approach. In: Proceedings of the 3rd European Conference of Computer Science, ECCS, pp. 110–114.

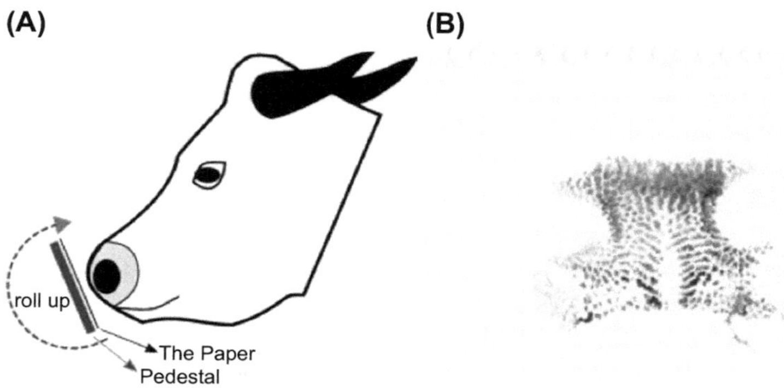

Figure 9.5 (A) The printing technique when taking the muzzle print and (B) the location and the example of the result of taking the muzzle print.
From Noviyanto, A., Arymurthy, A.M., November 2013. Beef cattle identification based on muzzle pattern using a matching refinement technique in the SIFT method. Computers and Electronics in Agriculture 99, 77–84.

A bioactive immunological labeling system, known as "ImmunoTrack," was developed in Germany by Responsif GmbH (Gonzales-Barron and Ward, 2005). This system uses highly antigenic peptide sequences with appropriate adjuvants to induce strong peptide-specific antibody responses in cattle or pigs. These antipeptide antibodies can be detected, using standard techniques, in the live animal's blood serum or the meat "juice" after slaughter. By varying the combinations of peptides, the origin (eg, country, region, farm), and other characteristics (eg, breed, organic farming) can be encoded (Gonzales-Barron and Ward, 2005).

The traceability system based on DNA analysis is currently the only technology that can give absolute truthfulness to the documentation that accompanies the meat up to the point of sale. DNA testing will not replace the traceability systems in place, but allows you to enhance and verify the validity of the information recorded in these systems, especially in the context of highly certified and special chains.

The genetic traceability is based on the concept that DNA is

- inalterable;
- a stable treatment suffered during transformations; and
- present in every cell in the body.

DNA analysis is carried out through the analysis of markers earphones. The DNA harvested from live animals can be extracted from the different matrices, including:

- ear cartilage, taken directly through the application of special ear tags;
- blood; and
- hairs with attached hair bulbs.

The analysis consists of a comparison, and then in the verification of the genetic equivalence, between the reference sample, taken from the animal in life (taken antemortem), registered and preserved, and the sample of meat (taking postmortem), taken at the point of sale or at any other point in the supply chain. Fig. 9.6 shows the process by which a genetic system can assist the ordinary traceability.

Allen et al. (2008) suggest that "DNA profiling, whilst being a powerful tool for scientifically verifying animal identity, is currently limited by the fact that verification results cannot be generated in real-time, ie, beside animal, for most applications. It is unlikely therefore to become the primary identifier for live animals. It can be used effectively in retrospective audits, meat tracing and parentage verification as a traceability technology and counter-fraud measure." Similar cautions could be issued about the other biometric techniques described, especially when considered in the context of on-farm operations involving large flocks or herds, but some at least have the potential to be generated in real time, and advances in technology are likely to facilitate their wider use. In addition, DNA profiling (and immunological labeling) could be used for carcasses, whereas muzzle patterns and retinal images suffer from the same problem as ear tags: they are lost when the animal's head is removed during processing at the abattoir.

For niche products or products with special quality awards, which are linked to a particular production specification or to a given territory (protected designation of origin (PDO), protected geographical indication (PGI), etc.), the geographical traceability is a

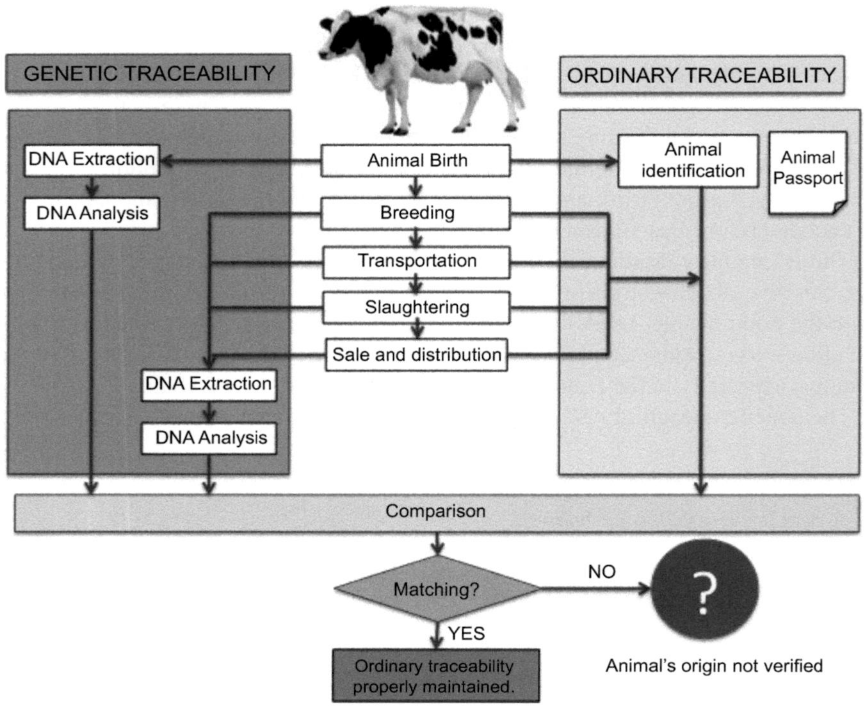

Figure 9.6 Genetic traceability supporting ordinary traceability.

useful alternative. Geographical traceability aims to protect the origin of a product through the study of "trace elements" as volatiles, microbial flora, special isotopes, or infrared spectroscopy (Franke et al., 2005; Mauriello et al., 2003; Pillonel et al., 2003; Schwägele, 2005).

In addition, this type of traceability is associated with good genetic analysis for the protection of products related to certain breeds (Dalvit et al., 2007). Traceability of race can assign a product to a specific race and is offered as a tool of protection and promotion of such products. The studies are conducted mainly in the Mediterranean countries (Italy, France, and Spain) that have a greater number of products protected by PDO and PGI.

3. Alternative and Future Technologies

There are many approaches that can be used to enhance traceability in the animal sectors. The individual identification of animals is a key point for the management, traceability, trade control, and disease eradication. Moreover, conventional identification used to identify animals is not sufficiently efficient to solve the traceability problem. As stated by Marchant (1991), in fact, an effective identification of animals and lots has to be individualized, permanent, simple to apply and read, welfare appropriate, and tamper-proof.

A series of national systems for animal traceability has been introduced in different countries at the mandatory and voluntary level. Several database systems for livestock tracking and tracing have been promoted and developed in Europe and worldwide. A simplified schema of the main requirements and regulations, along with the traceability systems imposed at the legislative level, is showed in Table 9.1.

European Countries are leaders in food traceability. Regulations and requirements in these countries address the traceability of a broad range of food and animal products of both domestic and imported origin. Specific regulations have been introduced, especially with regard to bovine identification and traceability of their movements and the resulting products (Reg. (EC) 17.7.2000 no. 1760/2000 as amended by Regulation Reg. (EC) 15.5.2014 n. 653/2014). In particular, the Regulation (EC) 1760/2000 provides for a system of registration and identification of cattle using ear tags, passports, and business records. Registered data must be constantly updated with the movements and deaths. It also lays down the mandatory labeling of meat at all stages of marketing and in all cuts, obtained though the definition of the identification code of the animal and the information about country of birth, country of fattening, the country and the slaughterhouse, and country sectioning.

There is also a voluntary labeling that contains any other information useful to characterize the animal and the derived meet products (race, age, holding of birth, date of slaughter, breeding system).

A comprehensive evaluation and comparison of the traceability regulations of 21 Organization for Economic Cooperation and Development (OECD) countries has been released in September 2014 from the Global Food Traceability Center (GFTC). According to the review carried out by Charlebois et al. (2014), some considerations can be made about the current regulatory framework of several countries. Canada is strengthening its traceability requirements through mandatory livestock identification, including a recently announced swine identification and movement-tracking program. However, the efforts to create a national traceability system have failed to produce anything beyond limited livestock tracking from birth to death. In the United States, while the new Food Safety Modernization Act (FSMA) is expected to improve food traceability capabilities for commodities, the development of regulations is still in the early stages. The United States does have robust identification and labeling requirements of packaged food products but is one of only two major beef producing countries without a national cattle identification or traceability system. On the other side, Australia, New Zealand, and Brazil have strong livestock identification and traceability systems. Finally, the traceability system in China is still under development, and traceability is largely unregulated. However, China has recently announced impending changes to its food traceability laws.

Although many countries lack specific legislation on food traceability, the global tracing and tracking of imported products is being achieved, often with difficulty, through record keeping, lot identification, labeling laws, and requirements for exporting countries to meet the standards of the domestic industries in those countries.

Nevertheless, legislation requires only minimal data to be retained, and any international standard has been introduced for the definition of data to collect in order to maintain the continuous traceability and monitoring of animals in the supply chain,

Table 9.1 **Traceability Requirements and Regulations**

Country	References	Description	Application Field	Technology	Type of Information
EU countries	Regulation (EC) 178/2002	The Regulation lays down the general principles governing food and feed in general, and food and feed safety in particular, at the community and national level. Art 18 states that food and any other substance intended to be, or expected to be, incorporated into a food or feed shall be established at all stages of production, processing, and distribution.	Food and feed in general	Not defined	Food and feed business operators shall be able to identify any person from whom they have been supplied with a food, a feed, a food-producing animal, or any substance intended to be, or expected to be, incorporated into a food or feed. To this end, such operators shall have in place systems and procedures which allow for this information to be made available to the competent authorities on demand.
EU countries	Regulation (UE) 653/2014, which amending Regulation (EC) 1760/2000	The Regulation establishes mandatory beef labeling. Beef labels must include precise information about where the animal was born and reared, as well as the place of fattening, slaughtering, and butchering. According to this regulation, each member state shall establish a system for the identification and registration of bovine animals.	Bovine animals and beef	All animals on a holding shall be identified by at least two means of identification. At least one of the means of identification shall be visible and bear a visible identification code.	The Regulation does not specify what type of information should be kept by the business operators. However, to fulfill the objective, any keeper of animals must maintain a computerized or manual register of animals on the holding. In addition, both ear tags bear the same unique identification code, which enables each animal to be individually identified, and the holding on which it was born.

		The system for the identification and registration of bovine animals shall comprise the following elements: a. ear tags to identify animals individually; b. computerized databases; c. animal passports; d. individual registers kept on each holding.			Any animal coming from an EU country 1 which has passed the veterinary checks laid down by Directive 91/496/EEC must be identified by an ear tag applied within 20 days of the veterinary checks and, at any rate, before it leaves the holding. Meat labeling must include the following information: • the reference number or code establishing the link between the meat and the animal (or group of animals) from which the meat was derived; • country where slaughter took place and licence number of the slaughterhouse; and • country where cutting was performed and licence number of the cutting plant. Moreover, since January 1, 2002, operators must also indicate: • the country where the animals were born; • the country where the animals were fattened/bred; and • the country where the animals were slaughtered.
EU countries	Regulation (EC) 1224/2009	The regulation requires that all lots of such products must conform to labeling requirements with identification numbers including the name and address of the suppliers.	Seafood	Not defined	Article 58 of EC 1224/2009 requires that "all lots of fisheries and aquaculture products shall be traceable at all stages of production, processing and distribution, from catching or harvesting to retail stage".

Continued

Table 9.1 **Traceability Requirements and Regulations—cont'd**

Country	References	Description	Application Field	Technology	Type of Information
Canada	National Livestock Traceability System defined by the Health and Animal Regulations	Mandatory livestock identification and tagging of cattle, sheep, and bison are regulated by the Health of Animals Regulations. The Canadian Cattle Identification Agency (CCIA) is the national administrator authorized by the Canadian Food Inspection Agency (CFIA) to issue and regulate approved RFID tags and monitor the reporting database—the Canadian Livestock Tracking System (CLTS). The CFIA is responsible for enforcing the identification requirements in Canada as per the Health of Animal Regulations. The national livestock traceability system is based on three pillars: animal identification, premise identification, and animal movement.	Livestock traceability	Tag including the official logo of the responsible administrator and a UID number that follows the ISO 11784	An approved CCIA radio frequency identification (RFID) tag is applied to the ear prior to leaving the farm of origin. All tags are visually and electronically imbedded with a unique identification number that is allocated from the national database and then distributed to producers through authorized tag dealers.

United States of America	National Animal Identification System (NAIS)	National Animal Identification System (NAIS) is an animal disease surveillance program implemented by the US Department of Agriculture, Animal and Plant Health Inspection Service (USDA/APHIS). This system involves tagging farm animals, providing a specific identification number to a specific farm and specific animal at that farm, and keeping the records of this identification at a USDA/APHIS office. The identification number is used to track the animal(s) and premise(s), when needed. Assigning a unique premises identification number (PIN) to each premises will provide select government officials the exact origin and location of an animal in the event of a disease investigation.	Livestock and some fish species	Not defined	Each individual animal, whether horse, cow, sheep, chicken, pigeon, etc., is given a unique identification code. The NAIS would require two types of mandatory registration. First, premises registration would require every person who owns even one horse, cow, pig, chicken, sheep, pigeon, or virtually any livestock animal to register their home, including owner's name, address, and telephone number, and keyed to Global Positioning System coordinates. Second, individual animal identification will require owners to obtain a 15-digit ID number, also to be kept in the federal database, for any animal that ever leaves the premises of its birth. The final part of the program is to track animal movement between the various premises.

Continued

Table 9.1 **Traceability Requirements and Regulations—cont'd**

Country	References	Description	Application Field	Technology	Type of Information
Australia	National Livestock Identification System (NLIS)—Animal Health Australia (2004)	The NLIS has been designed to improve traceability, to enhance food safety, to ensure beef product integrity, to allow and to sustain international market access, and to provide progressive livestock producers with enhanced management opportunities. The National Livestock Identification System (NILS) defines mandatory requirements for the tagging and identification of cattle, sheep, and goats. Through a national animal identification system and registration in the NLIS database, the program monitors the lifespan of individual animals, from birth to slaughter, and the movement of livestock.	Livestock traceability	NLIS-approved devices consist of a radio frequency identification device (RFID)	The system is based on a unique identification number, known as the property identification code (PIC). The tag identification number, which is the PIC, is eight digits in length, with the number accommodating a state identifier, a regional prefix, and an individual property number relating to that region. The National Livestock Identification System (NLIS) enables livestock to be traced from their property of birth to slaughter. Livestock must be tagged with an approved NLIS device (such as an ear tag or rumen bolus) before they leave their property of birth. The NLIS database records livestock movements.

New Zealand	National Animal Identification and Tracing Act 2012 (NAIT) (http://www.legislation.govt.nz/act/public/2012/0002/latest/DLM3430220.html)	The purpose of the Act is to establish an animal identification and tracing system that: a. provides for the rapid and accurate tracing of individual, or groups of, NAIT animals from birth to death or live export b. provides information on the current location and movement history of individual, or groups of, NAIT animals c. improves biosecurity management; and d. manages risks to human health arising from residues in food, food-borne diseases, and diseases that are transmissible between animals and humans; and e. supports improved animal productivity, market assurances, and trading requirements.	Cattle, buffalo, and deer	NAIT approved radio frequency identification device (RFID) ear tags	Tagged animals are registered in a national database and the details recorded include: • the animal's location, • movements in the animal's life, and • contact details for the person in charge of that animal.
Brazil	Brazilian System of Identification and Certification of Origin for Bovine and Buffalo	Brazil implemented a mandatory traceability and identification system for livestock. The SISBOV program is designed to identify animals born in Brazil, as well as animals imported into the country. The system, based on ear tags, matches the ear tags with individual animal certificates	Livestock traceability and identification	Ear tag	Animals registered with SISBOV have an identification number containing the origin property, month of birth, gender, raising system, feeding, and sanitary data.

Continued

Table 9.1 **Traceability Requirements and Regulations—cont'd**

Country	References	Description	Application Field	Technology	Type of Information
Japan	Beef Traceability Program (MAFF 2008)	Japan implemented, in July 2002, a set of bovine animal traceability and identification laws specifically designed as countermeasures to the spread of bovine spongiform encephalopathy. This law requires animal traceability from the feedlot to the abattoir. In 2003, Japan implemented the Beef Traceability Law that requires animal identification and traceability from "distribution to consumption."	Bovine animals	2 ear tags imprinted with a 10-digit number and a bar code.	At birth, each bovine animal is assigned an individual identification number, and the date of birth, gender, parents' individual identification numbers, and the breed of the animal are recorded. Animal movement records are required both for outgoing and incoming cattle, and the individual animal identification number, date of the transfer, and both parties involved in the transfer are required to be recorded in the database. Abattoirs are required to record individual animal identification numbers, the date of slaughter, and the party from which the cattle arrived at their facility.
China	Decree no. 67 (2006)	The regulation on animal labeling and their feeding documents establishment in farms has been promulgated by the Ministry of Agriculture in China (MOA) to encourage to keep recording within production.	Pig, cattle, and sheep/goat	Ear tag systems with a two-dimensional (2-D) barcode	Animals such as swine, cattle, and sheep/goats in the farms must be given a unique identity code around China, and it shall be labeled with a special tag embodying a unique code before moving from its region of origin. The feeding enterprises must establish their feeding information documents to record the inputs mainly used, such as feeds, feed additives, and veterinary drugs as treatment.

the format of those data, and the data standards to be used. As a consequence, each national system varies in terms of the data maintained, format of data, and standards used (Mirabelli et al., 2012). In addition, no existing national systems have a multilingual capability: data is stored in the national language only.

Little attention has been paid to the issue of interoperability of the different systems and efficient exchange of data with the European Commission or with other competent authorities. Data are generally stored in local databases that impose several restrictions to ensure efficient information access and management and hinder the interoperability between different systems (Pizzuti and Mirabelli, 2015). Hence, more expressive systems and techniques are required. These systems must be characterized by the ability to exchange relevant data between competent authorities in a timely, meaningful, coherent, and auditable manner. Rapid access to accurate data and the prompt and secure exchange of data are, in fact, essential prerequisites for the operation of the market, the detection of fraud, the prevention and containment of disease, and the monitoring and enforcement of agreed trade and veterinary protocols (McGrann and Wiseman, 2001).

In order to reach the goal of systems interoperability and easy communication between the different actors, the use of standards is necessary. The need to guarantee the external traceability requires the adoption of the same language by all the actors involved in the supply chain. In fact, when describing entities and activities, they should agree on the same terminology and adopt the same standard and notation (Pizzuti and Mirabelli, 2013).

Important technologies, such as the Internet and the new generation of communication infrastructure, have been developed for supporting new traceability applications.

In such a context, ontology-based representation systems are a new innovative technology that can be used to model the knowledge of interest and emulate the human logic by automated reasoning. Ontologies can be used to address many of the problems with data standards and data interchanging in the existing infrastructures.

In a typical supply chain, mass information and knowledge spread out in various formats among different enterprise systems. In addition, data related to traceability are generally stored in relational databases, and actors normally use the same terms with different meanings. Current enterprise informative systems usually do not contain information about the meaning of concepts and about the relations existing between different terms, and these conditions lead to semantic interoperability issues.

Pizzuti et al. (2014) proposed an ontological model for solving the above-mentioned problem, providing a structure for developing knowledge and unifying the metadata model of the current systems. They developed the Food Track and Trace Ontology (FTTO) for traceability control. FTTO is based on a language shared among all the actors involved in a generic food chain and supports the management of a single knowledge base through the integration of different concepts and terms coming from heterogeneous sources of information and users involved in the supply chain.

The main aim of containing all the information related to the food traceability domain in a unique ontology is directly related with the need of enabling information sharing along the food supply chain.

FTTO has been recently extended to the domain of the tracing of animals; in particular, the Meat Supply Chain Ontology (MESCO) has been developed, focusing on the domain area of traceability of pork and porcine products.

In particular, the ontology has been designed in order to model the whole domain of traceability of the animal and meat supply chain, from the agricultural producer to the final consumers, involving ingredients and service products as proposed in Pizzuti et al. (2014).

The methodological procedure followed to generate the ontology requires the preliminary analysis of the supply chain in order to identify actors, products, and processes involved in the traceability process. In the case of animals and meat products, the supply chain includes different steps: the primary production at the farm level, which refers to the phases of animals breeding and farming, the transportation of animals, the manipulating process with the relative slaughtering of animals and division in main cuts, the transformation process of the main cuts in meat products, and the final distribution of meat and meat products.

The analysis of the supply chain is a prerequisite for the identification of fundamental data to register for the maintenance of traceability of animals and processes. Information to record for traceability purposes must include, first of all, data imposed at the mandatory and, in addition, data able to provide further information about the animal and the environmental conditions in which the animal has grown. Data imposed at the mandatory level include the identification code of the animal, the identification of the animal's mother, the identification of the breeding company, the country of origin and the sex of the animal, and the identification of the owner.

With reference to the growing phase, as stated in previous sections, the complete traceability of animals should include the registration of feed, drugs, and medication used for the maintenance of the animals' welfare. In addition, every time that such a kind of operation is executed, information about the time of execution and the actors involved in the process should be recorded. From the mandatory point of view, data related with providers of feed; suppliers of drugs, medications, and other materials; and feeds and drugs used for animal breeding and growing are fundamental entities to trace.

At the time of the arrival of animal to the slaughterhouse, the phase of keeping track of animals, their documentation, and health records is a prerequisite for the execution of the successive operations.

It is clear that, in order to guarantee the complete and easy monitoring of the supply, a global standard for the definition of the information to record and the format to use for information encoding and information management is required. In this context, MESCO acts as a standard that enables interoperability and easy communication between the different actors involved in the supply chain. In order to guarantee external traceability, the actors involved in the whole supply chain should agree on the same terminology and adopt the same standard and notation when describing entities and activities of the various processes. This is straightforward to achieve by using an ontology-based representation system.

The structure of the ontology accesses all the information related to animals and the subsequent meat supply chain in a short time window to avoid delays in obtaining essential data, which is fundamental in case of a food-outbreak disease. The ontology can be easily queried and animals, animal cuts, and meat can be quickly traced to their origin.

In addition, in the case of nonconformal animals, the ontology can be queried for retrieving the supplier of the feed used for growing those animals.

4. Conclusions

Despite the benefits that can be achieved with the implementation of a traceability system, the current operating conditions show how the use of tracking systems is limited to compliance with the requirements imposed by law. Many barriers hinder the implementation of an efficient traceability system, including the need for high investments, the reluctance to change, the lack of trained staff, and the limitations of current systems (De Cindio et al., 2012). Nevertheless, some technologies have been introduced to solve the limitation of current systems. Animal ID technologies are evolving with the continuous advances in exiting technologies and the Internet of Things. In addition, these technologies could be supported by the introduction of biometric methods, such as the analysis of DNA markers, for the assessment of the identity of animals. Traceability systems may also be used in the future for the inclusion of additional information, not required at the mandatory level, with respect to the primary production circumstances (living conditions, use of medicines) of animals. Such additional information enhances product differentiation and branding. Therefore in order to solve the problem of systems interoperability and data uniformity, global standards should be introduced at the mandatory level. In such a context, the use of ontologies can be a valuable tool for information management: they can be proposed as global standards for information recording regarding all categories of animals.

References

Ahrendt, P., Gregersen, T., Karstoft, H., 2011. Development of a real-time computer vision system for tracking loose-housed pigs. Computers and Electronics in Agriculture 76 (2), 169–174.

Allen, A., Golden, B., Taylor, M., Patterson, D., Henriksen, D., Skuce, R., 2008. Evaluation of retinal imaging technology for the biometric identification of bovine animals in northern Ireland. Livestock Science 116 (1), 42–52.

Baranov, A., Graml, R., Pirchner, F., Schmid, D., 1993. Breed differences and intrabreed genetic variability of dermatoglyphic pattern of cattle. Journal of Animal Breeding and Genetics 110 (1–6), 385–392.

Bass, P.D., Pendell, D.L., Morris, D.L., Scanga, J.A., Belk, K.E., Field, T.G., Sofos, J.N., Tatum, J.D., Smith, G.C., 2008. Review: sheep traceability systems in selected countries outside of North America. Professional Animal Scientist 24, 302.

Bowling, M.B., Pendell, D.L., Morris, D.L., Yoon, Y., Katoh, K., Belk, K.E., Smith, G.C., 2008. Review: identification and traceability of cattle in selected countries outside of North America. The Professional Animal Scientist 24, 287–294.

Carné, S., Caja, G., Rojas-Olivares, M.A., Salama, A.A., 2010. Readability of visual and electronic leg tags versus rumen boluses and electronic ear tags for the permanent identification of dairy goats. Journal of Dairy Science 93 (11).

Charlebois, S., Sterling, B., Haratifar, S., Naing, S.K., 2014. Comparison of global food traceability. Regulations and requirements. Comprehensive Reviews in Food Science and Food Safety 13, 1104–1123.

De Cindio, B., Longo, F., Mirabelli, G., Pizzuti, T., 2011a. Modelling a traceability system for a food supply chain: standards, technologies and software tools. In: Proceeding of 10th International Conference on Modeling and Applied Simulation, MAS 2011, Held at the International Mediterranean and Latin American Modeling Multiconference, I3M 2011, pp. 488–494.

De Cindio, B., Longo, F., Mirabelli, G., Pizzuti, T., 2011b. Tracciabilità e rintracciabilità nel settore alimentare: analisi delle opportunità e vantaggi competitivi. Industrie Alimentari 7–19.

De Cindio, B., Seta, L., Mirabelli, G., Pizzuti, T., 2012. Nuovi sistemi informativi per la tracciabilità dei prodotti alimentari. Industrie Alimentari 51 (528), 47–53.

Dalvit, C., De Marchi, M., Cassandro, M., 2007. Genetic traceability of livestock products: a review. Meat Science 77, 437–449.

Fallon, R.J., Rogers, P.A.M., Earley, B., 2002. Electronic animal identification. Teagasc End of Project Report Beef Production Series No. 46.

Franke, B.M., Gremaud, G., Hadorn, R., Kreuzer, M., 2005. Geographic origin of meat elements of an analytical approach to its authentication. European Food Research and Technology 221, 493–503.

Geiger, R., Hayes, D., 2003. Identification and Traceability for the Pork Industry – a Proposal for a Feasibility Study. Report to the National Pork Board, NPB Project #01–158.

Grooms, D., January 2007. Radio Frequency Identification (RFID) Technology for Cattle. Published in Extension Bulletin E-2970. Michigan State University.

Gonzales-Barron, U., Ward, S., 2005. Review of Biometric and Electronic Systems of Livestock Identification. The BioTrack Project: Development of a Protocol for Biometric-based Animal Tracking and Tracing. Report to the Department of Agriculture and Food for Ireland, Dublin. Available at: www.ucd.ie/bioresources/biotrack/Downloads/Biometrics.doc.

Hogewerf, P., 2013. Current tools and technologies for the identification and traceability. In: Proceedings of the ICAR/FAO Workshop, Santiago, 5–7 December 2011. ICAR Technical Series 15(2013). ICAR, Rome, pp. 151–159. ISSN: 1563–2504.

Marchant, J., 2002. Secure Animal Identification and Source Verification. JM Communications, UK. Copyright Optibrand Ltd., LLC.

McMeekin, T.A., Baranyi, J., Bowman, J., Dalgaard, P., Kirk, M., Ross, T., Schmid, S., Zwietering, M.H., 2006. Information systems in food safety management. International Journal of Food Microbiology 112 (3), 181–194.

Mauriello, G., Moio, L., Genovese, A., Ercolini, D., 2003. Relationship between flavouring capabilities, bacterial composition, and geographical origin of natural whey cultures used for traditional water-buffalo mozzarella cheese manufacture. Journal of Dairy Science 86, 486–497.

Marchant, B.A., 1991. Identification requirements in the context of animal disease control in the community. In: Automatic Electronic Identification Systems for Farm Animals. Brussels, Belgium, pp. 122–126.

Mirabelli, G., Pizzuti, T., Goméz-González, F., Sanz-Bobi, M.A., 2012. Food traceability models: an overview of the state of the art. In: Proceeding of: 11th International Conference on Modeling and Applied Simulation, Vienna, Austria, pp. 268–277.

McGrann, J., Wiseman, H., 2001. Animal traceability across national frontiers in the European Union. Revue scientifique et technique (International Office of Epizootics) 20, 406–412.

Noviyanto, A., Arymurthy, A.M., 2012. Automatic cattle identification based on muzzle photo using speed-up robust features approach. In: Proceedings of the 3rd European Conference of Computer Science, ECCS, pp. 110–114.

Noviyanto, A., Arymurthy, A.M., November 2013. Beef cattle identification based on muzzle pattern using a matching refinement technique in the SIFT method. Computers and Electronics in Agriculture 99, 77–84.

Olivieri, M., 2007. Sistemi di tracciabilità delle carni bovine mediante identificazione elettronica e genetica. Biologi Italiani anno XXXVII n. 2.

Optibrand, 2005. Optibrand – a Universal and Secure Livestock ID System. Optibrand Ltd, LLC, Ft. Collins, Co.

Pillonel, L., Badertscher, R., Froidevaux, P., Haberhauer, G., Holzl Horn, P., Jakob, A., 2003. Stable isotope ratios, major, trace and radioactive elements in Emmental cheeses of different origins. Lebensmittel- Wissenschaft + Tecnologie 36, 615–623.

Pizzuti, T., Mirabelli, G., 2015. The global Track&Trace system for food: general framework and functioning principles. Journal of Food Engineering 159, 16–35. http://dx.doi.org/10.1016/j.jfoodeng.2015.03.001.

Pizzuti, T., Mirabelli, G., 2013. FTTO: an example of food ontology for traceability purpose. In: Proceedings of the IEEE 7th International Conference on Intelligent Data Acquisition and Advanced Computing Systems, p. 281–286. http://dx.doi.org/10.1109/IDAACS.2013.6662689.

Pizzuti, T., Mirabelli, G., Sanz-Bobi, M.A., Goméz-González, F., 2014. Food Track&Trace Ontology for helping the food traceability control. Journal of Food Engineering 120, 17–30.

Ribó, O., Korn, C., Meloni, U., Cropper, M., DeWinne, P., Cuypers, M., 2001. IDEA: a large-scale project on electronic identification of livestock. Revue scientifique et technique 20 (2), 426–436.

Rius, A., 2015. Mandatory Livestock Traceability as a Catalyst for Knowledge Intensive Services in Uruguay. See more at: http://publications.iadb.org/handle/11319/6953#sthash.KYZTPbn3.dpuf.

Rojas-Olivares, M.A., Caja, G., Carnè, S., Salama, A.A.K., Adell, N., Puig, P., 2011. Retinal image recognition for verifying the identity of fattening and replacement lambs. Journal of Animal Science 89, 2603–2613.

Schroeder, T.C., September 2011. Cattle Identification and Traceability: Implications for United States Beef Exports. Available at: www.agmanager.info.

Shanahan, C., Kernan, B., Ayalew, G., McDonnell, K., Butler, F., Ward, S., 2009. A framework for beef traceability from farm to slaughter using global standards: an Irish perspective. Computers and Electronics in Agriculture 66, 62–69.

Sheridan, M.K., 1991. Electronic identification systems: governmental considerations. pp. 109–113. EUR Report (13198).

Stanford, K., Stitt, J., Kellar, J.A., McAllister, T.A., 2001. Traceability in cattle and small ruminants in Canada. Revue scientifique et technique (International Office of Epizootics) 20, 510–522.

Schwagele, F., 2005. Traceability from a European perspective. Meat Science 71, 164–173.

Tharwat, A., Gaber, T., Ella Hassanien, A., Hassanien, H.A., Tolba, M.F., 2014. Cattle identification using muzzle print images based on texture features approach. In: Proceedings of the Fifth International Conference on Innovations in Bio-inspired Computing and Applications IBICA 2014 Volume 303 of the Series Advances in Intelligent Systems and Computing, pp. 217–227.

Voulodimos, A.S., Patrikakis, C.Z., Sideridis, A.B., Ntafis, V.A., Xylouri, E.M., 2010. A complete farm management system based on animal identification using RFID technology. Computers and Electronics in Agriculture 70, 380–388.

Wogerwerf, P., 2011. Current tools and technologies for the identification and traceability. In: Presentation at Workshop on Animal Identification and Recording Systems for Traceability and Livestock Development in Countries of Latin America and the Caribbean, 5–7 December. Santiago (Chile).

Whittier, J.C., Shadduck, J.A., Golden, B.L., 2003. Secure Identification, Source Verification of Livestock—the Value of Retinal Images and GPS., p. 167. Precision Livestock Farming. Wageningen Academic, Wageningen, The Netherlands.

Traceability Software for the Food Industry

10

A.K. Evizal
Islamic University of Riau, Pekanbaru, Riau, Indonesia; Universiti Teknologi Malaysia, Johor Bahru, Malaysia

S.K.A. Rahim
Universiti Teknologi Malaysia, Johor Bahru, Malaysia

T.A. Rahman
Universiti Teknologi Malaysia, Johor Bahru, Malaysia

S.L. Rosa
Islamic University of Riau, Pekanbaru, Riau, Indonesia

1. Introduction

The global economy in food products is struggling with the new threat of bioterrorism. People are very concerned about this issue and the implications of foodborne illness outbreaks. Establishing the security of the food supply chain requires control of food products in addition to regulations. In the United States, the Bioterrorism Act includes requirements in this area regarding the establishment of records to identify the immediate previous sources and immediate subsequent recipients of food, including its packaging; this is in effect for larger processors (US Food and Drug Administration, 2002). Food security regulations have been implemented in most countries worldwide; therefore every company must comply with food processing regulations. Enforcement depends on the size of the company. In response to these concerns, distributors and processors of food products are taking action by implementing traceability both of ingredients and finished goods products. Large business players in food products are demanding traceability from their suppliers as standard operation in running the business; most processors and suppliers have recently recognized that a traceability system that can be used to prove food safety must be implemented as a minimum operations standard.

Traceability is a technology that has the ability to track, trace, and locate an object (product). It can be used throughout a broad food product supply chain, from the farmer to the customer. The traceability process assists in tracking and tracing critical food products that might become contaminated and unsafe to human health at any stage during the delivery process, starting from the source, manufacturing, packaging, shipping, and distribution to a store or consumer. Which traceability system is suitable depends on the size of a business and the complexity of the food product; the supply chain may use

Advances in Food Traceability Techniques and Technologies. http://dx.doi.org/10.1016/B978-0-08-100310-7.00010-7

different tools, devices, and software for tracking and tracing, including using various levels of information and security, as appropriate. Some companies are required to do detailed tracking and tracing for hundreds or thousands of items from various suppliers through a multistep production process, whereas others are required simply to track and trace a food product only at key points. In some countries, traceability is not simply optional for the food product supply chain but is a law that every company must follow. Traceability systems have been obligatory for all businesses in the food chain in the European Union (EU) since January 2005 (European Commission, 2007). In the EU, all food product businesses are required to identify all suppliers and product recipients. This information must be properly managed, eg, in a central data center, to be available for inspection by authorities or on demand for traceability data.

A traceability solution in the product supply chain would be the implementation of traceability software to support the process that ties together all aspects of food product processing, including batches, tracking, and tracing. The use of software would have advantages in a complete system—it could identify all of the processes such as manufacturing, quality control, labeling, handling management, and accounting. With such an integrated system, the software would be an investment cost for a company; however, its benefits will likely outweigh the costs. A software system should exploit automated support technology in capturing food products, eg, using barcodes or radio frequency identification (RFID) with handheld scanners to avoid mistakes and errors in manual data entry. Automatically capturing each product batch yields much of the data associated with manufacturing, including the processing time, transportation, expiration, and storage time. Furthermore, a single information system and integrated software should result in benefits and efficiencies not only in processing but also in resource usage, planning, and warehouse management. Real-time traceability is required for all of these steps and to allow a quick response to determine which batch of food product is impacted so a warning or recall action can be fast. A streamlined system of food production using integrated software enables a business to establish and maintain quality and monitor business process controls that are critical to customers, such as the ability to maintain product traceability and safety. Merging information and communication technology (ICT) with the information software system enables the enhancement of food product quality and safety; a healthy product for the consumer is the highest priority for a food provider.

2. Traceability Software for the Consumer

A growing consumer interest in food products and the rising incidence of food fraud are forcing manufacturers with established production methods to adopt traceability systems. Consumers are now aware of the origins of their food because on occasion some processing may cause it to be unsafe for human consumption. Development of ICT through electronics such as the smartphone is advantageous for food traceability, eg, using mobility technology with software customized for the consumer. An advanced traceability system can be adapted to consumer solutions by developing application software to embed into mobile devices. Currently most mobile devices support RFID technology, near field communication (NFC) or the magnetic field system. Nasir Ali (2013) discusses the use of repository software for mining data and combining mined results with infrared (IR) techniques to improve accuracy (precision and recall). The proposed IR techniques such as the trust-race and trust-based traceability recovery approaches for medium-sized open-source systems can compare the accuracy of their traceability links with those recovered using state-of-the-art IR techniques based on the vector space model and Jensen–Shannon model. Thakur et al. (2011) introduced a new methodology for modeling traceability information using the electronic product code information services (EPCIS) framework, mapping food production processes to provide improved descriptions, and integration of traceability information. Rajo-Iglesias et al. developed a complete traceability system in the wine production sector by joining RFID technology with the use of wireless sensor networks (WSN) for monitoring at the vineyards (Rajo-Iglesias, 2013). The proposed system provides a manufacturing advantage by improving the visibility of the processes performed and the associated control over product quality. Parreño-Marchante et al. (2014) proposed a traceability system architecture based on Web services, which are used to integrate traceability data captured through RFID systems with environmental data collected with a WSN infrastructure. Fig. 10.1 shows an example architecture of a traceability system based on the EPC global standard network using RFID and WSN technology used by customers.

Implementation of a food traceability system enables consumers to check and trace product and processor histories through the use of application software installed in mobile devices that can detect food products. Some aspects that require attention to system design for interfacing with consumers include the following:

- *Simple*: the application software should show the product details and records and should be simple and user friendly. The information provided to the consumer will be more effective, efficient, and beneficial if it is only the required basic product information.
- *Speed*: accessing information and product details should be quick so customers can make their selections in a shopping complex or at a retailer. The use of mobile devices such as smartphones or tablets to access the information puts appropriate requirements on the network infrastructure.

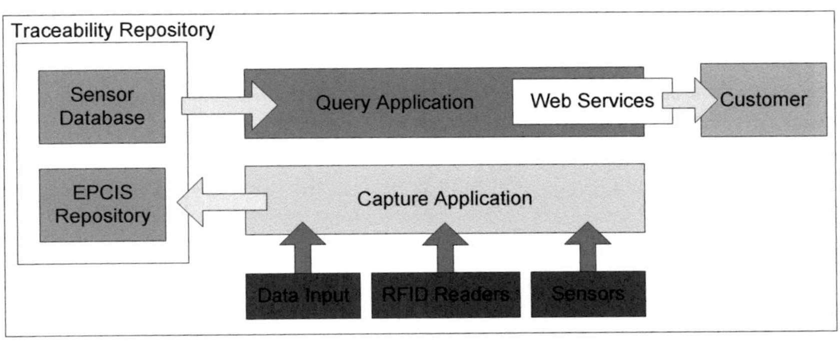

Figure 10.1 Traceability system architecture based on the EPC global standard network. Adapted from IBM DeveloperWorks., 2015. Traceability Solution Blueprint for Business Performance Optimization, Part 1: Understanding the Architecture of a Comprehensive Track-and-Trace Solution. http://www.ibm.com/developerworks/data/library/techarticle/dm-0902oberhofer/.

- *Accuracy*: because information on many products will be accessed by many people simultaneously, the system will be complex; the provisioning by various suppliers or manufacturers creates opportunities for miscommunication. Providing accurate information to the customer is very important; the customer depends on this information to determine whether a food product is good or safe to consume.
- *Support*: an integrated and multilevel system between the supply chain and the consumer application software means that the flow of data to the consumer is indirect. Various information is provided from different stages of the supply chain, so support for this level is important to get the correct data to the consumer.

A traceability system for the consumer application does not work alone. It depends on multiple stages in the supply chain to provide information and complete solutions to the consumer. Additionally, multilevel traceability processing must pair with a real-time system for data collection and then feed the consumer system. Furthermore, traceability software for the consumer can be integrated with other systems such as location services, online sales services, and retailer portals to have a complete solution for the consumer rather than only a single application.

2.1 Product Identification

Identification is defined as the act of determining or verifying that the object or material is correct. Many things need to be identified before use; the application could be in documentation, equipment, components, or material. Manufacturing and product maintenance require attention to ensure that items are correct because many objects are similar. The increase and spread of fraudulent and counterfeit food products has become an issue across all industries, and the globalization found in supply chain management systems puts consumers at risk. Food product fraud seriously impacts the performance, reliability, and reputation of the supplier and incurs financial losses. In a supply chain system, traceability can verify product details or locations by applying the identification of an item to what was previously documented (recorded). Implementation of a product identification and traceability system, particularly for food products, provides the ability to eliminate fraud and counterfeiting and provides safety, affordability, and improvement to the products consumed by humans. There are several techniques to identify food products; the most common systems are the barcode system and RFID. The evolution of mobile devices such as smartphones or tablets has enabled a built-in RFID system, the NFC reader. Identification of a food product can be simple once the application software has been developed for mobile devices; with one scan of a product, all of the previously recorded information appears on the device's display. Fig. 10.2 shows an example of identifying a food product in a shopping complex; the consumer just taps or scans a product to get the product details or processor history.

Item identification normally refers to a unique code or part number, a serial or batch number, or an additional technique to make identification easy. There are several methods or techniques for labeling an object that are currently used for item identification:

- RFID or NFC tag
- labels or tagging
- nameplates

Figure 10.2 Traceability system architecture based on the EPC global standard network. Adapted from New Zealand Trade and Enterprise.

- barcode
- permanent marker pen note
- dot peening
- etching
- laser jet or ink jet printing

However, the requirements for high-security identification techniques to prevent food product counterfeiting and fraud are gaining more attention; thus, advanced technology solutions are required to support the system. Details and the method of identification used, as well as where and how they were applied, must be indicated on drawings or otherwise made public so customers can use them. An advanced ICT solution assists in product identification, especially for food products whose detailed processes and sources should be known by the consumer. Barge et al. (2014) proposed a radio frequency (RF) traceability system at the item level; they performed experiments with methods of fixing tags to products and found solutions for automatic identification adapted to handling procedures in a manufacturing environment. Hua et al. (2013) discussed a system approach for developing a methodology to implement traceability for the vegetable supply chain. They considered various abstraction levels while elaborating traceability systems and introduced a unified modeling language (UML) model for traceability along with a set of suitable patterns. A series of UML class diagrams was developed to conceive a method for modeling the product, process, and quality information in the vegetable supply chain.

Gandino et al. (2009) published a framework for the evaluation of a traceability system for the agricultural food industry; the automation level in a traceability system based on RFID technology was analyzed and compared with traditional methods. An internal and external traceability system requires both techniques to be considered and formalized to classify various environments according to their level of automation.

Accurate product identification is a critical matter required for food products; in manufacturing, inefficient systems for providing information, identification, tracking, and tracing of food products will have a significant impact on manufacturing.

2.2 Product History Record

The rapid growth of food product industries worldwide through numerous brands, trademarks, and suppliers raises consumer concerns about accepting only safe and healthy products. A good food product manufacturing process provides a traceability system for the product and minimizes the production and distribution of unsafe or poor quality food products, thereby minimizing the potential for bad publicity, liability, and recalls. Another dimension is the ability to provide product details, not only in terms of ingredients but also product history, for example, supply chain history, traceability, and access at the consumer's fingertips. Product information, especially for food items that can be directly consumed, such as fruits, vegetables, and bread, is very important because some products come long distances from different countries or continents; for example, fruits such as apples and pears may be imported from African or Asian countries or from Australia. Thus, information such as the export date and time and the product storage history is needed by consumers. Some of the most common information that is important to publish for consumers using an advanced technology ICT system includes the following:

- date and time product was harvested
- processing information in manufacturing
- packing date and time
- exporting and importing companies
- storage information
- delivery to retailer
- food product expiration date
- halal certification (for Muslim countries)

Previous research and studies that discuss food product traceability for consumers, such as Aung and Chang (2014), discuss a traceability system for products as a tool to assist in the assurance of food safety and quality as well as to achieve consumer confidence. Comprehensive information about food products using traceability helps ensure safety and quality in the food supply chain. Aiello et al. (2015) discuss evaluating the expected value to consumers of implementation of traceability systems for perishable products such as fruits and vegetables. Their study presents a mathematical stochastic model and approach for optimizing this supply chain system and for establishing the optimal granularity level when an RFID solution is adopted. The supply chain system profit in the presence of RFID traceability was calculated and compared with the current expected profit in the absence of such a system; they obtained results that showed that the characteristics of the specific supply chain are important in determining the optimal configuration of the traceability system. Fig. 10.3 shows an example of a smart device screenshot and the detailed history of one food product.

An intelligent consumer is normally very selective in food products, both the types of products and their sources. Currently, most customers hold smart devices for communication, taking notes, and photos; thus, smart mobile devices can retrieve the full history and journey of an item. The consumer can make an informed decision about whether a product is safe to consume. In developed countries, a customer with inadequate knowledge about a food product may consume unsafe or unhealthy products

Figure 10.3 Example of screenshot display of product details on a consumer smart device. Reproduced from Evizal Abdul Kadir.

even after their expiration date, thus putting their health at risk. News reports of unsafe food products occur regularly. This may occur because someone in the supply chain gambles in food trading by selling or exporting a rejected product that should have been destroyed; the rejected product may then be bought by a customer who is not knowledgeable about food.

3. Traceability Software for Suppliers

Food suppliers and industries have built traceability systems not only to meet government regulations and legal requirements for customer standards but also to raise productivity through enhanced data and information management systems. Traceability software systems for suppliers assist food industries to manage data and information for food product safety, quality, and security throughout the food processing, manufacturing, and distribution supply chain to the consumer. If traceability in the supply chain is sufficiently robust, it can improve service, increase safety and quality control, and possibly gain a good market reputation relative to others, thus achieving a high market share. By investing in a traceability system and creating a safer food chain, the design will result in lower costs in manufacturing and distribution systems because the number of product recalls will fall and worldwide sales will expand for a high-value food product; thus, all of these lead to greater profitability for the company or industry.

Storøy et al. (2013) proposed a framework design for tracing food products and recommended good traceability practices using common principles of unique identification for food products, namely the common generic standard electronic exchange of traceability information (Trace Core XML). The use of a food-tracing framework has been applied in many countries funded by the EU funding project for the TRACE project. De las Morenas et al. (2014) proposed a new solution which is a customized and automated cooler for carrying samples and a smart sensor installed inside the cooler; data are collected during the process, and a memory stick is used to transfer the collected data to a computer for further analysis. Several technologies are used in this solution, such as a microcontroller, sensors, RFID, and the global positioning system (GPS).

There are some benefits in using integrated software in a traceability system beyond those specifically for the supplier or industries. Such a system enhances product traceability and has benefits beyond fast location of inventory; thus, it provides a justification of the asset for investors. The benefits of using such a traceability system include the following:

Brand Protection: the implementation of software for a traceability system enables customers to check product status and records. With 24-h Internet access to the product database, the online system can be active at all times; thus, product complaints related to any product recalls are minimized. Product recalls are an issue for the manufacturer; if they happen, they will likely cause significant impact and damage to a company's reputation and sales.

Error Prevention: human error at the distributor level is one of the issues of product delivery. Enhanced identification of items and tracking products at a more granular level improves accuracy in filling customer orders. In manufacturing industries, automating a database system by capturing product identities is more popular than is manual data entry by humans with an associated high possibility of error.

Better Inventory Visibility: real-time intelligence traceability software allows manufacturers to exploit the track-and-trace data to enhance analytics and product reports. Some products are kept in inventory for long periods because of lack of product rotation, which causes some products to expire. With "first-in and first-out" inventory rotation, a system is able to ensure that products with earlier expiration dates are shipped first to prevent waste. This additionally benefits companies in the level of insight for better inventory forecasting and more strategic operational inventory decisions.

Consistent Customer Experience: for some retailers, advanced traceability functionality can ensure shelf consistency at all premises. The ripest product can be delivered to the nearest premises or retailer, and greener products can be delivered to more distant locations. This increases the commonality of shoppers' experiences at all locations.

Quality Control: a good integrated system helps a manufacturer to better manage returns. If a product has multiple returns to the manufacturer, a root cause analysis must be made to find the problem. A traceability software system might be used to reveal data and quality issues with a specific ingredient or manufacturer. All of the information found can be used as a basis for discussions among vendors, suppliers, and the manufacturer to avoid more returns.

Waste Reduction: if a recall occurs, it will be targeted at the problem product. All affected products must be removed from the supply-and-distribution system. With integrated software, the affected products, particularly foods that may have a high risk for consumers, can be pinpointed, and manufacturers may avoid broader recalls of products that are not affected. Product recalls waste time, materials, and labor and all of their associated costs.

Reduced Insurance Premiums: with better traceability software system implemented in all manufacturing processes, it might be possible to demonstrate to insurers that the potential losses from a recall are well managed so the risk of losses is reduced.

Relationship Support: introducing a traceability software system for the supply chain and customer satisfaction improves relationships in the business process. Regardless of the manufacturer level within the food product supply chain, all traceability data are shared with all trading partners. Using an integrated software system in the supply chain ensures better support of those suppliers and customers and allows information to be more easily accessed by them.

The entire previous discussion is targeted at better food product processing and traceability in manufacturing. Communication with a central database allows validating and storing traceability information related to the production, operation, inventory, locations, and manufacturing ingredients. The central database with traceability software allows reporting to suppliers; it is a tool to allow the products to be tracked and recalled by the batch.

3.1 Product Tracking and Monitoring

Batch product tracking and monitoring is effectively implemented in the supply chain management system; this means that the manufacturer or industry knows in which products a batch of supplier product was used and who received those finished goods. The application of software to product tracking and monitoring is a way to follow product problems to the customer, particularly food products consumed by humans. Product tracking should extend through the entire distribution line, starting from the supplier who delivered the raw material and occurring in every subsequent step. Automatic information and communication technology provides several methods to track and monitor a product, such as GPS, RFID, NFC, and sensor networks; every transaction in any step of tracking must be recorded to enable retrieval of all data. Germani et al. (2015) proposed a system to improve the environmental sustainability and guarantee the traceability, including the tracking and monitoring of products along the entire product supply chain system for the food product. A proposed methodology has been developed and implemented in software in a distribution system to efficiently model complex supply chains, sharing data between every stage of supply chain processors and measuring the environmental sustainability. Several tracking systems for products in production or manufacturing processes include the following:

1. Tracking supplier batches at receiving
   The important aspect of batch tracking and monitoring in manufacturing is keeping track of the suppliers from which materials are received to ensure that the supplier is known in case of recall. Raw materials can be traced back to the source in case of a faulty batch to ensure that the supplier knows of the issue. Finally, batches of raw materials can be checked for the correct specifications and standards; if they are not met, substitution of a new batch can ensure the safety of finished goods.

2. Tracking raw materials through production
   Another important aspect of batch traceability is to keep track of which raw materials go into finished goods, thus ensuring that a finished good represents the correct collection of batch materials. This connects the gap between a batch of finished goods and supplier batches. This tracking of raw material can be challenging in small and medium-sized companies, especially for manufacturer processes with more than one product.
3. Tracking lots of finished goods to customers
   Another important aspect in batch tracking and monitoring is to ensure that all finished goods of each batch can be tracked to the customer to whom they have been shipped. This is urgent in case of a recall; the customer can be contacted and told to return the entire batch to prevent further distribution. Thus, with this system everything is monitored, and the data can be used to precisely trace each batch in case of a problem.
   If all of these levels of tracking and monitoring are implemented in the supply and distribution system, complete traceability is achieved for good product monitoring. In case of a product recall, the solution for taking action is easy because all batches of the product are known, as is the party that supplied it. Fig. 10.4 shows an example of a complete product tracking-and-monitoring system in the supply chain.

Product tracking and monitoring is not only for the suppliers and distributors in the supply chain but is applicable to the entire product batch, starting from raw materials until it reaches the customer. There are several benefits of using a tracking-and-traceability system for suppliers or customers:

- Ability to determine the real-time location and status of the products or batch items being delivered;
- Acquiring complete and accurate production counts at any point in the production process to help note any failures or bottlenecks;
- Immediate availability of failure rates of production items with precise tracking;
- More control for a complete range of effective product tracking-and-traceability systems, with application across all aspects of the industry;
- From initial sensing and detection, to barcode verification applications, to complete production line monitoring systems.

3.2 Product Traceability

Product traceability ensures that every product in the supply chain is traceable with all of its information, including expiration dates. Traceability is becoming more important and is required in manufacturing processes due to product safety concerns, particularly for food products consumed by humans. Badia-Melis et al. (2015) presented some of the conceptual advancements in the field of food traceability systems, such as the development of a common framework toward unifying the present technical regulations, the interconnectivity between agents, environment loggers and products, all in the form of the Internet of Things, and the development of intelligent traceability through which it is possible to retrieve the temperature of a product or its remaining shelf life.

Meimandi et al. (2014) investigated and identified contributions on traceability topics from the starting point to the end and characterized and analyzed the chosen contributions against three research questions by utilizing a comparative framework

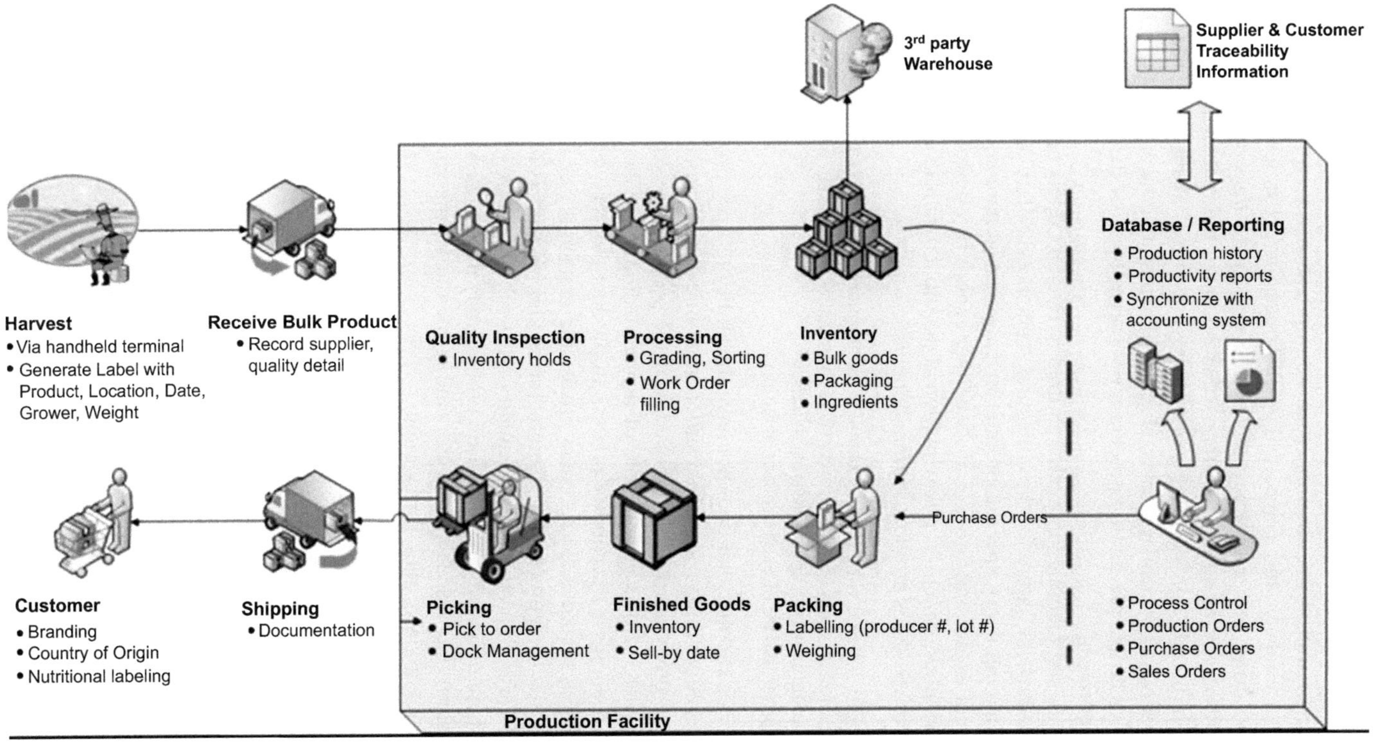

Figure 10.4 Complete product tracking-and-monitoring system.
Adapted from Infinium Solutionz.

including nine criteria. The study arrives at some interesting points and outlines a number of potential research directions, paving the way to facilitating and empowering traceability research in this domain to assist software engineers and testers in test management. Dabbenea and Gayb (2011) introduced novel criteria and methodologies for measuring and optimizing the performance of a traceability system. As opposed to previously introduced methods that optimize indirect measures, the proposed approach directly considers the worst-case (or average) quantity of product that should be recalled in the case of a crisis. Numerical examples concerning the mixing of batches in a sausage production process are reported to show the effectiveness of the proposed approach.

Food distributors and manufacturers are required to track products not only to sell the products but to provide safety for the consumer. "The Food Safety Modernization Act" requires managing every specific product inventory; it is paramount to ensure that products do not expire and become an issue that cannot be controlled in a non-food-specific distribution system. Implementation of potential recalls of food products and monitoring expired dates provides certain benefits:

- Providing the ability to trace with specific batch or lot numbers of products;
- Knowing the location of users and consumers of each traceable product;
- Providing product information enables easier recalls;
- Knowing the expiration date of each product or item.

A manufacturer or supplier that does not have the ability to trace products or does not know when a product is expiring might suffer economic consequences if a product becomes suspect and may need to be removed or if the product has expired and will be discarded. Fig. 10.5 shows an example of a manufacturer's product traceability system; Fig. 10.5A shows a traceability process in product delivery, and Fig. 10.5B shows a traceability system in use in product processing in a supply chain.

Traceability for the food processing manufacturing and industry is also utilized to identify key points of high production and quality areas of a business, versus those of low return, and where points in the production process may need to be improved. The challenge for food product traceability is its wide scope due to the variety of stakeholders in the supply chain system, for example:

- organized and nonorganized manufacturers and processors
- manual and automated packing systems
- merchandise of manufacturers in export systems
- world-class product testing laboratories
- more product control for tracking and traceability

To guarantee high levels of product quality and to comply with safety regulations, it is paramount that all of the products in the line can be effectively tracked and traced from their start at the farmer to finally reaching the consumer. Food and beverage products in particular require total product traceability in the supply chain system because of the high risks to the consumer. With modern product tracking and traceability tools, a production line can be accurately monitored through a tracking number or the unique identity of the product. A unique tracking identity also refers to an

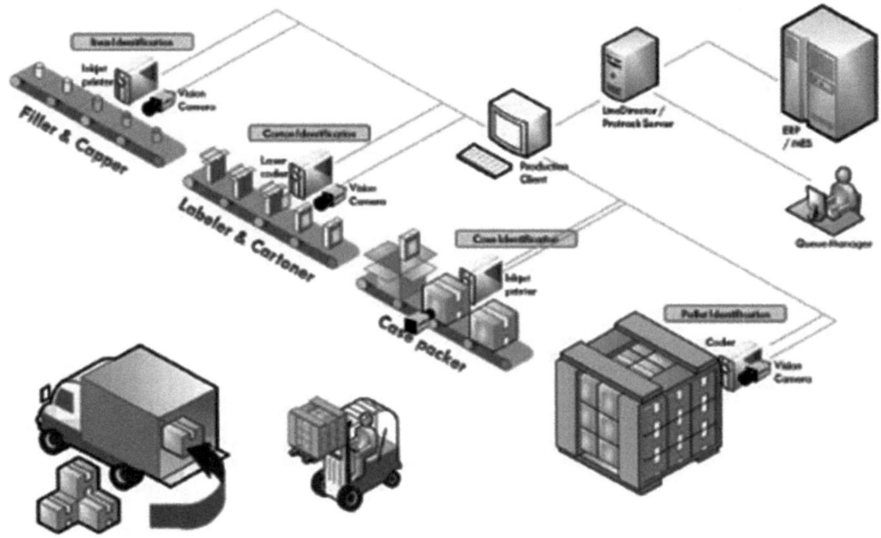

(A) (B)

Figure 10.5 Product traceability system: (A) traceability process in product delivery (B) traceability system in use in product processing in a supply chain.
Adapted from Food Safety Magazine.

individual production line; in the event of a potential error, an item can be accurately tracked back through the line from which it originally came.

3.3 Supply Chain Management System

Currently, many well-known food and beverage brands are trusted and employ online software for supply chain management and visibility because it supports the requirements of the food and beverage industries with advanced inventory and management capabilities as well as real-time product visibility, particularly for critical product information from manufacturers and suppliers to storage and all of the locations in the chain. A supply chain management system has the ability to control and monitor every step of product processes with the support of other technologies as components in an overall large system. Most large companies have already implemented a supply chain management system, starting from the source of material through the end of the process, which is the consumer. This is to ensure that the product is completely safe to use or consume. Fig. 10.6 shows an example of a product supply chain management system for a manufacturer, starting from an importer to its end, the retailer and consumer.

Introducing a supply chain management system that drives traceability of products, motivated by wide public interest and concerns, provides key benefits to the manufacturer or industries; these benefits include the desire to improve

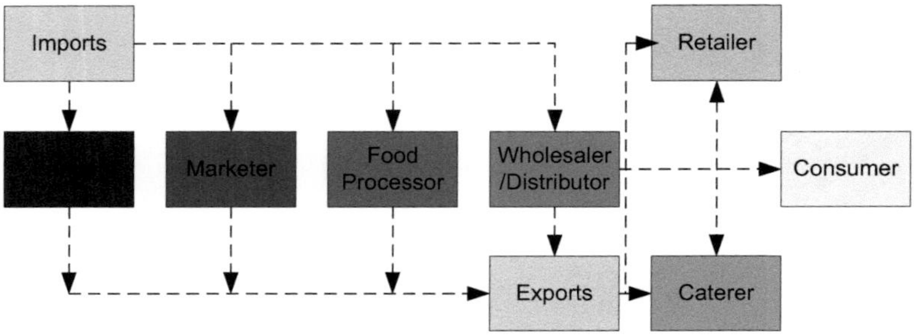

Figure 10.6 Example of product supply chain management system.
Reproduced from Evizal Abdul Kadir.

management, government regulations, and industrial standards. Responding to the social and economic impacts of major food safety incidents, many manufacturers and industries are implementing product supply chain management systems to ensure that all of their products are traceable. In case of a product recall, the system also assists in minimizing its extent and sustaining customer confidence to continue to consume the product; with such a system installed, the retailer becomes a more-trusted information source, and recalls are required only for contaminated products, thereby avoiding wastage of good products. A supply chain management system with product tracing ability additionally allows authorities to limit recalls to products related specifically to the food safety incident only; this helps mitigate the negative economic consequences to industries and is beneficial to public interest.

Supply chain management systems must be linked together throughout the supply process to ensure that food sources are safe for consumers and that food-related disease outbreaks can be detected quickly. The linkages depend on the sizes of processes; some may be larger, including many sources of material and processes, whereas others are relatively simple when their process needs only a few steps. Other potential solutions to improve supply chain management systems in providing product traceability include the following:

- Making a supply chain management system universal and comprehensive;
- Ensuring the supply chain management systems are compatible with the managed products;
- Mandating minimum requirements for suitable supply chain management systems;
- Making premises management systems mandatory for poultry and livestock producers;
- Requiring detailed product information to handle emergencies quickly;
- Cost-effective supply chain management systems for industries;
- Promoting the benefits of implementing a supply chain management system;
- Continuous evaluations to improve supply chain management system performance.

In some developed countries, food supply chain management systems for industries generate benefits for both public and private interests but also involve substantial costs for stakeholders. An efficient and effective supply chain management system provides optimum benefits with minimal costs involved. The ultimate prescription for good supply chain management for manufacturers or industries is to balance public and private costs in the supply chain with public and private benefits.

4. Future Trends

The global food industry must introduce traceability technology controls; progress will be technology driven as new technologies and applications are implemented. Several technologies implemented in the market can control product traceability and a supply chain management systems, such as RFID, barcodes, GPS, and infrared. All of these technologies have different applications, depending on the level at which they are applied. The barcode is widely used for product packaging, especially for food products, whereas RFID is coming into greater use because of its advantage of not requiring line-of-sight detection. Others, such as GPS and sensor networks, have not yet found significant use. The future trend of traceability system usage is one of growth at a healthy rate for all processes as the awareness of food safety increases among governments and consumers. Worldwide, governments are requiring traceability to track food products as a result of high concern for consumer health. Software is an important part in supporting traceability for products; steps at all levels require hardware, and the tool must have software to complete the system. There are two large classifications for the software part: support processors for data processing and database software for data storage. Cloud computing and a Web base linking to all parties in the supply chain will be required for the next major step of traceability systems. A sustainable food production system is another challenge for future trends of product traceability, including software parts for manufacturers and consumers. Food product businesses in the future will be more complex, encompassing such terms and entities as industrial agriculture, urban agriculture, bio-domes, small-scale farms, vertical growing spaces, backyard gardening, hydroponics, and community gardening.

References

Aiello, G., Enea, M., Muria, C., 2015. Production, manufacturing and logistics the expected value of the traceability information. European Journal of Operational Research 244, 176–186.

Ali, N., Gue´he´neu, Y.G., Antoniol, G., 2013. Trustrace: mining software repositories to improve the accuracy of requirement traceability links. IEEE Transaction on Software Engineering 39 (5), 725–741.

Aung, M.M., Chang, Y.S., 2014. Traceability in a food supply chain: safety and quality perspectives. Journal of Food Control 39, 172–184.

Badia-Melis, R., Mishra, P., Ruiz-García, L., 2015. Food traceability: new trends and recent advances. A review. Food Control 57, 393–401.

Barge, P., Gay, P., Merlino, V., Tortia, C., 2014. Item-level radio-frequency identification for the traceability of food products: application on a dairy product. Journal of Food Engineering 125, 119–130.

Dabbenea, F., Gayb, P., 2011. Food traceability systems: performance evaluation and optimization. Computers and Electronics in Agriculture 75, 139–146.

De las Morenas, J., García, A., Blanco, J., 2014. Prototype traceability system for the dairy industry. Computers and Electronics in Agriculture 101, 34–41.

European Communities, 2007. Food Traceability. Directorate-General for Health and Consumer Protection, B-1049 Brussels. http://ec.europa.eu/dgs/health_consumer/index_en.htm.

Gandino, F., Montrucchio, B., Rebaudengo, M., Sanchez, E.R., 2009. On improving automation by integrating RFID in the traceability management of the agri-food sector. IEEE Transaction on Industrial Electronics 56 (7), 2357–2365.

Germania, M., Mandolinia, M., Marconia, M., Marilungoa, E., Papettia, A., 2015. A system to increase the sustainability and traceability of supply chains. In: The 22nd CIRP Conference on Life Cycle Engineering (Procedia CIRP), vol. 29, pp. 227–232.

Hua, J., Zhang, X., Moga, L.M., Neculita, M., 2013. Modeling and implementation of the vegetable supply chain traceability system. Journal of Food Control 30, 341–353.

IBM DeveloperWorks, 2015. Traceability Solution Blueprint for Business Performance Optimization, Part 1: Understanding the Architecture of a Comprehensive Track-and-Trace Solution. http://www.ibm.com/developerworks/data/library/techarticle/dm-0902oberhofer/.

Meimandi Parizi, R., Lee, S.P., Dabbagh, M., 2014. Achievements and challenges in state-of-the-art software traceability between test and code artifacts. IEEE Transaction on Reliability 63 (4), 913–926.

New Zealand Trade and Enterprise. https://www.nzte.govt.nz/en/export/market-research/food-and-beverage/sustainability-market-intelligence-reporting-asia-february-2015-traceability/.

Parreño-Marchante, A., Alvarez-Melcon, A., Trebar, M., Filippin, P., 2014. Advanced traceability system in aquaculture supply chain. Journal of Food Engineering 122, 99–109.

Rajo-Iglesias, E., 2013. Wireless corner. IEEE Antennas and Propagation Magazine 55 (2), p. 255.

Storøy, J., Thakur, M., Olsen, P., 2013. The tracefood framework – principles and guidelines for implementing traceability in food value chains. Journal of Food Engineering 115, 41–48.

Thakur, M., Sørensen, C.F., Olav Bjørnson, F., Forås, E., Hurburgh, C.R., 2011. Managing food traceability information using EPCIS framework. Journal of Food Engineering 103, 417–433.

US Food and Drug Administration's, 2002. Implementation of the Bioterrorism Act: Applicability to Exporters of Products to the United States. Hogan & Hartson L.L.P. www.hhlaw.com.

Part Three

Legal Requirements and Regulation in Food Traceability

Real-Time Tracking and Remote Monitoring in Food Traceability

11

R. Badia-Melis, L. Ruiz-Garcia
Polytechnic University of Madrid, Madrid, Spain

1. Introduction

The long distance transport and the globalized markets make the development of the monitoring systems that guarantee the optimum conditions of the products essential. Exporting countries, such as Costa Rica, are sending their products to greater distances every time, as far as Russia, for example, and it implicates an increase in transportation time and, as a consequence, bigger risk for the products. These exported products spend some weeks traveling in refrigerated containers with no information until they reach their destination.

The transport includes various transitions and transferences of the product between companies, as well as administrative control in customs, borders, and seaports. Perishable food products have to maintain the temperature and relative humidity during the trip, so that they keep optimum conditions and freshness. Nonetheless, it is very common that those controlled conditions are not happening along the whole trip, especially in places such as seaports or transitions in airports, where the containers can be exposed to adverse conditions for long periods of time.

The UN FAO (2013) estimates that each year, approximately one-third of all food produced for human consumption in the world is lost or wasted. The United Nations Environment Program says that 222 million tons are annually wasted in developed countries, almost as much as the entire production of sub-Saharan Africa (230 million tons) (UNEP, 2013). A study performed in the United States, Canada, Australia, and New Zealand reveals that during production, 20% of the fruit and vegetables are lost, consumers waste 28%, and during distribution, 12% is wasted mainly because of the lack of refrigeration control (Gunders, 2012).

Adequate temperature monitoring in order to ensure the longest shelf life for perishable food products represents a difficult puzzle to solve due to the many factors involved. Thus studying and analyzing temperature gradient data inside refrigeration rooms, containers, and trucks is a primary concern for the industry. Any temperature disturbance can undermine the efforts of the whole chain (Mahajan et al., 2014). Perishable food products are very common, including goods that are poorly or excessive cooled. Jedermann et al. (2014a) reported many examples where the inadequate management during temperature control usually leads to losses in the food chain (at postharvest, distribution, and at home). It is possible to understand this variation by remote monitoring (Jedermann et al., 2015); this remote monitoring would require a high number of data loggers to identify weak points in specific spots through a container, cold room, or even a pallet.

Advances in Food Traceability Techniques and Technologies. http://dx.doi.org/10.1016/B978-0-08-100310-7.00011-9

The specialization of the agents in the current transportation and distribution chains must entail an improvement in the coordination; on the contrary, breaks in the cold chain would appear, resulting in the consequent loss of quality of the products.

Jedermann et al. (2009b) proposed the grouping of items with similar temperature requirements in separated batches in order to reduce the necessity of adding extra sensors; it could be carried out in his intelligent container. Similar studies propose the idea of an intelligent container where the sensors are positioned between the goods (Lütjen et al., 2012).

Approaches to better monitoring systems and real-time tracking are being assessed, as it is shown in this document. Those approaches would lead to a better understanding of food traceability and therefore a more effective traceability.

2. How Can We Track and Monitor in Real Time?

The integration of diverse technologies is required in the supervision of refrigerated transportation, and the communications and the sensors are especially important (see Fig. 11.1). This data collecting has to be done in critical points and in determinate moments; otherwise this data would not help. Other techniques will contribute to the adequate monitoring, such as the Internet of Things, the concept of intelligent traceability, and the temperature estimation methods.

2.1 Wireless Sensing Technologies

Wireless technologies have been considered key technological enablers, whereby intelligent tags are attached to an item within a network with wireless links, which

Figure 11.1 Monitoring system in refrigerated transport with wireless technology.

can transmit some physical parameters such as temperature or humidity in addition to other information like position or movement (Zou et al., 2014). Monitoring devices are used to ensure temperature integrity; however, resource limitations and cost factors severely limit their use to one-per-pallet or even one-per-container scenarios (Badia-Melis et al., 2013).

The potential for monitoring technologies, Wireless Sensing Network (WSN) and Radio Frequency Identification (RFID), has been suggested by several studies, and extensive literature has been published to address the issue (Badia-Melis et al., 2014; Georg et al., 2013; Ruiz-Garcia et al., 2010; Qi et al., 2014; Jedermann et al., 2014b). Nevertheless, due to characteristics of food and insulation of the packaging, the temperature is not always homogeneously distributed inside the pallet, and using a single sensor for the entire pallet does not provide a realistic representation of the temperature mapping since it only represents the temperature of one spot; meanwhile, the rest of the container or pallet does not have any temperature reference.

2.2 Remote Communications

Remote communications in long and middle ranges have been extraordinarily developed in the last few years since 2010. The usage of Wi-Fi, GSM, or GPRS is so common that they are barely noticed nowadays, but when it is about machine-to-machine (M2M) communications, these are not as common, for example, communications needed between a container traveling across the Atlantic Ocean and a server located thousands of kilometers away, even taking into account that there are hundreds of thousands of containers being moved daily in the world.

Jedermann et al. (2014b) present a detailed description of technical challenges and solutions for a wireless monitoring in transport, talking about pros and cons in device communication, protocols, energy consumption, and data accessibility. On the other hand, Badia-Melis et al. (2015) show how a warehouse with cold rooms of the size of $26 \times 6 \times 12 = 1848\,m^3$ can be instrumented by combining two different technologies, RFID and WSN, in a way that both can benefit from each other and give complementary data.

Fig. 11.1 represents a standard wireless monitoring system integrated into a refrigerated trailer. In every pallet, there are a certain number of devices that can communicate wirelessly with either a stand-alone gateway or a gateway connected to a computer, and at the same time, the gateway or the computer can be connected to the Internet and GPS, emitting to a remote database.

Short-range communications are not exempt of risks, since everything happens in the container with the size of 12 m length, 2.3 width, and 2.5 height (Table 11.1). The key is not only about transmitting data inside the container but making an adequate integration of the sensors. The fusion between both is called wireless sensing. Inside this field the technologies considered to have the greatest potential are RFID systems and WSN (Ruiz Garcia and Lunadei, 2011).

Currently, RFID is the most mature technology, and it is commercially available. On the other hand, WSN has not evolved as fast as expected in the agro-food sector, since important aspects are due to be solved, such as the encapsulation, the size of the devices, or the battery life, among others (Badia-Melis et al., 2014).

Table 11.1 Commercial Container Sizes

		Reefer 45′ High Cube (m)	Reefer 40′(m)	Reefer 20′(m)
Interior size	Length	13.10	11.57	5.44
	Width	2.29	2.29	2.29
	Height	2.50	2.21	2.14
Capacity		75.40m^3	58.70m^3	28.31m^3

The very first WSN nodes on the market held a whole variety of sensors able to control refrigerated logistics, such as temperature, relative humidity, and acceleration (vibration and shocks). Its drawback was that it was working in the microwave band (2.4 GHz), and its power was not higher than 1 mW (Fig. 11.2). The 2.4 GHz microwave band has the advantage of being available all around the world, but the main disadvantage is the coincidence with the water resonance frequency center (than is the reason of the food heating), therefore its propagation is difficult through the water and the air with high relative humidity. That is to say, in the typical conditions in fruit and vegetable transportation, the air has around 90% of RH and the product's water content is usually more than 90% (Ruiz Garcia and Lunadei, 2011).

2.3 Internet of Things

The Internet of Things (IoT) is a network that combines everyday objects with the ability to identify and interact with each other to reach cooperation goals (Giusto et al., 2010). The purpose of the IoT is that, all the items in the network can be perceived and controlled remotely, and also can be combined with the Internet to form a more intelligent production and create living systems. IoT describes a world where humans are surrounded by machines that communicate with each other and can allow people to interact with the digital world. The concept of the IoT breaks traditional ideas and starts a new technology field, and it will bring a new revolution of world information industry after the computer, the Internet, and mobile communication network.

The IoT is an infrastructure to facilitate the information exchange of products (Atzori et al., 2010; Leonardo et al., 2011). The system itself is composed of three dimensions which contain information items, independent networks, and intelligent applications. Information items are those that can identify and perceive their own message; independent networks have the capabilities of self-configuration, self-healing, self-optimizing, and self-protection; and intelligent applications mean the applications with the capabilities of intelligent control and processing.

RFID can enhance the efficiency of the food chain by storing the meaningful information such as producing area, the planting methods, the producing process, and other special information that the consumer will consider about them when they purchase products (Zhang and Li, 2012). The conceptual sensor data integration has been exploited by the classification approach by using the EPC in RFID (Theodorou, 2006).

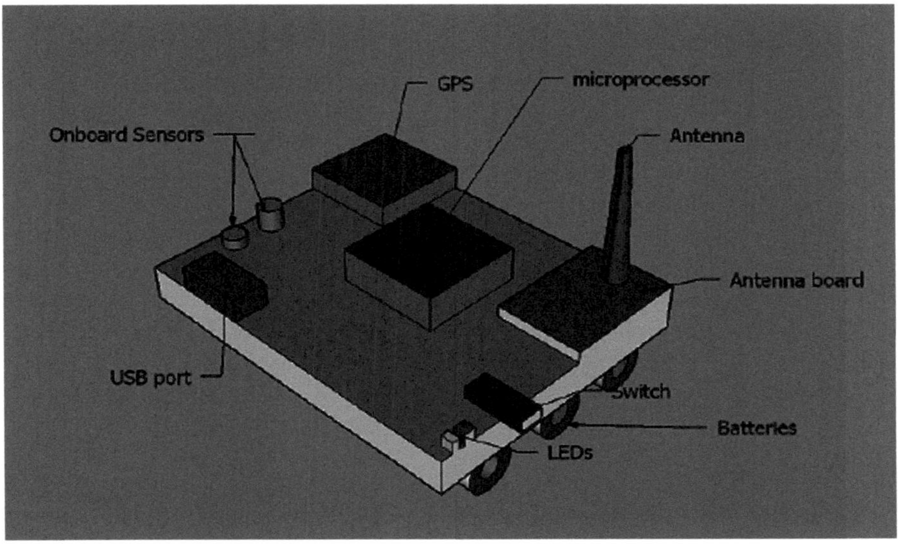

Figure 11.2 Standard mote structure.

Dimitris (2011) introduced the "intelligent products," which are meant to be used in the IoT.

Atzori et al. (2010) describe in their article how EPC is a mean to support the spreading use of RFID in worldwide modern trading networks and to create the industry-driven global standards for the EPCglobal Network. These standards are mainly designed to improve object visibility (ie, the traceability of an object and the awareness of its status, current location, etc.).

2.4 Intelligent Traceability

After an extensive review of many studies, Bosona and Gebresenbet (2013) concluded that food traceability should be considered as an important and integral part of logistics management, so the information is stored on the go and it can be retrieved at any moment.

The term "intelligent food logistics" is becoming an important topic about the food chain, meanwhile, as described by Jedermann et al. (2014a), intelligent food logistics are called to reduce the perishable waste along the food supply chain by means of reduction of the deviation from the optimal cold chain. To do that, it is necessary to quantify these deviations, making shelf life variation and remote monitoring. Ruiz Garcia et al. (2010) proposed a web-based system for data processing, storage, and transfer. It integrates all information along the food and feed chain, oriented for the needs of the consumer. These kinds of studies lead to a more intelligent food logistic chain.

On the other hand, traceability, as it is described in the introduction of this document, is an integral part of food safety, food quality, and food defense, which make it part of the food supply chain, pretending to be of high potential for the protection of consumers. Both intelligent logistics, like traceability, are happening during the supply chain.

As Scheen (2006) described in his article, the data obtained in the traditional tracking and tracing (T&T) systems are usually used only to recall management; nonetheless, the proposed quality-oriented tracking and tracing (QTT) system uses the information retrieved with what we could call intelligent logistic tools, such as wireless sensors (temperature and humidity for instance).

The QTT concept represents a perfect alliance between logistics and traceability, improving the supply chain. Jedermann et al. (2014b) published how the integration of QTT and FEFO (first expire first out), using shelf life models, can enhance the demand and supply chain. Van der Vorst et al. (2007) also emphasizes, in the implementation of quality-controlled logistics in the food supply chain networks, that if the product quality can be predicted in advance, it can lead to a better goods flow and better chain design.

2.5 New Advancements

The current advancements in RFID technology and the incorporation of integral parts, such as data logger capabilities and integrated sensors, have provided a new dimension to the application of RFID technology in food traceability systems. As it is gathered in the article by Ruiz-Garcia and Lunadei (2011), the applications of RFID to food traceability are many and varied. Since 2005, RFID has emerged as a lead actor in the development of traceability systems in the food supply chain, and their implementations are increasing at a fast rate (Costa et al., 2013). With the implementation of RFID technology, food traceability systems can become more reliable and efficient, since RFID enables a higher reading rate than traditional barcodes (Hong et al., 2011). With the ability to identify products without any physical contact, RFID technology provides effective information sharing with efficient customization and handling (Zhang and Li, 2012). Kelepouris et al. (2007) proposed an infrastructure using RFID, where it was compared with the traditional lot numbering and internal information systems; it has the advantages of automatic identification, uniform EPC for all partners, small investment in equipment, and easily drawn information.

Agro-food logistics and supply chain management processes for food traceability using RFID are discussed by several authors, ie, Jones et al. (2004), Angeles (2005), Twist (2005), Attaran (2007), Ngai et al. (2007), and Sugahara (2009). Amador and Emond (2010) developed a system of RFID temperature tracking for combat feeding logistics. The capability of hosting sensors of the tags allows the so-called "cold traceability," a concept where trace groups of temperature-sensitive products are transported in different atmosphere requirements (Ruiz-Garcia et al., 2010).

Feng et al. (2013) developed and evaluated a cattle/beef traceability system, which integrated RFID technology with PDA and barcode printers. They obtained real-time and accurate data acquisition and transmission and the high efficiency of information tracking and tracing across the cattle/beef supply chain.

An IoT platform is a concept used in this case by Zou et al. (2014). It consists of two-layer network architecture; it is an innovative way for implementing RFID for traceability in what is called "intelligent food logistics," consisting of an asymmetric tag-reader link (RFID layer) and an ad hoc link between readers (WSN layer), which are further connected to the Internet via cellular or Wi-Fi.

2.5.1 Near Field Communication

Near Field Communication (NFC) is actually an extension or a subcategory of RFID. NFC technology enables simple and safe two-way interactions between electronic devices; it complements many popular wireless technologies at the consumer level (in the existing standards for contactless card technologies ISO/IEC 14,443 A&B and JIS-X 6319-4). NFC enables devices to share information in less than 4 cm (NFC Forum, 2014). It operates at 13.56 MHz and currently supports data rates of 106, 212, 424, or 848 Kbit/s (Mainetti et al., 2013a). These tags are made very small so that they can fit inside products for various reasons such as security, antitheft and individual identification. NFC is the newer version of RFID that is typically for use in a very short distance range for making payments (Mainetti et al., 2012) and information retrieval. Like RFID, the main advantage of NFC technology, above barcode and Quick Response (QR) codes, is that it does not require a laser beam to have a solid path so that it can travel between two devices.

Massachusetts Institute of Technology (MIT) chemists have developed an NFC tag based in chemiresistors that is able to detect certain gases. They disrupted the electronic circuit by making a hole in it, after reconnecting the circuit with a linker made of carbon nanotubes. When the targeted gas is present the conductivity of the nanotubes change, and they are able to read the tag with a smartphone and detect the presence of the gas (Trafton, 2014). In this case, real-time tracking of gases can be done with NFC, and it can be recorded in the traceability of the product. At the same time, it is important to notice that the low cost of the NFC adds that benefit in front of other more expensive technologies.

2.5.2 Temperature Estimation

Some of the articles introduced the real possibility of temperature estimation (Jedermann and Lang, 2009a; Jedermann et al., 2009b, 2011), which means to estimate the temperature in critical places where there is no possibility of placing a sensor or there is no convenience for such a thing. The regular methods are based on the inverse of the distance weighting or kriging; however, it does not detect local variations unless it is used with a high density of local sensors, which represents a counter effect in reducing the number of sensors.

2.5.2.1 Capacitor

A temperature sensor is a capacitor/resistor system; therefore by taking the difference between the terminals, the signal estimation can be done. The temperature rise and fall rates within a food pallet fit quite accurately with the rise/fall time constants in a resistor-capacitor (RC) electrical circuit (Badia-Melis et al., 2013). Therefore pallet temperature is modeled by the potential differences between the terminals of a capacitor. Uysal et al. (2011) used this method to estimate the temperature profiles in First Strike Rations (FSR military US rations). The improvement was 90.11% using the estimated temperature by capacitor method as a reference versus the environmental temperature which is the traditional reference. Badia-Melis et al. (2013) achieved an 86.69% improvement in temperature estimation compared to no temperature estimation.

2.5.2.2 Kriging

The Kriging method is presented in his article as the second algorithm. There are various linear interpolation methods that multiply the measured values at the source points with a set of weighting factors in order to estimate the value at a destination point. The weighting factors can be set by a heuristic approach proportional to the inverse squared distance or by statistically more solid through the Kriging method. This method is based on an analysis of the spatial correlation of the measurements; it uses the Variogram, which gives the expected difference of the physical quantity between two points as a function of their distance (Jedermann and Lang, 2009a).

This method has previously been used in environmental estimation problems defined within a continuous feature space (such as temperature inside a homogenous container). Jedermann et al. (2011), Jedermann and Lang (2009a), and Badia-Melis et al. (2013) obtained, in the best cases, 20%, 79%, and 40% respective improvement of the errors in comparison with traditional methods.

2.5.2.3 Artificial Neural Network

Finally, as the third algorithm, it is assumed that the inherently nonlinear relationship between the product temperature inside a pallet and the air temperature can be modeled by an artificial neural network (ANN) (Zhang and Qi, 2008). Input to the network consists of time and temperature data provided by the sensor(s) placed outside the pallet, whereas output is the estimated time and temperature data for products placed inside the pallet. Hidden layers consist of artificial neurons which weigh and sum their inputs as they propagate their outputs to the next hidden layer (Fig. 11.3). ANNs need to be trained with part of the temperature data to learn how to estimate the nonlinear relationship between its input and target output.

3. Applications

As new advancements are showing the latest updates and trends in food traceability, the applications show how some of these advancements are applied to the real world, mostly in experimental approaches. Cold chain is an essential part of the traceability since many of the perishable goods need refrigeration during the supply chain, and it is important to record what is happening with the temperatures. On the other hand, the food processing needs a process registration for the record and for further recall in case it is needed due to a foodborne illness or information for the customer.

3.1 Cold Chain

It is important to take into account that the quality of perishable products cannot be enhanced during transportation, storing, and distribution. Each product has its own

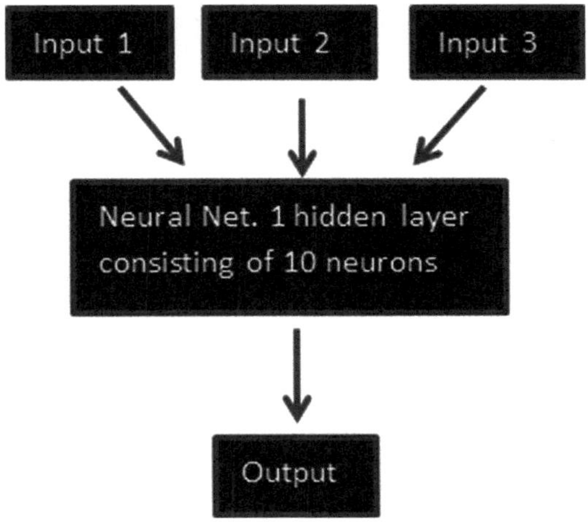

Figure 11.3 Structure of a Neural Network, one hidden layer with 10 neurons.

specific shelf life, limiting their good conditions for consumption. What is achieved with a proper cold chain is to delay the deterioration of the product (Ruiz-Garcia and Lunadei, 2011). But the loss of quality is accumulative, and it depends on numerous factors and interactions. The lack of control can occur at any moment. The temperature distribution is not homogeneous in a cold room and in a refrigerated container, shelf life is as well not homogeneous, having variations between pallets and even between cartons (Badia-Melis et al., 2013).

If the cold chain is broken at any moment the freight arrives in bad condition. The losses can go from a few cartons to an entire container. When the loss is significant, it is necessary to use insurance and experts, the responsibility assumption needs to be done, and it can fall into conflict. At this moment is when monitoring technologies become essential (Badia-Melis et al., 2015), since the information retrieved from the sensors determines the moments where the products were at risk.

3.2 Food Processing

Barge et al. (2014) automatically recorded cheese wheel movements during the production, handling in the maturing room and warehouse, delivery, packing, and selling phases in a dairy factory with the help different techniques by fixing RFID tags to the cheese. They considered factors such as tag type and shape, required power, antennas polarization and orientation, fixing method, and ripening duration to verify their effect on reading performance and system reliability.

Catarinucci et al. (2011) used a combination between WSN and RFID in order to enhance the traceability of the white wine from vineyard to consumer glass. WSN is a wireless technology like RFID and is tried by many other authors to improve the food quality in terms of traceability (Qi et al., 2014) and aquaculture (Qi et al., 2011).

A Wheat Flour Milling Traceability System (WFMTS) was developed by Qian et al. (2012), incorporating a 2-D barcode and RFID technology, to validate the system in a wheat flour mill in China. Labels with a QR Code were used to identify small wheat flour packages, and RFID tags were used to identify wheat flour bins and record logistics information automatically. They obtained that the total cost of the system increased by 17.2%; however, with the new system the sales income increased by 32.5% and proved the high potential of good application of WFMTS in medium and large wheat mill enterprises.

Recently, Mainetti et al. (2013b) described the potentialities from the combination of RFID and NFC; it allows the end consumer to know the complete history of the purchased product. They implemented the usage of NFC in a IV gamma products supply chain. NFC is proven to work together with a mobile app to allow the linking of plants and traceability information (Mainetti et al., 2013a).

Yu-Yi et al. (2014) proposed a scenario where the consumer can use the smartphone to read passive information and essential parameters and finally safely purchase the food.

4. Limitations and Challenges

The most important topic for food industries is to measure the temperature in the most possible points without a big investment. This achievement is possible by implementing RFID semipassive tags which incorporate temperature sensors. The major disadvantage is it operates in the 13.56 MHz frequency, and with this low emission power, the reading range does not go beyond 2–3 cm. Despite this, tags register date and time and record it in their internal memory, and it is necessary to wait until the end of the journey to download the information one by one. Retrieving this information takes time and the information is not live.

Are there any other systems able to provide live information? The answer is yes, but they are not completely developed due to its complexity. In order to achieve live information it is needed to implement active RFID or WSN, they can be configured to send data every certain periods of time, it is called the sample rate, and they are provided with the enough radio frequency power to send the signal through the pallets even they operate in 2.4 GHz band. Although it would be better to implement systems that use other frequencies, for example, 433 MHz. Currently this frequency band is considered the most appropriate for the refrigerated monitoring systems, since it is available all over the world and goes through the obstacles better than other high frequencies such as 915 MHz or 2.4 GHz (Laniel et al., 2010).

In regard to the cattle/beef sector, the main challenges for implementing RFID in their traceability systems are the inapplicable methods of inputting information, the inefficient sequence of data input and communication mechanism associated with RFID readers, and the high implementation cost (Feng et al., 2013). With many advantages of RFID technology, it is still not a preferable choice for most companies as it leads to additional cost. However, the balance between the benefit and safety requirement of the company is the major driving force for the adoption of technologies like RFID (Zhang and Li, 2012).

Another challenge is going around almost any obstacle, which these technologies have surpassed. The standard ZigBee protocol being used for some WSN and RFID class 4 are able to adopt net topology with data transference point to point (PtP).

It means that the data can be hopping from one node to the following node without having to go directly to the gateway; it allows the communication to go around any object, such as metals or liquids. It is necessary to take into account that this situation would need an increment on the number of nodes, increment of complexity, and the consequent increase in costs. Despite the sensing devices having a relatively low price (15–100 €), it is necessary to use routers, gateways, concentrators, and even sometimes specialized computers, and it increases the cost of the entire system.

Some of the Ultra High Frequency tags (bands 860 MHz Europe 915 MHz USA) are potentially useful and are currently used for cold chain monitoring and traceability control. They present the disadvantages of more interferences than other frequencies. These different bands are only available in the United States and Europe. Japan and China do not allow these transmissions (Ruiz-Garcia and Lunadei, 2011).

The reduced size of the wireless sensing devices allows them to be installed in any place inside the freight or the container. RFID tags have a similar size of a credit card, and WSN nodes are no bigger than a tobacco packet. This is a big advantage since it can be located either inside the cartons, the pallets, the container's walls, or in the direct airflow of the refrigerated system. Inside the container the temperatures are not homogeneous and different gradients can occur, warm air tends to stay in the upper part, and cold air stays at the bottom. From all the temperatures among the container, the most important to control is the product temperature itself (eg, the pulp of the pineapple or the leaf from the lettuce), because if the temperature is taken outside of the cartons or in the wall of the container, it cannot be possible to determine whether the product temperature is in the proper range or not. Therefore if the purpose is to avoid cold damage or excessive ripening, it is necessary for the best control over the product.

Is it mandatory to put a device in every box? It is not necessary, since it would dramatically increase the cost of the implementation. Determining the optimum number of sensors and their location is a topic not yet solved, and it is being investigated (Badia-Melis et al., 2013; Jedermann and Lang, 2009a; Jedermann et al., 2011) by using temperature estimation methods. It is a useful thing to know, since it will be known when a sensor is essential or can be substituted by an estimated point, and it is shown in this chapter in Section 2.5. A bigger number of nodes are providing not only more reliable data, but a rise in expenses, management, and maintenance.

Therefore the use of an estimation method can begin to consolidate, and so the use of ANN, which has been applied to forecasting for several authors (Zhang and Qi, 2008; Romeu et al., 2013; Liu et al., 2014; Yousefi et al., 2012). Badia-Melis et al. (2013), achieved an improvement of 95.3% in comparison with using only the ambient temperature as a reference, with errors of estimation always below 1°C.

It is known that while the average temperature throughout transportation might be similar for all cartons in a pallet, since the shelf life losses increase exponentially at higher temperatures, the remaining shelf life of products can be significantly different at the end of the transportation. Nascimento Nunes et al. (2014) showed, in their experiment, the cartons in the middle of the pallet possessed longer shelf lives compared to the cartons located near the outside. Therefore with the temperature estimation techniques, it is possible to assess temperature variations inside a pallet, hence they have a quantifiable impact on individual product qualities.

These techniques generate a large amount of data, which complicates the system and causes significant increments of daily data to manage. It is estimated that a regular RFID system generates between 10 and 100 times the information that a traditional 2-D barcode system does. The implementation of a management advanced system is necessary for the database that avoids collapses and guarantees an adequate speed for further consultation. Where some people see inconveniences, others see opportunities; for example, Walmart and other multinational companies work in data mining techniques that provide complementary information about performed operations (Ruiz Garcia and Lunadei, 2011).

In the case of fruit and vegetable logistics, the detection of volatile gases related with the ripening state of the freight involve a great challenge; the two most important gases are ethylene and ethanol. The development of specific sensors for those two gases is a problem which is being investigated, and an adequate solution has not been achieved. Certain sensors have been developed to detect both gases, but the detectors fail when it is necessary to differentiate them in the same air mass. This fact makes the use of these sensors with fruit impossible, since ethylene indicates the ripening in the fruit and the ethanol, on the other hand, indicates when the fruit is rotting. Hence, the usage of a nonspecific sensor could lead to serious mistakes in cargo management, such as discarding valid containers or accepting shipments that are rotting.

5. Conclusions and Future Trends

As it is shown in this chapter, there are numerous topics that need to be improved in the wireless sensing for real-time tracking and remote monitoring. That does not mean that these technologies are not valid. Many advances in the technologies have been done. There are new concepts that are created in order to complement the technological advancements, and the intelligent traceability will provide intelligence to the traceability chains. Combined with the IoT, it will be possible to train the systems, and those systems will learn and enhance the procedures. Temperature estimation can become key in these cases, saving money and reducing tool management operations. This forecasting makes the freight more predictable, and it is possible to build records of the chain and predict what is going to happen next.

Using real-time tracking and monitoring technologies, it is possible to reduce the risks of nondesired situations, the freight becomes a more controlled environment, and because the cargo is under surveillance, less people will be touching or opening up goods until the final destination.

Food processing, like the cold chain, is benefitting from the advancements that are really useful for the food industry. Usually the companies do not monitor the chain in all the stages or even in any of the stages; a simple sensor system would bring answers to the many problems that they face day-by-day and will reveal the unseen situations that need to be improved.

Although there are yet many limitations and challenges to face, the application of these technologies and advancements in refrigerated transport supervision is doable, under a technical and economical point of view, since their current state allows obtaining valuable information about transports that, until now, was not possible to achieve.

References

Amador, C., Emond, J.P., 2010. Development of RFID temperature tracking systems for combat feeding logistics. In: XVIIth World Congress of the International Commission of Agricultural and Biosystems Engineering (CIGR), CSBE/SCGAB. Quebec City, Canada, 2010.

Angeles, R., 2005. RFID technologies: supply-chain applications and implementation issues. Information Systems Management 22 (1), 51–65.

Attaran, M., 2007. RFID: an enabler of supply chain operations. Supply Chain Management: an International Journal 12 (4), 249–257.

Atzori, L., Iera, A., Morabito, G., 2010. The internet of things: a survey. The internet of things: a survey. Computer Networks 54, 2787–2805.

Badia-Melis, R., Ruiz-Garcia, L., Garcia-Hierro, J., Robla, J.I., 2015. Refrigerated fruit storage monitoring combining two different wireless sensing technologies: RFID and WSN. Sensors 15, 4781–4795.

Badia-Melis, R., Garcia-Hierro, J., Ruiz-Garcia, L., Jiménez-Ariza, T., Robla Villalba, J.I., Barreiro, P., 2014. Assessing the dynamic behavior of WSN motes and RFID semi-passive tags for temperature monitoring. Computers and Electronics in Agriculture 103, 11–16.

Badia Melis, R., Brecht, E., Lowe, A., Uysal, I., 2013. Pallet-wide temperature estimation and mapping for perishable food. In: American Society of Agricultural and Biological Engineers Annual Meeting, Kansas City, MO, 21–24 July 2013.

Barge, P., Gay, P., Merlino, V., Tortia, C., 2014. Item-level radio-frequency identification for the traceability of food products: application on a dairy product. Journal of Food Engineering 125 (0), 119–130.

Bosona, T., Gebresenbet, G., 2013. Food traceability as an integral part of logistics management in food and agricultural supply chain. Food Control 33.

Catarinucci, L., Cuiñas, I., Exposito, I., Colella, R., Fernández, J.A.G., Tarricone, L., 2011. RFID and WSNs for traceability of agricultural goods from Farm to Fork: electromagnetic and deployment aspects on wine test-cases. In: Paper Presented at the Software, Telecommunications and Computer Networks (SoftCOM), 2011.

Costa, C., Antonucci, F., Pallottino, F., Aguzzi, J., Sarriá, D., Menesatti, P., February 2013. A review on agri-food supply chain traceability by means of RFID technology. Food and Bioprocess Technology 6 (2), 353–366.

Dimitris, K., 2011. Closed-loop PLM for intelligent products in the era of the Internet of things. Computer-Aided Design 43, 479–501.

Feng, J., Fu, Z., Wang, Z., Xu, M., Zhang, X., 2013. Development and evaluation on a RFID-based traceability system for cattle/beef quality safety in China. Food Control 31 (2), 314–325.

Georg, H.H., Van Jos, H., Eelco, J., Klaas, S., 2013. Adoption of wireless sensors in supply chains: a process view analysis of a pharmaceutical cold chain. Journal of Theoretical and Applied Electronic Commerce Research 8, 138–154.

Giusto, D., Iera, A., Morabito, G., Atzori, L., 2010. The Internet of Things. Springer. ISBN: 978-1-4419-1673-0.

Gunders, D., 2012. Wasted: How America Is Losing up to 40 Percent of its Food from Farm to Fork to Landfill. Natural Resources Defense Council. Issue PAPER. August 2012 IP:12-06-B. http://www.nrdc.org/food/files/wasted-food-ip.pdf.

Hong, I., Dang, J., Tsai, Y., Liu, C., Lee, W., Wang, M., Chen, P., 2011. An RFID application in the food supply chain: a case study of convenience stores in Taiwan. Journal of Food Engineering 106 (2), 119–126.

Jedermann, R., Nicometo, M., Uysal, I., Opara, L., Janssen, S., Lang, W., 2015. Intelligent food logistics. Philosophical Transactions of the Royal Society A: Mathematical, Physical and Engineering Sciences 373.

Jedermann, R., Nicometo, M., Uysal, I., Lang, W., 2014a. Reducing food losses by intelligent food logistics. Philosophical Transactions of the Royal Society A: Mathematical, Physical and Engineering Sciences 372, 20130302.

Jedermann, R., Pötsch, T., Lloyd, C., 2014b. Communication techniques and challenges for wireless food quality monitoring. Philosophical Transactions of the Royal Society A: Mathematical, Physical and Engineering Sciences 372, 20130304.

Jedermann, R., Palafox-Albarrán, J., Barreiro, P., Robla, J.I., Ruiz-García, L., Lang, W., 2011. Interpolation of spatial temperature profiles by sensor networks. In: Proc. IEEE Sensors, Limerick, Ireland, 28–31 October 2011 778–781. http://dx.doi.org/10.1109/ICSENS.2011.6127148.

Jedermann, R., Lang, W., 2009a. The minimum number of sensors - interpolation of spatial temperature profiles in chilled transports. EWSN 2009. LNCS 5432, 232–246.

Jedermann, R., Ruiz-Garcia, L., Lang, W., 2009b. Spatial temperature profiling by semi-passive RFID loggers for perishable food transportation. Computers and Electronics in Agriculture 65, 145–154.

Jones, P., Clarke-Hill, C., Shears, P., Comfort, D., Hillier, D., 2004. Radio frequency identification in the UK: opportunities and challenges. International Journal of Retail & Distribution Management 32 (3), 164–171.

Kelepouris, T., Pramatari, K., Georgios Doukidis, G., 2007. RFID-enabled traceability in the food supply chain. Industrial Management & Data Systems 107 (2), 183–200.

Laniel, M., Uysal, I., Emond, J.P., 2010. Radio frequency interactions with air cargo container materials for real-time cold chain monitoring. In: XVIIthWorld Congress of the International Commission of Agricultural and Biosystems Engineering (CIGR). C.S.f.B. (CSBE/SCGAB), Québec City, Canada.

Leonardo, A., Fabiano, P., Eduardo, A., Jeronimo, C., Oliver, B., Thiago, F.O., 2011. eCloudRFID–A mobile software framework architecture for pervasive RFID-based applications. Journal of Network and Computer Applications 34, 972–979.

Liu, N., Babushkin, V., Afshari, A., October 2014. Short-term forecasting of temperature driven electricity load using time series and neural network model. Journal of Clean Energy Technologies 2 (4).

Lütjen, M., Dittmer, P., Veigt, M., 2012. New challenges for product and production engineering towards quality driven distribution of intelligent containers in cold chain logistics networks. In: 1st Joint International Symposium on System-integrated Intelligence 2012.

Mainetti, L., Patrono, L., Vergallo, R., 2012. IDA-Pay: a secure and efficient micropayment system based on Peer-to-Peer NFC technology for Android mobile devices. Journal of Communication Software and Systems 8 (4), 1–6.

Mainetti, L., Patrono, L., Stefanizzi, M.L., Vergallo, R., 2013a. An innovative and low-cost gapless traceability system of fresh vegetable products using RF technologies and EPCglobal standard. Computers and Electronics in Agriculture 98, 146–157.

Mainetti, L., Mele, F., Patrono, L., Simone, F., Stefanizzi, M.L., Vergallo, R., 2013b. An RFID-based tracing and tracking system for the fresh vegetables Supply chain. International Journal of Antennas and Propagation. http://dx.doi.org/10.1155/2013/531364 Article Number: 531364.

Mahajan, P.V., Caleb, O.J., Singh, Z., Watkins, C.B., Geyer, M., 2014. Postharvest treatments of fresh produce. Philosophical Transactions of the Royal Society A: Mathematical, Physical and Engineering Sciences. http://dx.doi.org/10.1098/rsta.2013.0309.

Nascimento Nunes, M.C., Nicometo, M., Emond, J.P., Badia-Melis, R., Uysal, I., 2014. Improvement in fresh fruit and vegetable logistics quality: Berry logistics field studies. Philosophical Transactions of the Royal Society A: Mathematical, Physical and Engineering Sciences 372, 20140212.

Ngai, E.W.T., Cheng, T.C.E., Au, S., Lai, K.H., 2007. Mobile commerce integrated with RFID technology in a container depot. Decision Support Systems 43 (1), 62–76.

Qian, J., Yang, X., Wu, X., Zhao, L., Fan, B., Xing, B., 2012. A traceability system incorporating 2D barcode and RFID technology for wheat flour mills. Computers and Electronics in Agriculture 89 (0), 76–85.

Qi, L., Zhang, J., Xu, M., Fu, Z., Chen, W., Zhang, X., 2011. Developing WSN-based traceability system for recirculation aquaculture. Mathematical and Computer Modelling 53 (11), 2162e2172.

Qi, L., Xu, M., Fu, Z., Mira, T., Zhang, X., 2014. C2SLDS: a WSN-based perishable food shelf-life prediction and LSFO. Food Control 38, 19e29.

Romeu, P., Zamora-Martínez, F., Botella-Rocamora, P., Pardo, J., 2013. Time-series forecasting of indoor temperature using pre-trained deep neural networks. In: Artificial Neural Networks and Machine Learning – ICANN 2013Lecture Notes in Computer Science, vol. 8131, pp. 451–458.

Ruiz-Garcia, L., Steinberger, G., Rothmund, M., 2010. A model and prototype implementation for tracking and tracing agricultural batch products along the food chain. Food Control 21, 112–121.

Ruiz-Garcia, L., Lunadei, L., 2011. The role of RFID in agriculture: applications, limitations and challenges. Computers and Electronics in Agriculture 79, 42–50.

Scheen, F.P., 2006. Optimizing supply chains using traceability systems. In: Smith, I., Furness, A. (Eds.), Improvising Traceability in Food Processing and Distribution. Woolhead publishing Ltd, Cambridge, UK, pp. 52–64.

Sugahara, K., 2009. Traceability system for agricultural products based on RFID and mobile technology. In: Zhao, C., Li, D. (Eds.), Computer and Computing Technologies in Agriculture II, vol. 3. Springer, Boston.

Theodorou, S., 2006. A Classification Approach for RFID-based Sensor Integration (MPhil thesis). University of Cambridge.

Trafton, A., 2014. Detecting Gases Wirelessly and Cheaply. MIT News Office. http://newsoffice.mit.edu/2014/wireless-chemical-sensor-for-smartphone-1208 (08.12.14.).

Twist, D.C., 2005. The impact of radio frequency identification on supply chain facilities. Journal of Facilities Management 3 (3), 226–239.

UN/FAO, 2013. Food Wastage Footprint Impacts on Natural Resources: Summary Report, 2013. Available online: http://www.fao.org/docrep/018/i3347e/i3347e.pdf (Last consultation 12.09.14.).

UNEP, 2013. Food Waste Facts. http://www.unep.org/wed/2013/quickfacts/ (Last consultation 09.04.15.).

Uysal, I., Emond, J.P., Bennett, G., 2011. Tag testing methodology for RFID enabled temperature tracking and shelf life estimation. In: IEEE International Conference on RFID Technology and Applications, 2011.

Van der Vorst, G.A.J., Van Kooten, O., Marcelis, W., Luning, P., Beulens, A.J.M., 2007. Quality controlled logistics in food supply chain networks: integrated decision-making on quality and logistics to meet advanced customer demands. In: Proceedings of the Euroma 14th international annual conference, 18e20 June, Ankara, Turkey, 2007.

Yousefi, A., Asadi, V., Nassiri, S.M., Niakousari, M., Aghdam, S.K., September 2012. Comparison of mathematical and neural network models in the estimation of Papaya fruit moisture content. Philippine Agricultural Scientist ISSN 0031-7454 95 (3), 246–251.

Yu-Yi, C., Yao-Jen, W., Jinn-Ke, J., 2014. A novel deployment of smart cold chain system using 2G-RFID-Sys. Journal of Food Engineering 141, 113–121.

Zhang, M., Li, P., 2012. RFID application strategy in agri-food supply chain based on safety and benefit analysis. Physics Procedia 25 (0), 636–642.

Zhang, P., Qi, M., May 2008. Trend Time–Series modeling and forecasting with neural networks. IEEE Transactions on Neural Networks 19 (5).

Zou, Z., Chen, Q., Uysal, I., Zheng, L., 2014. Radio frequency identification enabled wireless sensing for intelligent food logistics. Philosophical Transactions of the Royal Society A 372, 20130302 (published 05.05.14).

Legal Requirements for Food Traceability in the European Union

12

M. Borit
University of Tromsø (UiT) – The Arctic University of Norway, Tromsø, Norway

1. Introduction

Since the mid 90s, traceability has become a popular concept in industrial logistics (Opara, 2003), regardless of the production regime and type of product (Flapper et al., 2002; Jansen-Vullers et al., 2003). At the same time, traceability has been both a politically and strategically controversial issue and has acted as a major deterrent for multidisciplinary cooperation and understanding (Vorst, 2004). Transparency lies at the core of the decision to regulate traceability at the European Union (EU) level, as at the time that the first EU legal rule implementing traceability in the European food chains, it was hoped that this risk management tool would increase food products' safety through increasing the transparency of the food chain. As a consequence of the multiple food scandals in Europe during the last two decades of the 20th Century, such as the Bovine Spongiform Encephalopathy (BSE, or mad cow disease) and the dioxin contamination of chicken feed in Belgium, the European Commission (EC) drew the first rules requiring the implementation of traceability in food chains at the Union level. This chapter looks closer at how the EU body of law related to food traceability came into being, what the legal text entails, and what are the consequences of this text. It is important to perform such an analysis of the legal text because the legislature sets minimum compliance requirements that businesses have to follow.

Besides describing the EU food traceability legislation, one objective of this chapter is to assess whether the wording of the legal text is consistent with the declared objective of the law, which is to implement traceability throughout the entire product supply chain. The focus is on the words of the law articles and not on the general policy arguments, usually formulated in the preamble of the legal rule. If the legal text has weaknesses, traceability systems implemented as minimum requirements of this text can be ineffective, and supply chains may thus become vulnerable to fraud, as exemplified by the food scandals such as the beef-horsemeat one (NTB, 2013). In such cases, it is most likely that consumers might question their trust in a legislature that puts them at risk by using the notion of traceability inconsistently.

Advances in Food Traceability Techniques and Technologies. http://dx.doi.org/10.1016/B978-0-08-100310-7.00012-0

2. Background

2.1 The European Union Law

2.1.1 Difference Between Directives and Regulations

As explained by the European Commission itself (European Union, 2015), "there are three basic types of EU legislation: regulations, directives and decisions. A regulation is similar to a national law with the difference that it is applicable in all EU countries. Directives set out general rules to be transferred into national law by each country as they deem appropriate. A decision only deals with a particular issue and specifically mentioned persons or organizations."

2.1.2 Principles of Law

The principles of consistency and coherence have a prominent place in the EU law (Barnard et al., 2013). Consistency refers to the absence of contradiction, whereas coherence relates to "positive connections" or "the construction of a united whole" (Hillion, 2014). The former refers to ideas of compatibility and of making good sense, whereas the latter relates more to synergy and added value. "Hence, coherence in law would be a matter of degree, whereas consistency would be a static notion in the sense that concepts of law can be more or less coherent but cannot be more or less consistent. They are either consistent or not" (Hillion, 2014). At a general level, the Treaty on the Functioning of the European Union stipulates that the EU shall ensure consistency between its policies and activities (European Commission, 2012).

From a theoretical point of view, the meaning of consistency has been refined as horizontal consistency and vertical consistency. The latter applies to the relations between the Member States and the EU, while the former focuses on implementation and refers to consistency at the EU level, either interinstitutional or interpolicy (Minkova, 2011). In terms of this second aspect, consistency can be defined as "the way in which the substance of different policies generated by the EU forms part of whole" (Portela and Raube, 2008).

2.2 The Concept of Traceability

2.2.1 What Is Traceability?

Answering this question seems to be a difficult task, and the food industry is the one where the most varied and contradictory definitions of traceability can be found. This situation is true, although traceability has general common requirements and drivers across industries (Jansen-Vullers et al., 2003). To ascertain which definitions are most commonly used, Olsen and Borit (2013) performed a comprehensive systematic literature review of scientific articles relating to food traceability, giving an overview of relevant traceability definitions and outlining similarities, differences, and the consequences of choosing one definition over another.

This study reveals that one-third of the reviewed scientific articles did not mention a traceability definition, possibly meaning that their authors took the definition of traceability for granted. Out of the articles referring to a definition, two-thirds used a single definition. The remaining articles referred to at least two definitions, possibly meaning that maybe a traceability definition should not be taken for granted. The most common definition used was the one provided by the EU Regulation 178/2002 (European Commisssion, 2002), which was followed, in order, by the definitions given by the international standards ISO 8402 (ISO, 1994) and ISO 9000/ISO 22005 (ISO, 2005). Moving further with their analysis, the authors of this review show that all the definitions commonly referred to in these articles have weaknesses, including the mostly used definition. However, the definition provided by ISO 8402 is the one incorporating all the critical properties of a traceability system as described in the scientific literature. Nevertheless, considering that this standard has been superseded, the authors conclude that an ideal definition does not currently exist. By combining the best parts of the existing definitions, Olsen and Borit (2013) offer a new possible definition of traceability that can potentially be applied to the traceability of any products, not only food related: traceability (*n.*) is the ability to access any or all information relating to that which is under consideration, throughout its entire life cycle, by means of recorded identifications.

Moving beyond the definition of traceability, a review of the specialty literature indicates that the theoretical disagreement continues with the breadth of traceability (ie, how much information should the traceability system record), its depth and granularity (ie, how small should the unit identified by the system be), its coverage (ie, the entire supply chain or only specific parts of it), and the regulatory body who should be responsible for implementing traceability (ie, the law-making bodies or the industry).

2.2.2 Traceability and Data Validation and Verification

Contrary to popular belief, traceability is not a method to ensure that information about a certain product is true. Borit and Santos (2015) argue that traceability is an infrastructure that can be used by control agencies for retrieving different data for various reasons (such as quality problems) or for verifying these data with their specific means. As previously mentioned in the specialty literature, "a traceability system is quite similar to a filing cabinet in that they both deal with systematic storing and retrieving of data. Importantly, neither a traceability system nor a filing cabinet care about what types of data are being stored" (Olsen and Borit, 2013). Thus error and fraud can make that the recordings are neither true nor complete. Therefore the need to verify these claims is obvious, and here, alongside thorough inspections and certification, analytical methods play an important role (Kelly et al., 2011).

2.2.3 Traceability and Certification

When analyzing food products' traceability, it is important to notice that documenting traceability and documenting an eco-label type chain of custody are two different concepts. While being two different processes, traceability can be used as a tool in the certification process. Documentation of chain of custody is part of what is required for certification, especially certification related to the use of eco-labels. As explained in the specialty literature, there are crucial differences between the two processes

(Borit and Olsen, 2012, 2016). For example, while traceability allows random splitting and joining of products as long as one documents the fact that the traceable units have been split up or joined together, certification has stricter rules for the chain of custody requirements, and traceable units from different suppliers should not be mixed.

2.2.4 Traceability and Transparency

Transparency is a critical element in risk communication because it is directly linked to trust building among stakeholders (Hofstede, 2004a; Renn, 2008). Transparency of a supply chain is the degree of shared understanding of and access to product-related information as requested by supply chain's stakeholders without loss, noise, delay, or distortion (Hofstede, 2004a). Transparency does not equate to traceability because the latter only sets the framework for the former (Egels-Zandén et al., 2014). Depending on whether transparency is aimed at the past, present, or future, it can be divided into three types: history-, operations-, and strategy transparency (Hofstede, 2004a). When it comes to products, it can be said that traceability is a means of acquiring the first two types of transparency, since it addresses the past and the present of the product.

3. The EU Food Legislation

Until the outbreak of the BSE crisis (late 1980s), many of the Community's rules relating to food safety regulation were mainly created on an ad hoc basis and/or developed in the jurisprudence of the European Court of Justice (Vos, 2000). However, this crisis was a major catalyst for change in the EU regulatory regime for food. After one green paper on food, one white paper on the same matter, and the dioxin contamination scandal, it was decided to establish general principles of food law at the EU level, to establish common definitions and to lay down overarching guiding principles and legitimate objectives for food law in order to ensure a high level of health protection (O'Rourke, 2005). All these were included in the EC Regulation 178/2002, laying down the general principles and requirements of food law, establishing the European Food Safety Authority, and laying down procedures in matters of food safety, usually referred to as the General Food Law (GFL). As a regulation, this act was directly applicable in the legislation of the Member States, without transposition. Separation of legislation from scientific consultation and from inspection, thus leading to greater transparency, are the general principles of this more coherent approach to food safety (Martinez and Epelbaum, 2011; O'Rourke, 2005). The GFL is the backbone of the new European food law, and it is backed up by numerous other regulations and directives dealing with a multitude of food-related topics, ranging from materials and articles intended to come into contact with food to food-and-feed products produced from genetically modified organisms (GMOs). For the untrained eye, the entire EU body of food law, at first sight, might be perceived as an endless amount of provisions of a very technical and detailed nature (Meulen, 2014).

4. Traceability in the EU Food Legislation

4.1 The General Food Law (GFL)

Regulatory institutions in the EU attempted to restore consumer confidence in policy making and industry practices by introducing food and food ingredient traceability systems through the provisions of the GFL (Martinez and Epelbaum, 2011). Recitals 28 and 29 of the GFL refer to the main driver of implementing traceability at the EU level, ie, enabling the identification of food safety problems at the source and across the food chain and the removal of unsafe food from the market:

> *(28) Experience has shown that the functioning of the internal market in food or feed can be jeopardised where it is impossible to trace food and feed. It is therefore necessary to establish a comprehensive system of traceability within food and feed businesses so that targeted and accurate withdrawals can be undertaken or information given to consumers or control officials, thereby avoiding the potential for unnecessary wider disruption in the event of food safety problems.*
>
> *(29) It is necessary to ensure that a food or feed business including an importer can identify at least the business from which the food, feed, animal or substance that may be incorporated into a food or feed has been supplied, to ensure that on investigation, traceability can be assured at all stages.*

The specific GFL provisions broaden the scope of traceability to all foods and food ingredients. Article 3 Paragraph 15 defines traceability as follows:

> *"Traceability" means the ability to trace and follow a food, feed, food-producing animal or substance intended to be, or expected to be incorporated into a food or feed, through all stages of production, processing and distribution.*

It is worth noting that food crises in the past have shown that tracing the commercial flow of a product by keeping invoices was not sufficient to follow the physical flow of that product. Therefore it is considered essential that the traceability system of each food business operator is designed to follow the physical flow of the products (EC, 2010). Article 18 provides further details on how traceability can be achieved.

> *Traceability*
>
> 1. *The traceability of food, feed, food-producing animals, and any other substance intended to be, or expected to be, incorporated into a food or feed shall be established at all stages of production, processing, and distribution.*
> 2. *Food and feed business operators shall be able to identify any person from whom they have been supplied with a food, a feed, a food-producing animal, or any substance intended to be, or expected to be, incorporated into a food or feed.*

To this end, such operators shall have in place systems and procedures which allow for this information to be made available to the competent authorities on demand.

3. *Food and feed business operators shall have in place systems and procedures to identify the other businesses to which their products have been supplied. This information shall be made available to the competent authorities on demand.*
4. *Food or feed which is placed on the market or is likely to be placed on the market in the Community shall be adequately labeled or identified to facilitate its traceability, through relevant documentation or information in accordance with the relevant requirements of more specific provisions.*
5. *Provisions for the purpose of applying the requirements of this Article in respect of specific sectors may be adopted in accordance with the procedure laid down in Article 58(2).*

This article refers to "any substance intended to be, or expected to be, incorporated into a food or feed" as a part of a food or feed during its manufacture, preparation, or treatment. Furthermore, the article applies to food business operators at all stages of the food/feed chain, from primary production (food-producing animals, harvests) and food/feed processing to distribution and supply, including brokers, regardless of whether they take physical possession of the food/feed in question. However, the traceability provisions of the Regulation do not apply outside the EU (EC, 2010).

According to the text of Article 18, the legal requirements rely on the "one-step-back" and "one-step-forward" approach to traceability. This approach implies that food business operators have to establish two types of links: a link "supplier-product" (which products supplied from which suppliers) and a link "customer-product" (which products supplied to which customers). Nevertheless, operators do not have to identify the immediate customers when they are the final consumers. The Regulation does not expressly constrain operators to establish a link "incoming and outgoing products" (so-called "internal traceability"). Thus there is no requirement for records to be kept identifying how product batches are split and combined within a business to create particular products or new batches (EC, 2010). In other words, businesses have to know where the ingredients came from and where the products went, but not necessarily which ingredients went into which products (Meulen and Velde, 2008). This specific approach is considered to be not effective by the specialty literature, and it was proven so by the beef-horsemeat scandal in Europe, when agricultural experts declared that "the lack of transparency and traceability is flagrant in regard to poultry, pork, and horsemeat" (Priscille, 2013). The Regulation (EU) No 931/2011 of September 19, 2011 on the traceability requirements set by GFL (Europan Commission, 2011) tries to improve this situation by adding several details to the list of information that has to be kept by food operators. However, following the methodological approach for evaluation of regulatory requirements related to data recording and traceability(Borit and Olsen, 2012; Borit and Santos, 2015), this present study considers that even with this new addition, the GFL approach to traceability is still not effective.

Arguing that Article 18 of GFL is worded in terms of its goal and intended result, rather than in terms of prescribing how that result is to be achieved, the European Commission is nevertheless acknowledging that "an internal traceability system would contribute to more targeted and accurate withdrawals. Food business operators are likely to save costs in terms of time of a withdrawal and in avoiding unnecessary wider disruption. This in turn would help maintain consumer confidence. [...] It would also be very helpful if details are kept of any reference or batch number enabling the product to be identified" (EC, 2010). The legislature considers that "this more general approach allows industry greater flexibility in the implementation of the requirement and is thus likely to reduce compliance costs. However, it requires both food businesses and the control authorities to take an active role in ensuring effective implementation" (EC, 2010). In this case, the question that arises is whether the businesses and the control authorities are indeed taking this active role.

In conclusion, the main aim of implementing traceability though GFL is facilitating rapid identification of the source of a food safety issue in order to conduct aimed recalls that take faulty products out of the market. However, in practice, traceability systems are used for other purposes as well. As indicated in Section 2.2.2, these systems can be used as an infrastructure enabling retrieval of data for various purposes, such as assessing certain invisible qualities related to their origin or the way the products have been handled (eg, organic). The traceability requirements of GFL are general requirements and are therefore always applicable (EC, 2010).

4.2 Other Legislation

Several other EU rules, among which are the ones regulating the trade in fish and fish products, or sprouts and seeds intended for the production of sprouts, impose traceability systems similar to the GFL one. However, for some specific products that proved to pose safety problems in the past, such as beef, eggs, and items produced from GMOs, more specific requirements apply, which go beyond the "one-step-back" and"one-step-forward" approach and implement, at least from a legislative point of view, an effective traceability system (Borit and Santos, 2015). The same holds true for the new rules for identification of Equidae (Regulation 2015/262 (European Commission, 2015)). This new rule was drawn in the aftermath of the beef-horsemeat scandal, when it was observed that the mislabeled beef was traced back to its original supplier due to the effective traceability implemented in the beef sector, but because of the noneffective traceability of the horsemeat sector, the origin of this meat could not be identified (Priscille, 2013). In the case of products that make very specific claims, such as organic products, the legislative approach is somewhere in between the GFL one and this very strict one (Borit and Santos, 2015).

Besides food safety, these rules might have different drivers for the implementation of traceability in the specific food chains. One example is rules having environmental sustainability as the main driver, such as Regulation 1005/2008 on deterring and eliminating illegal, unreported, and unregulated fishing (European Commission, 2008). Another example is rules that reassure that traceability is also meant to facilitate attribution of responsibility (eg, liability), such as Regulation 1935/2004 on materials and articles intended to come in contact with food (European Commission, 2004). Whether this possibility is considered an advantage or a disadvantage depends on the position of the actor in the food chain. As noted in the literature, for business at the end of the food chain, compared with those further upstream, it seems advantageous for liability to be passed to the actors where the problem originated (Meulen and Velde, 2008).

5. Consistency of the EU Body of Law Related to Traceability and Future Trends

As it is clear from the analysis performed in Section 4, the EU food law related to traceability lacks interpolicy horizontal consistency. This conclusion is in line with the results of a systematic review of legislation that analyzed the EU body of law related to traceability in the supply chain of 16 groups of products (including nonfood ones) (Borit and Santos, 2015).

Following the same line of thought, according to Lavelli (2013), the EU legislature works with two different models when it comes to food traceability. The one implemented through the application of the GFL leads to a generic (nonspecific) low-warranty traceability of the food supply chain. A second, more complex model is followed in rules regulating products such as those derived from GMO. The application of this second model leads to a specific, high-warranty traceability system for any product unit in the food supply chain. Naming both these opposite models as traceability can have negative consequences, as both the legislature and the consumers could develop a false sense of security and thus hinder further action in the form of research and effective new legislation in the field (Borit and Santos, 2015). In addition, the public trust can be shaken if the authorities cannot rely on the current rules in order to perform quick and accurate recalls of faulty products, as in the beef-horsemeat scandal.

It is interesting to analyze if there is a trend in how the EU legislature approaches traceability, in order to understand the rationale behind the inconsistent choice between the two models, and to be able to forecast how future legislation will tackle the matter. A recent study concluded that these approaches are strongly correlated with the original driver for risk management (eg, human safety or concern for environmental sustainability), moderately correlated with the type of goods involved (eg, low-technology or high-technology products), and uncorrelated with their trade value (Borit and Santos, 2015). These results are validated by the preference of the legislature for the high-warranty effective model when implementing the Tobacco Products Directive 2014/40/EU. Safeguarding human safety was the focus when formulating this rule, as annually, around 700,000 people die in the EU due to illness related to tobacco usage (Europan Commission, 2014). Thus it is expected that food-related legislation implementing traceability because of environmental issues would fall short of effectiveness.

6. Conclusions

In the aftermath of several food scandals, the EU decided to implement new legal rules related to food supply chains, and traceability was chosen as a management tool to safeguard human safety and other claims such as environmental sustainability and product quality. However, the body of law dealing with this matter is considered to be inconsistent, as the approach to traceability implementation varies from product to product, depending on the answers to the questions "why traceability?" and "what kind of product is it?". On a more profound level, the multiple perceptions about the meaning and applicability of traceability held by different people and different legislatures may be related to idiosyncratic cultural backgrounds and therefore to their basic notions of trust and transparency requirements (Hofstede, 2004b). However, regardless these aspects, a traceability system is considered good when it delivers accurate information in a fast manner (EC, 2010). This study asks whether the current food legislation uses the best available models in order to set the implementation basis of good traceability systems, regardless of the food product and traceability drivers.

Acknowledgments

This research was facilitated by the EWMA project (Norwegian Research Council project number 195160).

References

Barnard, C., Albors-Llorens, A., Gehring, M., Schütze, R. (Eds.), 2013. Cambridge Yearbook of European Legal Studies, vol. 15 2012–2013. Bloomsbury Publishing.

Borit, M., Olsen, P., 2012. Evaluation framework for regulatory requirements related to data recording and traceability designed to prevent illegal, unreported and unregulated fishing. Marine Policy 36 (1), 96–102. Retrieved from: http://www.sciencedirect.com/science/article/pii/S0308597X11000716.

Borit, M., Olsen, P., 2016. Analysis of gaps and inconsistencies in the seafood traceability standards and norms, FAO COFI:FT/XV/2016/Inf.8.

Borit, M., Santos, J., 2015. Getting traceability right, from fish to advanced bio-technological products: a review of legislation. Journal of Cleaner Production. 104, 13–22. http://doi.org/10.1016/j.jclepro.2015.05.003.

EC, 2010. Guidance on the Implementation of Articles 11, 12, 14, 17, 18, 19 and 20 of Regulation (EC) N° 178/2002 on General Food Law. Conclusions of the Standing Committee on the Food Chain and Animal Health.

Egels-Zandén, N., Hulthén, K., Wulff, G., 2014. Trade-offs in supply chain transparency: the case of Nudie Jeans Co. Journal of Cleaner Production. http://doi.org/10.1016/j.jclepro.2014.04.074.

Europan Commission, 2011. Commission Implementing Regulation (EU) No 931/2011 of 19 September 2011 on the Traceability Requirements Set by Regulation (EC) No 178/2002 of the European Parliament and of the Council for Food of Animal Origin Text with EEA Relevance. Retrieved from: http://eur-lex.europa.eu/legal-content/EN/ALL/?uri=CELEX:32011R0931.

European Commission, 2014. Tobacco Policy. Retrieved from: http://ec.europa.eu/health/tobacco/policy/index_en.htm.

European Commission, 2004. Regulation (EC) No 1935/2004 of the European Parliament and of the Council of 27 October 2004 on Materials and Articles Intended to Come into Contact with Food and Repealing Directives 80/590/EEC and 89/109/EEC. Retrieved from: http://eur-lex.europa.eu/legal-content/EN/ALL/?uri=CELEX:32004R1935.

European Commission, 2008. Council Regulation (EC) No 1005/2008 of 29 September 2008 Establishing a Community System to Prevent, Deter and Eliminate Illegal, Unreported and Unregulated Fishing, Amending Regulations (EEC) No 2847/93, (EC) No 1936/2001 and (EC) No 601/2004 and Repeal. Retrieved from: http://eur-lex.europa.eu/legal-content/EN/TXT/?qid=1408984470270&uri=CELEX:02008R1005-20110309.

European Commission, 2012. Consolidated Version of the Treaty on the Functioning of the European Union. EC.

European Commission, 2015. Commission Implementing Regulation (EU) 2015/262 of 17 February 2015 Laying Down Rules Pursuant to Council Directives 90/427/EEC and 2009/156/EC as Regards the Methods for the Identification of Equidae (Equine Passport Regulation). Retrieved May 20, 2008, from Commission Implementing Regulation (EU) 2015/262 of 17 February 2015 Laying Down Rules Pursuant to Council Directives 90/427/EEC and 2009/156/EC as Regards the Methods for the Identification of Equidae (Equine Passport Regulation).

European Commission, 2002. Regulation (EC) No 178/2002 of the European Parliament and of the Council of 28 January 2002 Laying Down the General Principles and Requirements of Food Law, Establishing the European Food Safety Authority and Laying Down Procedures in Matters of Food Safe. Retrieved from: http://eur-lex.europa.eu/legal-content/EN/TXT/?uri=celex:32002R0178.

European Union, 2015. Regulations, Directives and Other Acts. Retrieved from: http://europa.eu/eu-law/decision-making/legal-acts/index_en.htm.

Flapper, S.D.P., Fransoo, J.C., Broekmeulen, R.A.C.M., Inderfurth, K., 2002. Planning and control of rework in the process industries: a review. Production Planning & Control 13 (1), 26–34. http://doi.org/10.1080/09537280110061548.

Hillion, C., 2014. Tous pour un, un pour tous! Coherence in the external relations of the European Union. SSRN Electronic Journal. http://doi.org/10.2139/ssrn.2405364.

Hofstede, G.J., 2004a. Hide or Confide? The Dilemma of Transparency. Emerging World of Chains and Networks. 's-Gravenhage: Reed Business Information. Retrieved from: http://catalog.hathitrust.org/Record/007261712.

Hofstede, G.J., 2004b. The dynamics of trust and transparency. In: Hofstede, G.J., Schepers, H., Spaans-Dijkstra, L., Trienekens, J., Beulens, A. (Eds.), Hide or Confide? The Dilemma of Transparency. 's Gravenhage: Reed Business Information, pp. 201–224.

ISO, 1994. ISO 8402:1994 Quality Management and Quality Assurance. Retrieved from: http://www.iso.org/iso/catalogue_detail.htm?csnumber=20115.

ISO, 2005. IISO 9000/ISO 22005 Food Safety Management Systems – Requirements for Any Organization in the Food Chain. Retrieved from: http://www.iso.org/iso/home/store/catalogue_tc/catalogue_detail.htm?csnumber=35466.

Jansen-Vullers, M., van Dorp, C., Beulens, A.J., 2003. Managing traceability information in manufacture. International Journal of Information Management 23 (5), 395–413. Retrieved from: http://www.sciencedirect.com/science/article/pii/S0268401203000665.

Kelly, S., Brereton, P., Guillou, C., Broll, H., Laube, I., Downey, G., et al., 2011. New approaches to determining the origin of food. In: Food Chain Integrity. A Holistic Approach to Food Traceability, Safety, Quality and Authenticity.

Lavelli, V., 2013. High-warranty traceability system in the poultry meat supply chain: a medium-sized enterprise case study. Food Control 33 (1), 148–156. http://doi.org/10.1016/j.foodcont.2013.02.022.

Martinez, M.G., Epelbaum, F.M.B., 2011. The role of traceability in restoring consumer trust in food chains. In: Food Chain Integrity. A Holistic Approach to Food Traceability, Safety, Quality and Authenticity. Woodhead Publishing Limited, Cambridge.

Minkova, V., 2011. Horizontal Consistency in the Areas of the European Union S External Action. Diplom.de.
NTB, 2013. Mattilsynet: Problemet er usporbart kjøtt. Retrieved from: http://www.aftenposten.no/nyheter/iriks/Mattilsynet-Problemet-er-usporbart-kjott-7122093.html#.Uc337tgUmJm.
O'Rourke, R., 2005. European Food Law. Thomson Sweet & Maxwell, London.
Olsen, P., Borit, M., 2013. How to define traceability. Trends in Food Science & Technology. Retrieved from: http://www.sciencedirect.com/science/article/pii/S0924224412002117.
Opara, L.U., 2003. Traceability in agriculture and food supply chain: a review of basic concepts, technological implications, and future prospects. Journal of Food, Agriculture and Environment 1, 101–106.
Portela, C., Raube, K., 2008. Six authors in search of a notion: (In) coherence in EU foreign policy and its causes. Hamburg Review of Social Sciences 3 (1), 1–10.
Priscille, L., 2013. Horsemeat Scandal Highlights Lack of Traceability. Retrieved from: http://www.france24.com/en/20130212-france-findus-beef-horse-meat-lasagna/.
Renn, O., 2008. Risk Governance: Coping with Uncertainty in a Complex World. Earthscan.
Van Der Meulen, B., Van Der Velde, M., 2008. European Food Law Handbook. Wageningen Academic Publishers.
Van Der Meulen, B. (Ed.), 2014. EU Food Law Handbook. Wageningen Academic Publishers.
Vorst, J.V.D., 2004. Traceability: state of the art in seven countries. In: Hofstede, G.J., Schepers, H., Spaans-Dijkstra, L., Trienekens, J., Beulens, A. (Eds.), Hide or Confide? The Dilemma of Transparency. 's Gravenhage: Reed Business Information, pp. 73–80.
Vos, E., 2000. EU food safety regulation in the aftermath of the BSE crisis. Journal of Consumer Policy 23 (3), 227–255. http://doi.org/10.1023/A:1007123502914.

Legal Requirements and Regulation for Food Traceability in the United States

13

L. Zach
Food System Solutions, Mt Horeb, Wisconsin, WI, United States; Food System Solutions, Christchurch, New Zealand

1. Introduction

As this global market grows, food supply chains are becoming longer and more complex. As a result, the tools of traceability become necessary in order to offer accountability to both consumers and food manufacturers. Such accountability is helpful should the foods result in illness, allergic reactions, or fraud. Traceability offers authorities the ability to quickly find the source and to stop further disease spread, toxic reaction, or fraud. Traceability also offers a respite to those industries or producers who can use their traceability systems to show that a harmful product is not theirs.

Each year about 48 million people in the United States (one in six) get sick from foodborne diseases, 128,000 are hospitalized, and 3000 die, according to 2011 data from the Centers for Disease Control and Prevention (2011). This is a significant public health burden that is largely preventable. The economic burden from such illnesses was estimated by Hoffman et al. (2015) for the USDA Economic Research Service to be over $15.5 billion. Prompt responses from government entities and the food industry can play a vital role in stopping the spread of illnesses and deaths.

This chapter seeks to summarize the current legislation in force in the United States, as well as the ongoing development of legal requirements, guidance, and rule-making at the US Food and Drug Administration (FDA) to meet the legislative requirements in the Food Safety Modernization Act of 2011 (FSMA). The chapter also seeks to give the reader some perspective on future developments in this area by giving an overview of the standards under development by the US food science and technology professional body, the Institute of Food Technologists (IFT). IFT has collected information and input from a variety of experts, including industry and consumer advocates, and as such has completed a pilot project on traceability for the US FDA. It has published standards for traceability for regulators and many sectors of the food industry. It is currently working to develop further standards in a global context.

1.1 Definitions

English terms sometimes have different meanings in the United States compared to England. Therefore a few definitions are in order:

Advances in Food Traceability Techniques and Technologies. http://dx.doi.org/10.1016/B978-0-08-100310-7.00013-2

"Animal food" applies to food for both livestock and companion animals.

"Business size": The regulations for Federal Legislation apply differently to the various sizes of business and types of business. For example, a farm is differentiated from a covered processing facility near a field. Also note that exact definitions for "small" and "very small" businesses may vary depending on the particular rule in question. In general, the following guidelines apply:

Very Small Businesses: defined as having less than $1 million in total annual sales, adjusted for inflation;
Small businesses: defined as employing fewer than 500 persons; and
Other Businesses: defined in the negative, as not small or very small.

"Center for Disease Control and Prevention" (CDC): The CDC is the national public health institute of the United States. It is a Federal Agency under the Department of Health and Human Services.

"*Escherichia coli* O157:H7," also called Hemolytic *E. Coli*, may cause hemolytic uremic syndrome, especially in children under five years of age and the elderly, often resulting in death.

"Food" is defined in Section 201(f) of the Federal Food, Drug, and Cosmetic Act as "(1) articles used for food or drink for man or other animals, (2) chewing gum, and (3) articles used for components of any such article."

"Food defense" refers to the protection of food from intentional contamination.

"Food protection" is a more general term referring to all aspects of making food safe from contamination, which might be intentional, accidental, chemical, or microbial.

"FSMA" refers to the Food Safety Modernization Act of 2011. The US FDA is currently working, often in consultation with the USDA and State regulators, toward developing the regulations to allow implementation of this Act.

"Food security" refers to populations having enough nutritious food to eat. It does not address fraud, food safety, or food protection.

"High-risk food" (as used in Section 204 of the US FSMA of 2011) applies to food with a high-potential risk for microbial or chemical contamination or food which would support growth of pathogenic microorganisms due to the nature of the food or processes used to produce the food. Such a food is judged to be high-risk because it has a historical public health record with respect to outbreaks, causes of foodborne disease, and other factors. The US FDA is developing the exact guidelines to designate food as "high-risk" as part of Section 204 of the FSMA. Spices, eggs, and sprouts are examples of foods which have a propensity to be contaminated with the pathogen *Salmonella* and would therefore be considered "high-risk." The FSMA focuses regulation on tracing systems for high-risk foods (US FDA, 2014b).

"IFT": Institute of Food Technologists. This is a nonprofit scientific society of professionals engaged in food science, food technology, and related professions. IFT has collected information and input from a variety of experts, including industry and consumer advocates, on the subject of traceability.

"Law" commonly refers to the entire body of law on a subject but is also a synonym for statute. A statute is a law enacted by a legislative body (Fortin, 2009).

"Rule," also called a regulation, is a law promulgated by an administrative agency and involves the details necessary for enforcement of the law (Fortin, 2009).

"US FDA": Food and Drug Administration. An agency under the US Department of Health and Human Services. The US FDA has authority over all food, as defined earlier, except for those foods for which the USDA has jurisdiction.

"USDA": United States Department of Agriculture. The agency that does the regulation and inspection for food safety in the USDA is the Food Safety and Inspection Service (FSIS). In general, this agency is responsible for food safety of meat, poultry, and processed eggs. However whether the USDA or the FDA has jurisdiction over these foods is complicated and can be uncertain (Fortin, 2009).

US FDA versus USDA jurisdiction over foods:

The US FDA has authority over all food, as defined earlier, except those foods which the US Department of Agriculture (USDA) has jurisdiction for. The category of "processed eggs" include liquid or powdered eggs that is, all eggs except those in-the-shell sold at retail. For meat-containing food products, the percentage of meat determines which agency has jurisdiction over the food. If a food product contains 3% or less raw meat, it falls under US FDA jurisdiction, but there are intricacies based on internal decisions which should be checked (Fortin, 2009). The US FDA has jurisdiction over live meat animals intended for food, particularly animal food and drugs. However, the Food Safety and Inspection Service is the regulatory and inspection arm of the USDA and, as such, has exclusive jurisdiction over the slaughter and processing of meat animals. Further, while the US FDA and USDA have joint jurisdiction over transport of meat products after processing, only the US FDA has direct authority over retail establishments. Please note that the new FSMA legislation for tracing, preventive controls, and so on, is written to include the USDA in a cooperative or consultative manner with specifics to be addressed in published guidance. However, the Bioterrorism Act, requiring "one-step-forward and one-step-back" traceability, applies to all food (Fortin, 2009).

1.2 Background

Numerous well-published foodborne disease outbreaks in the 1990s and 2000s pushed the need for legislative action for improved food safety regulations (GAO, 2001, 2007). The September 2001 terrorist incident, involving the destruction of New York's World Trade Center and the loss of many lives, initiated an examination of security in many areas, from the border, to disease spread, to food supply. Therefore the need for some degree of traceback in the food supply crystallized into law in the *US BioTerrorism Preparedness and Response Act of 2002*.

In spite of this, foodborne disease outbreaks involving hundreds of people in states located diversely around the country continued to occur in the United States. The CDC improved identification of disease outbreaks and tracing of the source of these illnesses with tools of "DNA" fingerprinting and the PulseNet database (CDC, 2014). These tools allowed public health authorities to investigate a foodborne illness when two or more people were found to be ill from an organism with the same genetic fingerprint. PulseNet enabled comparison of the "DNA fingerprints" of bacteria from ill patients to those in its database, allowing identification of diverse clusters of disease which would otherwise remain unrecognized and unlinked outbreaks. The use of PulseNet allowed public health authorities to attribute many illnesses in diverse locations to the same organism (CDC, 2014).

Public health authorities were able to close the loop when they used the genetic fingerprint to link the organism from the sick person with the organism isolated from their food (if it was still available), or from the premises of the manufacturer or producer. In this way, the public could see that one organism in one food or food ingredient, distributed widely across the country might be responsible for several deaths and numerous hospitalizations. Therefore it became obvious that more attention needed to be focused on prevention, food safety, and improved traceability at all steps in the food supply chain.

While improved illness tracing was helpful, tracking the product through the food supply chain needed to be improved in order to speed efforts to find the source and remove contaminated products from shelves. Furthermore, recalls of food products that were mistakenly identified or mistakenly traced triggered wrongful economic losses for the food industry. Three major foodborne disease outbreaks of a national scale illustrated the need for improved traceability in the United States.

i. The Peanut Corporation of America (PCA) *Salmonella enterica* serovar typhimurium outbreak of 2009:
The Peanut Corporation of America (PCA) *Salmonella enterica* serovar typhimurium outbreak of 2009 caused at least 714 people in 47 states, half of them children, to be sickened. Nine deaths were thought to be attributed to bacteria from PCA peanut products (CDC, 2009). This infection triggered the most extensive food recall in US history. This was because PCA sold peanut products as ingredients to other food manufacturers. Its products included peanut butter, peanut paste, peanut meal, and whole and chopped peanuts used in a variety of baked goods, snacks, and food service meals. As of April 2009, the recall involved at least 361 companies and 3913 different products manufactured using PCA ingredients (Goetz, 2013). The losses resulted in the shutdown and bankruptcy of the company in 2009 with criminal charges filed against the company executives.
Although the recall did not involve major brand peanut butters, many consumers reacted by avoiding peanut products altogether, driving down the sales of all brands of peanut butter by nearly 25%. This caused harm to the peanut industry and farmers, already suffering from low prices due to the 2008 bumper crop and the deepening economic crisis. Estimated losses to the US peanut industry because of this outbreak were on the order of $1 billion (Doering, 2009). This is a noteworthy recall because of the sheer number of food brands and food service outlets (including nursing homes) affected. The complexity and interconnectedness of the supply chain defined it as more of a supply "web," and this resulted in a lengthy and difficult recall.

ii. *Salmonella enterica* serovar Saintpaul in raw "tomato" outbreak of 2008:
This large foodborne disease outbreak started slowly in April of 2008 but, over several months, resulted in 1442 persons infected in 43 states, the District of Columbia, and Canada, with the relatively rare strain of a genetically identical *Salmonella enterica* serovar Saintpaul. At least 286 persons were hospitalized, and the infection might have contributed to two deaths (CDC, 2008). The source of the microbial contaminant was not found for several months, although government investigators identified a statistically significant association between consumption of certain types of tomatoes and illness. The US FDA legally but mistakenly warned people in a press briefing that tomatoes from Mexico or Florida were likely responsible for the multistate outbreak (Flynn, 2013). However, the lack of traceback records in the supply chain of the initially implicated raw tomatoes complicated the investigation, lengthening the period of uncertainty. The FDA's investigation

subsequently confirmed the pathogen on samples of capsicum and in irrigation water from a farm in Mexico. A recall of tomatoes was never associated with this outbreak; however, FDA warnings about tomatoes had led consumers to stop purchasing them, causing an estimated $145 million loss to tomato growers and shippers (Flynn, 2013).

iii. *Escherichia coli* O157:H7 outbreak in spinach in 2006:
In 2006, 205 confirmed illnesses and three deaths in 26 states resulted from infections with the outbreak strain of *E. coli* O157:H7. Among the ill persons, 102 were hospitalized and 31 developed a type of kidney failure called Hemolytic-Uremic Syndrome (HUS). Of the victims, 22 were children under five years old, a highly vulnerable age group for this disease (CDC, 2006). The FDA and the California Department of Public Health reported that the outbreak was linked to raw spinach produced in California's Salinas Valley. Perhaps due to the serious health problems and bad press, consumers stopped buying all fresh leafy greens, including spinach. The market did not recover for several years, and as such, the outbreak caused an estimated $100 million loss to the spinach industry (US GAO, 2012). As a result, The California Leafy Greens Producers took the lead on food safety in this sector to prevent this from happening again and developed the "California Leafy Green Products Handler Marketing Agreement" to require certain food safety and traceability practices for all members (http://www.lgma.ca.gov/).

The next sections address the legislative reaction to protect public health then briefly describes the US regulatory system.

2. Current Legislation for Food Traceability in the United States

Laws and regulations in the United States governing safe food for humans and animals may be found at the federal, tribal (Native American jurisdiction), state, and local level. It should be noted that the US food safety system is based on strong, self-enforced compliance by food companies, distributors, wholesalers, and retailers. This chapter discusses traceability, and because this generally applies to food in interstate and global commerce, federal laws apply.

2.1 General US Federal Structure Governing Food Traceability

In addition to the US FDA and the USDA oversight of food, specifically with respect to traceability, there are a number of other federal agencies involved. These include (in part, from Fortin, 2009, pp. 23–27)

- Centers for Disease Control and Prevention: Investigates with local, state, and other federal officials sources of foodborne disease outbreaks. Maintains a nationwide system of foodborne disease surveillance (for more information, www.cdc.gov).
- US Environmental Protection Agency: Overseas drinking water and pesticide safety. Regulates toxic substances and wastes to prevent their entry into the environment and food chain. Determines safety of new pesticides, sets tolerances for pesticide residues in foods, and publishes directions on the safe use of pesticides (for more information, www.epa.gov).
- National Marine Fisheries Service: Overseas fish and seafood products through voluntary, fee-for-service system. This agency inspects and certifies fishing vessels, seafood processing

plants, and retail facilities for federal sanitation standards which are written by the FDA (FDA, 2014f). The National Oceanic and Atmospheric Administration (NOAA) Seafood Inspection Program is often referred to as the US Department of Commerce (USDC) Seafood Inspection Program and uses marks and documents bearing the "USDC" moniker. Further, this agency prevents the illegal, unregulated, and unreported harvesting and trafficking of fish and wildlife and tackles seafood fraud (for more information: www.nmfs.noaa.gov/index.html. For specific information on seafood inspection, and laboratory analysis: http://www.seafood.nmfs.noaa.gov/).

- Alcohol and Tobacco Tax and Trade Bureau: Oversees alcoholic beverages except wine beverages containing less than 7% alcohol. This agency enforces food safety laws governing alcoholic beverages and investigates adulteration of alcoholic products, sometimes with help from the FDA (for more information: www.ttb.gov).
- US Customs and Border Protection Service: Oversees imported foods. This agency works with federal regulatory agencies to ensure that all goods entering and exiting the United States do so according to US laws and regulations. They inspect shipments at the borders using, for example, agricultural protection standards and the Bioterrorism Act, among others (for more information: www.cbp.gov).
- US Department of Justice: Prosecutes companies and individuals suspected of violating food safety laws. The US Marshals Service in this department seizes unsafe food products not yet in the marketplace, as ordered by courts (for more information: www.justice.gov).

Currently in the United States, food industry traceability standards are governed under federal laws imposed in 2002 under the Bioterrorism Act, although a few exceptions to this include finalized piecemeal regulations under the FSMA of 2011.

2.2 Public Health Security and Bioterrorism Preparedness and Response Act of 2002

In response to the attacks of September 11, 2001, Congress wanted to keep the nation's food supply safe. As a result, the Public Health Security and Bioterrorism Preparedness and Response Act was passed in 2002, which is commonly called the Bioterrorism Act. This law was a major step forward in the use of product tracing systems and gave authorities in the FDA and the USDA new requirements intended to protect the food supply from an attack or from other food-related emergencies. The Act requires a paper trail documenting food distribution in order to allow the source of contamination to be determined in the event of a foodborne disease outbreak. These new authorities given to the US FDA and USDA were not restricted to problems that were caused by bioterrorism. In particular, and among other requirements, the law required that all food handlers keep records showing the path of food and food ingredients from farm to fork, "one step forward and one step back."

The Bioterrorism Act directed the US FDA in consultation with the USDA to establish the regulations and guidance summarized below (US FDA, 2009), which provide additional authority and provide a framework for a developing food traceability system.

i. Food facilities register with the US FDA. This includes all facilities, domestic or foreign, that manufacture, process, pack, or hold food for human consumption, unless they are exempt. Exempt facilities include farms, food service (cafes and restaurants), food retail, transport vehicles, food establishments which are not-for-profit, fishing vessels which do not engage in processing, and facilities regulated by the USDA. This may soon be superseded by the FSMA regulation on Registration of Food Facilities (Section 102 of FSMA) when it becomes finalized (US FDA, 2015b).

ii. The US FDA has the authority to detain food. The US FDA was given authority of temporary detention of food shipments, if the food was considered to pose a threat of serious adverse health consequences.

iii. The US FDA has the authority to require records be established and maintained relating to the manufacture, importation, and distribution of food. These requirements specify the recording of immediate sources and destinations of incoming and outgoing food products and must include: full contact details (including emergency contact information) of sources, destinations, and transporters, description of food (brand name and variety), date received and released, lot code number (if relevant), and quantity and packaging (for example, 50 pound bags or 12-oz bottles). This was intended to enable traceability along the food supply chain in the event of detecting contaminated food.

iv. The US FDA has the authority to inspect these records if there is reasonable belief that the food is adulterated and could pose serious health consequences. This has now been superseded by the FSMA Final Rule on Inspection of Records, Section 101 of the FSMA (US FDA, 2015f).

v. The US FDA must be given advance notice on shipments of imported food. Prior notice regulations of food imports (human or animal) must be submitted electronically no more than five days prior to arrival.

Other exemptions from record-keeping requirements, but subject to records access requirements include food retailers with 10 or fewer employees nonprocessing fishing vessels and nonprofit food establishments. Further, direct-to-consumer distributors are not required to keep records of the people to whom they sell. Food transfers within a company are not subject to record keeping (US FDA, 2009).

Records must also include information identifying specific sources of each ingredient for each lot of finished products. In cases where food processors commingle ingredients from different suppliers, such as flour or milk, the US FDA accepts that manufacturers may not be able to identify one specific source. These regulations became effective in December of 2003 (Fortin, 2009; McEntire, 2010).

2.2.1 Discussion of the Bioterrorism Act

The intent of the bioterrorism legislation was to protect the public from a variety of food-related emergencies by requiring these relatively straightforward strategies. However, implementation of these regulations to a reasonably high operational standard was much more challenging. At the time the Act was passed, the majority of facilities in the US stored their records in paper files, which were time-consuming to access and even more time-consuming for officials to correlate with records from other facilities in the food chain. As a result, a common time frame for a traceback to occur might be two weeks or more and recall might take even longer.

The traceback attempted by the FDA during *Salmonella enterica* serovar Saintpaul in the raw "tomato" outbreak of 2008 (described earlier, with illnesses occurring mid-April through late August) demonstrated one of the major difficulties for authorities working with this regulation. In July of 2008, according to Venkataraman, CDC and

FDA officials were not able to find the source of the fresh tomato and pepper contamination causing the *S. enterica* serovar Saintpaul outbreak (Venkataraman, 2008a).

The issue was that processors frequently repack boxes of raw tomatoes to meet a buyer's demands, termed an "internal tracking event" not required by the regulation. They were not required to record the tomatoes' farm, state, or even country of origin. The rule required only that produce handlers keep track of "external tracking events," or food, "one step back and one step forward" in the supply chain. In addition, the rule did not apply to retailers or growers, two of the types of exempt facilities in the rule (see item "i" above). Finally, because the rule did not specify the format for record keeping, investigators took many weeks sifting through a hodgepodge of paper trails in order to identify the source of the contaminated produce. This time not only allowed the traceback trail to grow cold but also was time during which consumers stopped buying tomatoes, believing they were the cause of the outbreak. A better traceback regulation could have shortened the traceback process, preventing both illnesses and the major economic losses of an entire industry (Venkataraman, 2008b).

2.3 Food Safety Modernization Act of 2011 (FSMA)

FSMA was passed by the US Congress (US Congress, 2011; US FDA, 2014d) and signed into law by President Obama on January 4, 2011. A brief and easily readable general summary of the Act and its sections was published by the law firm Morgan Lewis (Morgan Lewis, 2011). It enables the US FDA to better protect public health by helping to ensure the safety of the food supply (both human and animal) and is the most sweeping reform of the FDA's food safety authority in 70 years. It enables the FDA to focus more on preventing food safety problems rather than relying primarily on reacting to problems after they occur. The law also provides the FDA with new enforcement authorities to help it achieve higher rates of compliance with risk-based and prevention-based food safety standards. It allows the FDA to better respond to and contain problems when they do occur. In addition, the law gives the FDA important new tools to better ensure the safety of imported foods and directs the FDA to build an integrated national food safety system in partnership with state and local authorities. In regards to this chapter, Section 204 of the Act requires the FDA to develop additional record-keeping requirements for high-risk food to improve their traceability.

The FSMA is a law that was passed by Congress and signed into law in 2011, but "rule-making" for this law is incomplete. Rule-making involves the development of detailed regulations for enforcement. The regulations specify the technical details that are necessary to comply with the law's broad requirements. For example, the Preventive Controls Rule is a newly finalized regulation in Section 103 of FSMA, as of September 10, 2015, and goes into effect in 2016 and 2017 (by facility type and size). US law requires the agencies making the rules to follow minimum procedural safeguards. For example, notice of a proposed rule must be published in the "Federal Register" (https://www.federalregister.gov/). In addition, the agencies must allow interested parties time to submit comments on the rule, and in some instances, public hearings are conducted. Public comments must be reviewed and considered by the agency before final adoption of a regulation. Some regulations require several reviews.

In the case of the FSMA, for example, one of several contentious areas is that the new rules require inspections, with the cost borne by the facility being inspected. Small and very small businesses and farms, which may not have dealt with such issues before, are affected by these rules and inspections. Moreover, costs of following the regulations and paying for inspections to these smaller enterprises might be a significant portion of their modest operating budget. Therefore the issue became politicized and there have been several versions of the same rule. In addition, the FDA has been operating with a tight budget that has not seen expansion in many years, in spite of its expanded authorities. Congress has been slow to approve budgets and caused a government shutdown in 2013. Therefore the regulatory structure regarding food traceability in the US, which is discussed in this chapter, is only partially complete.

The author will discuss the current situation, which will include a few of the building blocks of a traceability standards system and which are currently finalized rules of the FSMA. These building blocks include subsections of the Preventive Controls Rule addressing responsibility for the supply chain (upstream and downstream) and a section requiring recall, as well as a rule on Inspection of Records (Section 101). Also addressed in this section will be the legal language of FSMA Section 204, *Enhanced Tracking and Tracing of Food and Record Keeping*, "although there is no regulation, proposed or otherwise, for this section and no timetable for when it will be addressed". One portion of Section 204 has been completed, and that is the report for the pilot project to evaluate the current tracing system. This is briefly discussed. Section 3 of this chapter will describe proposed regulations affecting food traceability and Section 4, *Future Trends*, will describe industry standards in food traceability which are widely accepted by food professionals and which might become traceability rules and/or guidance under Section 204 of the FSMA.

The major elements of the new FSMA law can be divided into five key areas (US FDA, 2014d):

i. **Preventive controls**: For the first time, the FDA has a legislative mandate to require comprehensive, prevention-based controls across the food supply.
ii. **Inspection and Compliance**: The legislation recognizes that inspection is an important means of holding the industry accountable for its responsibility to produce safe food; thus the law specifies how often the US FDA should inspect food producers. The US FDA is applying its inspection resources in a risk-based manner and, given current budget constraints, may adopt innovative inspection approaches.
iii. **Imported Food Safety**: The US FDA has new tools to ensure that those imported foods meet US standards and are safe for consumers. For example, for the first time, importers must verify that their foreign suppliers have adequate preventive controls in place to ensure safety, and the US FDA will be able to accredit qualified third party auditors to certify that foreign food facilities are complying with US food safety standards.
iv. **Response**: For the first time, the US FDA will have mandatory recall authority for all food products. The US FDA expects that it will only need to invoke this authority infrequently since the food industry largely honors its requests for voluntary recalls. "This portion of FSMA incorporates the requirements for traceability and record keeping for high-risk foods".
v. **Enhanced Partnerships**: The legislation recognizes the importance of strengthening existing collaboration among all food safety agencies—US federal, state, local, territorial, tribal, and foreign—to achieve the FDA's public health goals. For example, it directs the FDA to improve training of state, local, territorial, and tribal food safety officials.

Note: In general, the FSMA legislation does not cover seafood, except raw oysters, nor does it address grain storage. Seafood sanitary processing and importing was addressed in its own 1995 food safety regulations (US GPO, 1995). There have been reports of seafood in the US being labeled with an incorrect market name. In response to these reports, the FDA began conducting DNA testing on fish that have a history of being misidentified in an effort to determine the accuracy of the market names on their labels. The FDA found that fish species at the wholesale level were correctly labeled 85% of the time. The FDA has devoted a website to this issue with information about specific regulations and guidance on labeling, as well as specific tips for retailers and others in the supply chain to discourage this fraud (US FDA, 2014e).

2.3.1 Enhanced Tracking and Tracing of Food and Record Keeping

This is also known as Section 204 of the FSMA, the Traceability Rule. At the time of this writing, the US FDA has a directive (from the FSMA legislation) to develop regulations for the traceability for high-risk foods in consultation with the USDA. It has not developed the regulations, nor does it have a timeline for when a draft of such rules will be written. Before this can be addressed, the high-risk food methodology must be developed (see Section 3 of this chapter). Traceability regulations are therefore piecemeal until this rule is drafted and finalized, but the section of the law is reviewed below.

2.3.1.1 General Language of Section 204 of the FSMA on Traceability

Section 204 of the FSMA legislation requires a product tracing system for the purpose of improving the capacity to detect and respond to food safety problems because the rapid tracing of food products prevents illness. In general, Section 204 makes the following points (US FDA, 2014d):

- **i.** Establish a pilot project to explore and evaluate methods for rapidly and effectively tracking and tracing food, taking into account recommendations from the USDA, FDA, and State Departments of Health, in coordination with the food industry.
- **ii.** Improve tracking and tracing of processed foods and fruits and vegetables that are raw agricultural commodities in the event of a foodborne illness outbreak in coordination with the USDA. Further requirements include establishment of standards for the type of information, format, and time frame for people to submit records to aid the FDA in such tracking and tracing.
- **iii.** Require additional record keeping for high-risk foods: the FDA must designate and publish a list of high-risk foods and propose regulations for record keeping specific to high-risk foods.
- **iv.** Exclude food that is produced and packaged on a farm from the requirements of this section and limit the requirements on food obtained from fishing vessels and commingled raw agricultural and other foods of facilities as the FDA, in consultation with the USDA, may exempt.
- **v.** In spite of this, farms are required to respond to requests for tracing information.
- **vi.** The *Small-Entity Compliance Guide* is to be issued by FDA.
- **vii.** Failure to comply.

In general, a product tracing system involves documenting the production and distribution chain of products so that in the case of an outbreak or evidence of contaminated food, a product can be traced back to a common source or forward through distribution channels. Product tracing systems enable government agencies and those who produce and sell food to take action more quickly when an outbreak of foodborne illness occurs or contaminated product is identified, thus preventing illnesses. Actions include removing a product from the marketplace and alerting the public if a product has already been distributed.

Many producers, manufacturers, and retailers have product tracing systems in place, but they vary depending on the amount of information the system records, how far forward or backwards in the supply chain the system tracks, the technologies used to maintain records, and the precision with which a system can pinpoint a product's movement. It is for this reason that item "i" was written above, where the FDA was tasked to work with the USDA and state agencies through a pilot project to evaluate and characterize the current tracing system and make recommendations for improvements. This task has been completed and is discussed in item "b" below.

2.3.1.2 Pilot Project to Evaluate the Current Tracing System Project

The FDA requested assistance from The Institute of Food Technologists (IFT) in 2011 to work with the food industry, regulatory authorities, and consumer advocates in order to execute the legislative requirement of a pilot project for traceability. The pilot projects were designed to explore and demonstrate methods for the rapid and effective tracking and tracing of food, including types of data that are useful, ways to connect the various points in the supply chain, and how quickly data can be made available to the US FDA. The final report was released in 2013 (IFT, 2013).

Information from the pilots will help to determine what data are most needed to trace a product that has been distributed widely in the marketplace back to a common source. Tracing a product forward, such as in the case of an ingredient known to be contaminated, was also tested. Two pilots were required by the FDA Food Safety Modernization Act (FSMA)—one with the processed food sector and one with processors or distributors of raw fruits and vegetables. FSMA required that the pilots reflect the diversity of the food supply, taking into account practicality for small businesses and including at least three different types of foods that have been the subject of significant outbreaks. International product tracing practices were to be considered as well.

With input from a variety of stakeholders, the following types of foods were selected for the pilot projects:

- Tomatoes, grown in fields and greenhouses; whole and sliced; and distributed to restaurants and other institutions like hospitals, schools, nursing homes, and grocery stores. Tomatoes were chosen because they have been involved in a number of significant and repeated outbreaks. Tomatoes represent a complex food supply chain and were identified by most industry associations as a top candidate for the produce-related pilot.
- Frozen Kung Pao-style dishes that contain peanut products, red pepper spice, and chicken were chosen because they contain multiple ingredients involved in significant outbreaks. They also offered a variety of supply chain distribution channels and, like tomatoes, involved both domestic and imported products.
- Jarred peanut butter and dry, packaged peanut/spice were added to the pilot projects to enhance the complexity of the pilots.

The IFT report covered baseline studies, pilot execution using trace scenarios, use of technology, cost and benefits, domestic and global practices, and recommendations. Within the context of the pilot projects, the report estimated the costs associated with improved record keeping and product tracing. Costs varied greatly based on a number of factors, such as using a paper-based or electronic system, scanning labels, nomenclature standardization, and electronically linking supply chain partners. The pilot projects found that technology was no substitute for quality data. Once challenges with data quality could be overcome, significant improvements in the speed and accuracy were achieved with the aid of technology.

A variety of technology solutions were explored for data capture, visualization, and analysis. The benefits of implementing a product tracing system and recommendations from the pilot project study, from IFT (2013):

- **i.** Product tracing systems enable government agencies and those who produce and sell food to take action more quickly when an outbreak of foodborne illness occurs or contaminated product is identified, thus preventing illnesses.
- **ii.** As part of the overall project, cost savings, driven by reductions in illness, were estimated for reducing traceback duration by 25, 50, 75, and 100%. The range of public health benefit per outbreak ranged from $18,000 to $14 million, depending on the characteristics of the outbreak.
- **iii.** In addition to quickly discovering which food(s) may be causing an outbreak, tracing systems might also be able to determine what foods can be eliminated from consideration, reducing economic impact on the industry.
- **iv.** Product tracing also is helpful for other functions, such as inventory control.

Finally, the IFT report (IFT, 2013) had 10 recommendations:

- **i.** From an overarching perspective, IFT recommends that the FDA establishes a uniform set of record-keeping requirements for all FDA-regulated foods and not permit exemptions to record-keeping requirements based on risk classification.
- **ii.** FDA should require firms that manufacture, process, pack, transport, distribute, receive, hold, or import food to identify and maintain records of Critical Tracking Events (CTEs) and Key Data Elements (KDEs) as determined by the FDA.
- **iii.** Each member of the food supply chain should be required to develop, document, and exercise a product-tracing plan.
- **iv.** The FDA should encourage current industry-led initiatives and issue an advance notice of proposed rule-making or use other similar mechanisms to seek stakeholder input.
- **v.** The FDA should clearly and more consistently articulate and communicate to the industry the information it needs to conduct product tracing investigations.
- **vi.** The FDA should develop standardized electronic mechanisms for the reporting and acquiring of CTEs and KDEs during product tracing investigations.
- **vii.** The FDA should accept summarized CTEs and KDEs that are submitted through standardized reporting mechanisms and initiate investigations based on such data.
- **viii.** If available, the FDA should request more than one level of tracing data.
- **ix.** The FDA should consider adopting a technology platform that would allow efficient aggregation and analysis of data submitted in response to a request from regulatory officials. The technology platform should be accessible to other regulatory entities.

x. The FDA should coordinate traceback investigations and develop response protocols between state and local health and regulatory agencies, using existing commissioning and credentialing processes. In addition, the FDA should formalize the use of industry subject matter experts in product tracing investigations.

In addition to providing the findings of the pilot projects, the report contains IFT's recommendations to the FDA for improving the tracking and tracing of food. It is interesting to note that recommendation "i," above, recommends that the FDA establish a uniform set of record-keeping requirements and not permit exemptions based on risk classification. This makes sense because a low-risk food can quickly become a high-risk food, as did peanut butter in the 2009 *Salmonella* outbreak from peanut products. However, this recommendation contradicts the language of the legislation passed by Congress, which specifies tracing only for high-risk foods. In this situation, the FDA may find a voluntary alternative, but the language of Congress takes precedence.

The FDA has sought public comment on this report. After public comments and additional data are gathered, the FDA will initiate rule-making on traceability for high-risk foods. The FDA must first define high-risk foods, considering such factors as the known risks of a food based on foodborne illness data, the likelihood that a particular food has a high potential risk for contamination, and the likely severity of an illness attributed to a particular food. There is no current time frame.

2.3.2 The Preventive Controls Rule

Also known as the rule for Current Good Manufacturing Practice and Hazard Analysis and Risk-Based Preventive Controls, this regulation (finalized September 10, 2015) describes current good manufacturing practice as well as hazard analysis and risk-based preventive controls for human food. There is a similar regulation for animal food (US FDA, 2015j), but this chapter will focus on human food. The legal requirements may be found in Section 103 of FSMA (US FDA, 2015a), but the detailed rules and guidance for human food are better explained on the US FDA's web page (US FDA, 2015d). The rule requires all facilities that manufacture, process, pack, or hold human food to have a written preventive control plan, with exceptions of farms and retail food establishments. It applies to domestic and imported food. In addition to its other requirements, this regulation requires companies to have control over hazards in their supply chain and have a plan to recall their product. In effect, this is the closest regulation requiring traceability that is currently approved and published because in order to recall food, there must a form of traceability in place.

Some key requirements of the Preventive Controls Rule, with a bias toward traceability in supply chains for human food (US FDA, 2015d):

a. Covered facilities must establish and implement a food safety system that includes an analysis of hazards and risk-based preventive controls. The rule sets requirements for a written food safety plan that includes:
   i. Hazard analysis
   ii. Preventive controls (These measures are required to ensure that hazards requiring a preventive control will be minimized or prevented. They include controls for

processes, food allergens, and sanitation, as well as controls for a supply chain program and a recall plan.)

iii. Oversight and management procedures for preventive controls issued above (including procedures for monitoring, corrective action procedures, and verification procedures: this includes such procedures for supply chain management)

b. The facility must have a risk-based supply chain program for raw materials and other ingredients. The program should specify how the hazards are being controlled, either by the supplier or by the receiving facility, and include appropriate verification activities.

The recall plan in this regulation is quite general. However, the FDA developed guidance on policy, procedures, and industry responsibilities in 1978 and updated them in April 2015 (FDA, 2015e). The recall strategy is specified for a variety of situations, depending on the health hazard evaluation, whether the recall is initiated by the manufacturer or by the FDA, the depth into the distribution chain, the need for effectiveness checks, and many other factors. The FDA classifies the recall according to Class I, Class II, or Class III to indicate the relative degree of health hazard of the product. A Class I recall is a situation in which there is a reasonable probability that the product will cause serious adverse health consequences or death. A Class II recall is a situation in which the product may cause temporary adverse health consequences or where the probability of serious adverse health consequences is remote. A Class III recall is a situation in which the product is not likely to cause adverse health consequences. If the FDA must take over the recall situation for any reason, their charges apply (see the fee schedule, Federal Register, 2015).

In regards to compliance dates, very small businesses have until September 2018, small businesses have until September 2017, businesses subject to the Pasteurized Milk Ordinance have until September 2018, and all other businesses have until September 2016 to comply with this regulation. Separate compliance dates have been established for the supply chain program provisions to accommodate compliance dates for suppliers of different sizes and subject to different rules, such as Produce Safety Standards and the Foreign Supplier verification program.

Note: The Preventive Controls Rule does not cover seafood, juice, or low-acid canned foods because these areas have their own preventive controls regulations from previous legislation. The seafood HACCP was legislated in 1995 (US GPO, 1995), with rules and guidance at these US FDA sites (FDA, 2014f; FDA, 2014i). The "Juice HACCP Regulations of 2001" can be found at US FDA (2014g). Low-Acid canned foods were regulated in 1974 with an HACCP-like set of regulations (US FDA, 2015c).

2.3.3 Inspection of Records: Requirements and Establishment, Maintenance, and Availability of Records

Also known as Section 101 of FSMA, the final version of the regulation requiring the inspections of records was published in the Federal Register on April 4, 2014 (Federal Register, 2014b). A more understandable version with guidance is available at FDA (2015f). This regulation expands the FDA's former records access authority beyond access to records relating to a specific suspect article of adulterated food that will cause serious adverse health consequences or death to humans or animals. The authority is expanded to records that the FDA reasonably believes likely to be affected in a similar manner. That is, the FDA shall have access to records that are needed to assist it in determining whether a food is adulterated. This new regulation is applicable to food traceability because detailed records must be kept of all food received into the facility and released out of it.

The records that must be established include the details of the food or food ingredient received into the facility (name and contact details of originating firm whether foreign or domestic, type of food, date received, quantity, and type of packaging) and the details of the transporter. Similarly, records must be kept including the details of the food sent from the facility to a recipient (name of firm and contact details, type of food, date released, quantity, and type of packaging) and the details of the transporter. In addition, transporters must maintain records with details including the immediate subsequent recipient, origin and destination points, date shipment was received and date released, number of packages, description of freight, route of movement during the time the food was transported, and transfer point(s) through which the shipment moved. This applies specifically to food facilities and transporters within the United States but can include transporters who are foreign. Failure to establish and maintain records carries both civil and criminal penalties.

The records, whether paper or electronic, must be available to the FDA as soon as possible, not to exceed 24h. These records do not allow full traceability as they do not include a traceability step (such as transformation), where multiple ingredients are mixed together to produce a different product. Instead, the regulation goes part of the way and is focused on following all items into and out of the facility and through a transporter.

2.3.4 Sanitary Transportation

Sanitary transportation is the final rule under FSMA Section 111. It requires those who transport food for humans and animals to use sanitary transportation practices in order to ensure the safety of food. Specifically, the FDA establishes conditions and practices, training, and record keeping, for the sanitary transportation of food (FDA, 2016).

With some exceptions, this proposed rule applies to shippers, receivers, and carriers who transport food in the United States by motor or rail vehicle, whether or not the food enters interstate commerce. It does not apply to an exporter, who ships food to the United States in an international freight container and arranges for the transport of the intact container through the United States to Canada or Mexico so that it does not enter distribution within the United States. Companies involved in the transportation of food intended for export are covered by the rule until the shipment reaches a port or US border.

The rule establishes requirements for the following:

- **Vehicles and transportation equipment**: The design and maintenance of vehicles and transportation equipment are specified to ensure that it does not cause the food that it transports to become contaminated.
- **Transportation operations**: The measures taken during transportation are specified to ensure food is not contaminated, such as adequate temperature controls and separation of food from nonfood items in the same load.
- **Information exchange**: Procedures are required for the exchange of information about prior cargos, cleaning of transportation equipment, and temperature control between the shipper,

carrier, and receiver, as appropriate to the situation. For example, a carrier transporting bulk liquid nondairy foods would want to ensure that vehicles that have previously hauled milk will not introduce allergens into nondairy foods through cross-contact.
- **Training**: Training is required of carrier personnel in sanitary transportation practices and documentation of the training is required.
- **Records**: Maintenance of written procedures and records is required by carriers and shippers related to transportation equipment cleaning, prior cargos, and temperature control. This contributes to traceability records.
- **Waivers**: Procedures are specified by which the FDA will waive any of these requirements if it determines that the waiver will not result in the transportation of food under conditions that would be unsafe for human or animal health and that it is in the public interest.

The following are exempted from coverage:

- Shippers, receivers, or carriers engaged in food transportation operations that have less than $500,000 in total annual sales
- Transportation activities for raw agricultural commodities performed by a farm
- Transportation of shelf-stable food that is completely enclosed by a container, except for food that requires temperature control
- Transportation of compressed food gases
- Transportation of live food animals, except molluscan shellfish

3. Pending Rules and Guidance

This section includes rules and guidance proposed and drafted by the FDA for the FSMA. However, these draft rules and guidance are not final; that is, they are open for comments or pending approval. As such, the Traceability rule (Section 204 of FSMA) has not yet been addressed, except for the pilot project reports (IFT, 2013 and Section 2.3.1.2 of this chapter), but the proposed method for identifying high-risk foods is addressed below.

3.1 *Identifying High-Risk Foods*

Under Section 204 of the FSMA (the traceability rule), the FDA is required to designate high-risk foods for human consumption for which additional record-keeping requirements are appropriate (Federal Register, 2014c). These additional record-keeping requirements are important to enable rapid tracing of these high-risk foods during a foodborne disease outbreak or other event. As of October 2015, the FDA has published only a proposed methodology for designating high-risk foods. No list has been published. In February 2014, the FDA published a "Request for Comments and for Scientific Data and Information" in the Federal Register (2014c) to seek comments and scientific data that will help refine a methodological approach for identifying high-risk foods (see below). After a method is finalized, perhaps sometime in the year 2016, the FDA will develop the list of high-risk foods. The public will again have the opportunity to comment on this list before it is finalized.

The FSMA specifies that the method FDA develops to designate high-risk foods for traceability must be based on the following factors:

a. the known safety risks of a particular food, including the history and severity of foodborne illness outbreaks attributed to it, taking into consideration foodborne illness data collected by the Centers for Disease Control and Prevention
b. the likelihood that a particular food has a high potential risk for microbiological or chemical contamination or would support the growth of pathogenic microorganisms due to the nature of the food or the processes used to produce such food
c. the point in the manufacturing process of the food where contamination is most likely to occur
d. the likelihood of contamination and steps taken during the manufacturing process to reduce the possibility of contamination
e. the likelihood that consuming a particular food will result in a foodborne illness due to contamination of the food
f. the likely or known severity, including health and economic impacts, of a foodborne illness attributed to a particular food
g. the factors to designate high-risk foods must be science-based and must have the public health benefits outweigh the cost of complying with the requirements
h. specific technologies to maintain records may not be prescribed and must be practical for facilities of varying sizes and capabilities
i. maintenance of records at a reasonably accessible location must be allowed, provided that the records can be made available to the FDA within 24 h of a request
j. a record of the complete previous distribution history of the food from the point of origin may not be required (Federal Register, 2014c)

Taking these congressionally specified factors into account, the FDA published a draft methodological approach for designating high-risk foods (US FDA, 2014b). This report reviews a number of methods, qualitative decision trees, semiquantitative risk scoring, and quantitative risk assessment, but appears to favor a semiquantitative risk-scoring method published by Anderson et al. (2011). Further, this report proposes high-risk food hazards to include microbial hazards (including bacterial, amebic, and mold), chemical hazards (including allergens, mycotoxins, pesticides, metals, and marine biotoxins), and industrial chemicals, as well as those formed during processing. However, each aspect of risk is proposed to be classified as a criteria and is scored based on a variety of risk factors.

Although FSMA specifies that additional record-keeping requirements must apply only to high-risk foods, the FDA will be seeking input from stakeholders in order to evaluate whether to develop voluntary guidance for foods beyond those designated as high risk, which could enhance product tracing in the supply chain. The IFT pilot projects report for improving traceability (IFT, 2013) recommended traceability for all foods. Therefore the call for voluntary guidance would be appropriate and would perhaps allow the market to select for those companies and foods who can offer consumers better accountability in their food.

3.2 Registration of Food Facilities

Section 102 of the FSMA is focused on creating a biennial registration requirement for food facilities, which is important to a traceability system. For more information about laws and rules that have been finalized regarding the registration of food facilities, see US FDA (2015g), and for the latest amendments and guidance to the proposed regulations for facility registration, see US FDA (2015h).

The latest proposed regulation would require biennial registration of food facilities using electronic submission (ie, email) (US FDA, 2015h). The term "food facility" is redefined in the amendment (US FDA, 2015i). In particular, "retail food establishments," farms, restaurants, and certain other entities are exempt from the requirement to register. The definition of a farm has not changed, although farms may be subject to another rule in the FSMA, the produce safety rule. The amended definition for a retail food establishment would exempt additional facilities from the requirement to register. An additional requirement of this regulation is that registered food facilities must assure the FDA that they will be permitted to inspect the facility when a request is made. Further, the registration comes with the stipulation that the FDA may suspend registration if it determines that the food facility was responsible for or knew of the likelihood of adverse health consequences or death to humans or animals from a food manufactured, processed, packed, received, or held by that registered food facility.

4. Future Trends and Business Governance in Traceability

4.1 What to Expect From the US FDA for Traceability

The US FDA acknowledges that the central force driving change is the dramatic expansion in the global scale and complexity of the food system. It further understands that the foundation for reducing the risk of preventable foodborne illness lies with action by industry, with top-level management commitment (US FDA, 2014j). In addition to its regulatory role, Mike Taylor, Deputy Commissioner for Foods and Veterinary Medicine for the US FDA, suggested that his agency had an important role as an "influencer" (US FDA, 2015k).

Regarding the FSMA traceability regulation, it is likely the US FDA will selectively adopt standards being developed by IFT's Global Food Traceability Center. However, it should be emphasized that it is unlikely the US FDA will recommend specific product tracing technologies as they move forward with the implementation of this rule. Rather, it is likely that the agency will focus on the elements of a product tracing system that will enable rapid and effective tracing of foods, such as what data should be kept and ways to link food along the supply chain. Furthermore, it is likely that the US FDA will focus on how quickly the data can be provided, especially when there is an emergency situation and the risk of illness exists. In addition, while the US FDA is charged with not endorsing any particular system or technology in the traceability regulations it develops, it is expected that it may specify general characteristics of commercial collaboration platforms that enhance the speed and accuracy of tracing (US FDA, 2014j).

4.2 The IFT's Global Food Traceability Center

IFT has been involved in and has led the discussion on traceability since 2006. In September 2013, IFT launched the Global Food Traceability Center (GFTC), a not-for-profit collaborative, public–private partnership. The GFTC brought together stakeholders in the agricultural and food system to collaborate on product tracing and has provided leadership as a scientific and unbiased source for food traceability. Its mission is to assist companies and government agencies to better understand the nature of food traceability requirements, how to use technologies to improve responsiveness and reliability in the event of food-related emergencies, and the value and commercial benefits of food traceability (IFT, 2015). This organization has and continues to work as an intermediary between the government, consumers, and the industry.

The Global Food Traceability Center (GFTC) worked to further understand food traceability regulations on a global scale and to provide regulators and the food industry with the current best practices. Zhang and Bhatt (2014) published *A Guidance Document on the Best Practices in Food Traceability* to address the lack of uniform requirements across food sectors and supply chains. The report presents food traceability best practices in terms of a generalized framework involving critical tracking events and key data elements, although these concepts were first generalized by IFT before 2010 (McEntire, 2010). The developed framework was applied to six food sectors: baked goods, dairy, meat and poultry, processed foods, produce, and seafood, using the expertise of 55 subject matter experts from 11 countries. This collaborative team was involved in developing the guidance, and the criteria proposed are the minimum necessary to manage a proper food traceability system. Food sectors yet to be addressed include oils, seeds, and nuts.

In further work by IFT's Global Food Traceability Center, Charlebois et al. compare traceability regulations among 21 OECD countries using a questionnaire (Charlebois et al., 2014). The questions scored information based on (1) whether mandatory traceability regulations exist in a country, (2) whether regulations include imported products and the nature of required documentation, (3) whether an electronic database for traceability exists, and, if present, its accessibility, and (4) whether labeling regulations allow consumer access and understandability. Results of the survey educate the reader regarding the state of traceability on the global stage and provide a helpful background for government agencies constructing a traceability framework expected to perform in the global context. Survey results were unsurprising, showing that, among other conclusions, EU countries adopting EU legislation were strong leaders in global food traceability.

4.3 Businesses Interested in Traceability for Their Own Benefits

Many businesses are interested in traceability for the variety of internal benefits that collecting such data may provide, beyond complying with legislation. The data can provide production and manufacturing management information through enterprise resource planning (ERP) software that, while meeting the needs of the Bioterrorism Act traceability requirements, can also improve scheduling, efficiency, and quality assurance checks and balances. This includes generating the following information:

- identification of specific data throughout the process, including dates, times, personnel, machinery, and individual efficiencies;
- incorporation of quality assurance checks and balances throughout the production process;
- provision of customized multiple reports to assist in the scheduling process; and
- provision of reporting at user-definable intervals for productivity analysis, trend analysis, and incentive programs.

In addition to the internal benefits, companies can be interested in traceability record keeping for reasons beyond food safety or productivity, including documentation for a sustainability label or simply addressing increased consumer interest of traceability in the food supply chain. Finally, there is a growing demand by food suppliers' customers for improved traceability and recalls, which is proven through certification to Global Food Safety Initiative (GFSI)-recognized schemes.

4.4 Industry Initiatives in the United States

Because a good traceability scheme is dependent on its weakest link, businesses have relied on their trade associations to develop appropriate standards for all supply chain partners. Unlike the government regulations which constrain recommendations of a specific technology, these industry initiatives are based on specific technologies. The GS1 technology appears to be popular and is used by the following traceability schemes:

i. The seafood industry in association with GS1 has put together a document to address standards for traceability in the food supply chain (National Fisheries Institute, 2011).
ii. The US meat and poultry companies (mpXML) have put together a comprehensive tracing standard using IFT's concept of critical tracking events (mpXML, 2014; Angus-Lee, 2014) and have published a document, *A Model for the Adoption of Critical Tracking Events (CTEs) in the Meat and Poultry Supply Chain* (mpXML, 2013). The members of this industry trade organization are composed of GS1 US, the National Chicken Council, the American Meat Institute, the National Turkey Federation, the American Lamb Association, Pork Checkoff, Beef Cattlemen of America, and industry leaders including Walmart, Costco, Safeway, Farmland Foods, Wegmans Food Markets, Smithfield Foods, Supervalu, and Tyson Foods.
iii. GS1 holds the US implementation guideline for traceability for the dairy, deli, and bakery associations (GS1, 2013). The Innovation Center for US Dairy (2013) has produced an industry guidance for enhanced traceability.
iv. GS1 also holds the US implementation guideline for the traceability for fresh fruits and vegetables (GS1, 2010).

4.5 For Questions or More Information

The US FDA encourages questions, because later they will assist with writing guidance. Eventually a "Frequently Asked Questions" page will be added to their website. In the meantime, submit your questions to http://www.fda.gov/Food/GuidanceRegulation/FSMA/ucm459719.htm.

US FDA rule-making will continue. To keep up to date on FSMA rule-making and guidance and to see the historical sequence of the FDA's rule-making regarding this Act, see http://www.fda.gov/Food/GuidanceRegulation/FSMA/ucm257986.htm.

5. Conclusion

Food traceability is a cornerstone of the increasingly complex, industrialized, and global food system. In the event of a recall, traceability is critical to respond rapidly and reliably to protect consumer, or animal or plant, health, especially when large quantities of contaminated products have been distributed across widespread markets. Traceability is also essential for producers and manufacturers who need to track items for supply chain management and for clients who want seamless data about the products they buy.

Current US Government regulations specify a minimum requirement of one step forward and one step back. However, there is a need for standardized and harmonized requirements across all food sectors and between countries. Developing specialized rules and mandates, including exceptions for specific foods, will be complicated to implement and difficult to use.

The ability to ascertain product origin and attributes from the farm through food processing, retail, and food service to consumers is growing in importance. Increasingly, public health concerns are requiring traceability. However, economic competition, which will reward those who can more effectively and reliably track and trace products back and forth through each step of the chain, will drive long-term adoption.

It is hoped that this chapter and book are a step forward toward consistent global baseline requirements for food traceability.

References

Anderson, M., Jaykus, L.-A., Beaulieu, S., Dennis, S., 2011. Pathogen-produce pair attribution risk ranking tool to prioritize fresh produce commodity and pathogen combinations for further evaluation (P^{3}ARRT). Food Control 22, 1865–1872.

Angus-Lee, H., December 2014/January 2015. The Evolution of Traceability in the Meat & Poultry Industry. Food Safety Magazine. http://www.foodsafetymagazine.com/magazine-archive1/december-2014january-2015/the-evolution-of-traceability-in-the-meat-poultry-industry/ (accessed 16.09.15.).

Centers for Disease Control and Prevention, 2014. Food Safety: Foodborne Illness: Outbreaks and Surveillance (Tracking). http://www.cdc.gov/foodsafety/facts.html (accessed 16.09.15.).

Centers for Disease Control and Prevention, 2011. CDC Estimates of Foodborne Illness in the United States. http://www.cdc.gov/foodborneburden/2011-foodborne-estimates.html (accessed 20.08.15.).

Centers for Disease Control and Prevention, 2009. Multistate Outbreak of *Salmonella* Typhimurium Infections Linked to Peanut Butter, 2008–2009. (Final Update) http://www.cdc.gov/salmonella/2009/peanut-butter-2008-2009.html (accessed 15.09.15.).

Centers for Disease Control and Prevention, 2008. Multistate Outbreak of *Salmonella* Saintpaul Infections Linked to Raw Produce. (Final Update) http://www.cdc.gov/salmonella/2008/raw-produce-8-28-2008.html (accessed 15.09.15.).

Centers for Disease Control and Prevention, 2006. Multistate Outbreak of *E. coli* O157:H7 Infections Linked to Fresh Spinach. (Final Update) http://www.cdc.gov/ecoli/2006/spinach-10-2006.html (accessed 15.09.15.).

Charlebois, S., Sterling, B., Haratifar, S., Kyaw Naing, S., 2014. Comparison of global food traceability regulations and requirements. Comprehensive Reviews in Food Science and Food Safety 13, 1104–1123. or http://onlinelibrary.wiley.com/enhanced/doi/10.1111/1541 4337.12101/ (accessed 18.09.15.).

Doering, C., 2009. Peanut Recall Having Big Impact on Small Firms, US Edition. Reuters, Washington, DC. http://www.reuters.com/article/2009/03/11/us-salmonella-peanuts-idUSTRE52A4AT20090311 (accessed 15.09.15.).

Flynn, D., 2013. Tomato Growers Want Compensation for Losses in 2008 Outbreak. Food Safety News. http://www.foodsafetynews.com/2013/08/tomato-growers-want-to-be-compensated-for-losses-in-2008-outbreak/#.VfhUKhF4qUk (accessed 15.09.15.).

Fortin, N.D., 2009. Food Regulation: Law, Science, Policy, and Practice. John Wiley & Sons, Hoboken, NJ, USA.

GS1 US, 2013. Traceability for Dairy, Deli, Bakery US Implementation Guideline. Available from: http://www.gs1us.org/industries/fresh-foods/dairy-deli-bakery/guide (accessed 30.09.15.).

GS1 US, 2010. Traceability for Fresh Fruits and Vegetables Implementation Guide. Available from: http://www.producetraceability.org/documents/Global_Traceability_Implementation_Fresh_Fruit_Veg.pdf (accessed 30.09.15.).

Goetz, G., February 22, 2013. Peanut Corporation of America from Inception to Indictment: A Timeline. Published in "*Food Safety News*". http://www.foodsafetynews.com/2013/02/peanut-corporation-of-america-from-inception-to-indictment-a-timeline/#.Vfg_7hF4qUk (accessed 15.09.15.).

Hoffman, S., Maculloch, B., Batz, M., 2015. Economic Burden of Major Foodborne Illnesses Acquired in the United States. Economic Research Service, of the US Dept. of Agriculture. http://www.ers.usda.gov/publications/eib-economic-information-bulletin/eib140.aspx (accessed 15.09.15.).

Innovation Center for US Dairy, 2013. Guidance for Dairy Product Enhanced Traceability. Available from: http://www.usdairy.com/ (accessed 30.09.15.).

Institute of Food Technologists, 2013. Pilot Projects for Improving Product Tracing Along the Food Supply System. Final Report by Jennifer McEntire & Tejas Bhatt http://www.fda.gov/Food/GuidanceRegulation/FSMA/ucm270851.htm (accessed 15.09.15.).

Institute of Food Technologists, 2015. Global Food Traceability Center. http://www.ift.org/gftc.aspx (accessed 25.09.15.).

McEntire, J.C., 2010. Traceability (product tracing) in food systems: an IFT report submitted to the FDA, Volume 1: technical aspects and recommendation. Comprehensive Reviews in Food Science and Food Safety 9 (1), 92–158.

Morgan Lewis, 2011. Summary of the Food Safety Modernization Act. http://www.morganlewis.com/pubs/foodsafetymodernizationact (accessed 15.08.15.).

mpXML, 2014. Meat and Poultry Data Standards. Available from: http://www.mpxml.org/I_DataSync.html (accessed 01.10.15.).

mpXML, 2013. A Model for the Adoption of Critical Tracking Events (CTEs) in the Meat & Poultry Supply Chain. Available from: https://www.gs1us.org/industries/fresh-foods/meat-and-poultry/cte (accessed 30.09.15.).

National Fisheries Institute, 2011. Traceability for Seafood US Implementation Guide. Available from: http://www.aboutseafood.com/sites/all/files/FINAL%20Seafood%20Trace%20Guide_v1.1.pdf (accessed 28.09.15.).

US Congress, 2011. Public Law 111–353. The Food Safety Modernization Act. U.S. Government Printing Office, Washington, DC. http://www.gpo.gov/fdsys/pkg/PLAW-111publ353/html/PLAW-111publ353.htm (accessed 18.09.15.).

US Federal register, August 3, 2015. Food Safety Modernization Act Domestic and Foreign Facility Reinspection, Recall, and Importer Reinspection Fee Rates for Fiscal Year 2016. https://www.federalregister.gov/articles/2015/08/03/2015-18906/food-safety-modernization-act-domestic-and-foreign-facility-reinspection-recall-and-importer (accessed 20.09.15.).

US Federal Register, April 4, 2014b. Establishment, Maintenance, and Availability of Records: Amendment to Record Availability Requirements. https://www.federalregister.gov/articles/2014/04/04/2014-07550/establishment-maintenance-and-availability-of-records-amendment-to-record-availability-requirements.

US Federal Register, April 2, 2014c. Designation of High-Risk Foods for Tracing; Request for Comments and for Scientific Data and Information. https://www.federalregister.gov/articles/2014/02/04/2014-02255/designation-of-high-risk-foods-for-tracing-request-for-comments-and-for-scientific-data-and (accessed 21.09.15.).

US Food and Drug Administration, 2015a. FDA Food Safety Modernization Act. http://www.fda.gov/Food/GuidanceRegulation/FSMA/ (accessed 12.08.15.).

US Food and Drug Administration, May 12, 2015b. Registration of Food Facilities. http://www.fda.gov/food/guidanceregulation/foodfacilityregistration/default.htm (accessed 18.09.15.).

US FDA, 2015c. FDA Low-acid Canned Foods Regulation of 1974. http://extension.psu.edu/food/entrepreneurs/regulations/fda-canning-regulations/FDA-21CFR-Part-113.pdf/view.

US FDA, September 10, 2015d. FSMA Final Rule for Preventive Controls for Human Food. http://www.fda.gov/food/guidanceregulation/fsma/ucm334115.htm (accessed 18.09.15.).

US FDA, 2015e. Recalls – Guidance on Policy, Procedures, and Industry Responsibilities. http://www.accessdata.fda.gov/scripts/cdrh/cfdocs/cfcfr/cfrsearch.cfm?cfrpart=7&showfr=1&subpartnode=21:1.0.1.1.6.3.

US FDA, 2015f. Guidance for Industry: What You Need to Know about Establishment and Maintenance of Records; Small Entity Compliance Guide. April 3, 2014 http://www.fda.gov/Food/GuidanceRegulation/GuidanceDocumentsRegulatoryInformation/ucm391329.htm (accessed 17.09.15.).

US FDA, 2015g. Registration of Food Facilities. http://www.fda.gov/Food/GuidanceRegulation/FoodFacilityRegistration/default.htm (accessed 21.09.15.).

US FDA, 2015h. FSMA Proposed Rule: Amendments to Registration of Food Facilities. http://www.fda.gov/Food/GuidanceRegulation/FSMA/ucm440988.htm (accessed 21.09.15.).

US FDA, 2015i. Guidance for Industry: Questions and Answers Regarding Food Facility Registration, sixth ed. http://www.fda.gov/Food/GuidanceRegulation/GuidanceDocumentsRegulatoryInformation/ucm331959.htm (accessed 21.09.15.).

US FDA, 2015j. FSMA Final Rule for Preventive Controls for Animal Food. http://www.fda.gov/Food/GuidanceRegulation/FSMA/ucm366510.htm (accessed 28.09.15.).

US FDA, 2015k. Influencing Outcomes in the Global Food System: FDA's Evolving and Expanding Food Safety Role. http://www.fda.gov/food/guidanceregulation/fsma/ucm443731.htm (accessed 01.10.15.).

US FDA, February 2014b. FDA's Draft Approach for Designating High-risk Foods as Required by Section 204 of FSMA (pdf document) http://www.fda.gov/food/guidanceregulation/fsma/ucm380210.htm (accessed 15.09.15.).

US FDA, 2014d. Full Text of the Food Safety Modernization Act (FSMA) [same text as US Congress. 2011. Public Law 111-353 above but in More Readable Format]. http://www.fda.gov/Food/GuidanceRegulation/FSMA/ucm247548.htm (accessed 19.09.15.).

US FDA, October 23, 2014e. DNA Testing at Wholesale Level to Evaluate Proper Labeling of Seafood Species. http://www.fda.gov/Food/GuidanceRegulation/GuidanceDocuments-RegulatoryInformation/Seafood/ucm419982.htm (accessed 20.09.15.).

US FDA, 2014f. Seafood HACCP. http://www.fda.gov/Food/GuidanceRegulation/HACCP/ucm2006764.htm (accessed 20.09.15.).

US FDA, 2014g. Juice HACCP. http://www.fda.gov/Food/GuidanceRegulation/HACCP/ucm2006803.htm (accessed 20.09.15.).

US FDA, October 26, 2014i. Fish and Fishery Products Hazards and Controls Guidance, fourth ed. http://www.fda.gov/Food/GuidanceRegulation/GuidanceDocumentsRegulatoryInformation/Seafood/ucm2018426.htm (accessed 24.09.15.).

US FDA, 2014j. Operational Strategy for Implementing the FDA Food Safety Modernization Act (FSMA). http://www.fda.gov/food/guidanceregulation/fsma/ucm395105.htm (accessed 30.09.15.).

US FDA, 2016. FSMA Final Rule on Sanitary Transportation of Human and Animal Food. http://www.fda.gov/Food/GuidanceRegulation/FSMA/ucm383763.htm (accessed 6. 5. 2016).

US Food and Drug Administration, 2009. The Bioterrorism Act of 2002: Title 3-Protecting the Safety and Security of the Food and Drug Supply. http://www.fda.gov/RegulatoryInformation/Legislation/ucm155769.htm (accessed 18.09.15.).

U.S. General Accountability Office (GAO), 2012. Food Safety: FDA's Food Advisory and Recall Process Needs Strengthening. GAO-12-589. http://www.gao.gov/products/GAO-12-589 (accessed 02.10.15.).

U.S. General Accountability Office (GAO), 2007. High Risk Series: An Update. GAO-07-310 http://www.gao.gov/new.items/d07310.pdf (accessed 15.09.15.).

U.S. General Accountability Office (GAO), 2001. Food Safety and Security: Fundamental Changes Needed to Ensure Safe Food. GAO-02-47T. http://www.gao.gov/new.items/d0247t.pdf (accessed 15.09.15.).

US GPO, December 1995. Procedures for the safe and sanitary processing and importing of fish and fishery products. Published in the Federal Register 60 (242). Final Rule effective December 1997. http://www.gpo.gov/fdsys/granule/FR-1995-12-18/95-30332/content-detail.html (accessed 25.10.15.).

Venkataraman, B., 2008a. As Outbreak Affects 1,000, Experts See Flaws in Law. July 10. The New York Times. http://www.nytimes.com/2008/07/10/health/policy/10tomato.html?fta=y (accessed 18.09.15.).

Venkataraman, B., 2008b. Amid Salmonella Case, Food Industry Seems Set to Back Greater Regulation. July 31. The New York Times. http://www.nytimes.com/2008/07/31/health/policy/31outbreak.html?ref=us (accessed 18.09.15.).

Zhang, J., Bhatt, T., 2014. A guidance document on the best practices in food traceability. Comprehensive Reviews in Food Science and Food Safety 13, 1074–1103. or http://bit.ly/1pGr1py.

Part Four

Case Studies in Food Traceability

Effective Use of Food Traceability in Product Recall

14

T.M.L. Diallo, S. Henry
University of Lyon, University Lyon 1, Villeurbanne, France

Y. Ouzrout
University of Lyon, University Lyon 2, Lyon, France

1. Introduction

Due to the globalization of exchanges (several suppliers and customers across all continents), product complexity (several ingredients and complex manufacturing processes), and regulations (accountability of the manufacturer on its product), the product recall is a challenge that is facing more and more industries (Kumar, 2014).

Product recall is "any measure aimed at achieving the return of a dangerous product that has already been supplied or made available to consumers by the producer or distributor." The difference between product recall and product withdrawal should be noted. The latter is defined as "any measure aimed at preventing the distribution, display and offer of a product dangerous to the consumer" (EU, 2001). According to the degree of dangerousness of the product, three classes of recall are distinguished (Kumar and Budin, 2006):

- Class 1: This is the more stringent class. It is advocated when the use or exposure to the product can cause serious and lasting health problems or death.
- Class 2: The product may cause temporary health problems but can lead to serious problems in the long-term.
- Class 3: With the lowest severity, it concerns cases where there is no health risk.

The strategy and the impact of the recall obviously vary according to the class concerned. A company may decide to issue a voluntary recall or be forced to do so by authorities.

Although product recall may concern any type of product, it is especially more frequent and critical in the field of food products (see reports (Commission, 2013b) and (Commission, 2013a)).

The causes of these recalls mainly come from raw materials (contamination, wrong dosage, foreign objects, etc.), equipment (failure, configuration mistakes, etc.), or processes (design errors, manufacturing, labeling, marking, handling, etc.) (Commission, 2013a,b; Kumar and Budin, 2006; Potter et al., 2012; Hora et al., 2011; Berman, 1999).

As the causes of a recall may arise from the materials or the process itself, traceability shall therefore bear on these two aspects, namely material and process traceability. The determination of the causes of a nonconformity will depend in large part on the quality of the traceability system.

Advances in Food Traceability Techniques and Technologies. http://dx.doi.org/10.1016/B978-0-08-100310-7.00014-4

First, in the next section, we describe the key tasks of any recall procedure. In Section 3, we explain how food traceability can be effectively used in product recall. The issue of reducing the size of the recall through the reduction of the dispersion and root cause analysis is addressed in Section 4. The contribution of internal and external traceability to the product recall process is presented in Section 5, and finally, we make a distinction between unitary traceability and batch traceability and their uses in the recall process in Section 6.

2. Product Recall

The product recall scenarios are quite varied depending on the type of product, the type and scope of the supply chain, and the regulations that are in force. These scenarios also vary depending on the lifetime of the product, its manufacturing process, and the actors involved in its life cycle (manufacturer, distributor, and retailer) (Wynn et al., 2011; Hora et al., 2011; Kumar, 2014; Malickson, 1983).

Because of the unforeseen nature of such incidents and the need to react quickly in the event of product recalls, one should be prepared in advance. The actions to be implemented to cope with this type of incident should be well defined in advance. It is in this spirit that the authorities of some countries provide product recall guidelines. That is the case, for instance, for the United States (US Food and Drugs Administration [FDA], Consumer Products Safety Commission [CPSC], National Highway Traffic Safety Administration [NHTSA]) (CPSC, 2014; FDA, 2014), the EU (PROSAFE, 2011; Services, 2014), and Australia (*Australian Competition and Consumer Commission*) (Branch and Commission, 2014). Standardization bodies such as GS1 (GS1, 2012) or industrial federations have also developed product recall standards. The GS1 product recall standard, for example, defines and standardizes critical attributes to be collected and exchanged between partners in the supply chain and authorities during a recall. More or less, similar recall procedures have been published in the literature. Analysis of these procedures identifies the main actions carried out during a recall. In Wynn et al. (2011), the authors propose a generic workflow to coordinate the product recall process and communication with stakeholders. The main identified activities were:

- in-depth assessment of the situation in order to investigate the causes and involved products and entities
- risk analysis to decide whether or not to make a recall and to determine the scope of the recall
- when a recall is decided, perform the call by following a preestablished guide
- correct the causes of the recall
- communicate and document the incident

Kumar relates the recall procedure of a large international group that includes the following steps: (1) notification to the relevant authorities of the intention of the company to undertake a recall, (2) publication of the recall notice on the media (websites and/or newspapers), (3) incentive for customers to return products, and

(4) compensation to customers on the basis of existing laws and the causes of the recall (Kumar, 2014). Berman (1999) divided the recall activities into three groups depending on whether they are carried out before, during, or after the recall. With product recall being by definition unpredictable, it should be anticipated. Among preventive actions, B. Barry cites the designation of a product recall committee and a coordinator, the establishment of a planning of security operations, the development and maintenance of effective channels of communication, and the development and maintenance of efficient products and customer databases. Safety analysis, budget estimation, information of the intermediaries and end customers, the recovery of recalled products, and repair or replacement of the recalled products are the activities promoted during the recall. After the recall, the author suggests to tackle the restoration of the reputation of the firm and to evaluate the effectiveness of the recall process. Hamory and Duffy identified six actions to be taken following the detection of the problem requiring a recall, which are verification and characterization of the problem, determination of the scope of the problem, information for in-house staff, warning the other stakeholders, implementation of corrective action (recall, replacement, repair, destruction), and documentation of the recall process (Riggs Duffy and Hamory, 1987).

By analyzing these different procedures, we find common actions to all of these procedures expressed differently by the authors or sometimes in an implicit way. These common actions that contribute to achieve an efficient recall are root cause analysis, risk analysis, the recall itself, communication and documentation, and post-crisis management.

2.1 Cause Analysis

Analysis of the causes should take place immediately after the detection or reporting of noncompliance. Noncompliance may be detected by the manufacturer, distributor, retailer, consumer, or authorities (Magno, 2012; Kumar, 2014; Riggs Duffy and Hamory, 1987). The determination of the causes may be obvious or may require further investigation. The causes can be determined by analyzing the traceability data related to manufacturing and distribution processes. Thanks to the determination of the causes, one can situate accountability for the detected noncompliance. The causes may result from the manufacturing process, transportation, storage, or use (Riggs Duffy and Hamory, 1987). The information collected during this cause analysis phase will serve in all the other actions of the recall process (Kramer et al., 2005).

2.2 Risk Analysis

Risk analysis can be performed after the reporting and confirmation of noncompliance or after the search of the causes of the noncompliance. One of the risk analysis objectives is to decide whether or not a recall is necessary. This decision may be taken by authorities and imposed on the manufacturer. The recall decision can also be taken voluntarily by producers when it is aware of the danger or as a precautionary measure. The risk analysis also helps to define the products to be recalled. If the causes of the noncompliance are not known, all manufactured lots are generally recalled.

A SWOT (Strengths, Weaknesses, Opportunities, and Threats) analysis can be used to decide whether it is appropriate to make a recall in the case of a voluntary recall. FMEA (Failure Mode and Effects Analysis), HACCP (Hazard Analysis Critical Control Points), and fault-tree methods can also be used for risk assessment.

Risk analysis can be used as a preventive measure to avoid the recall or as corrective action to better manage the crisis related to the product recall by choosing the most appropriate corrective action (Berman, 1999). It assists in determining the criticality class of the recall and thus to make appropriate decisions. Among the possible strategies, the company may decide to make a full or a selective recall to issue a warning or to modify the instruction manual and/or the labeling (Berman, 1999).

2.3 The Recall Itself

When a recall is decided, two cases can be distinguished. In the first case, the product is still within the supply chain in a warehouse, stored by a distributor, or at a retail outlet. In the second case, the product has reached the end customers. The first case is relatively easier to manage. If a good traceability exists, it will be easier to determine which items to recall and their locations. In the second case, the task is more delicate and requires the cooperation of the retailer and the media to inform potential holders of the products of their dangerousness and induce them to return the noncompliant products. In the latter case, it is necessary to master the communication with the public. To encourage consumers to return defective products, awards, in addition to reimbursement of the defective product, can be granted to them (Berman, 1999). To manage the process, the product recall can be treated as a reverse logistics process (Hora et al., 2011). The recalled faulty products are either repaired or destroyed.

2.4 Communication and Documentation

The management of the incident related to the recall includes internal communication (legal department, management, employees, etc.) and external communication with other stakeholders in the supply chain (suppliers, subcontractors, etc.) with authorities, media, customers, and the general public. The constraints and means vary depending on the interlocutor. It is appropriate to be prepared in advance and not to improvise, because a communication error could have significant financial and legal consequences. The messages delivered should be consistent and comply with the legal obligations. External communication is often about the causes of the recall and corrective actions taken or to be taken (Hora et al., 2011). The concerned products, the nature of the problem, and the primary contact person must be correctly indicated (Berman, 1999). The company responsible for the recall must inform its employees and partners (intermediaries, distributors) about the nature of the risk and the procedure put in place to make the recall (Berman, 1999). The media can help in the management of the crisis by informing the public concerned about the crisis and the procedure to be followed to reduce risk. They can also complicate the management of the crisis by creating panic among consumers, blurring the message, or by affecting the company's brand image (Malickson, 1983). A clear communication policy must be set to take advantage of the media and minimize the effects of negative publicity.

2.5 Postcrisis Management

All actions undertaken as part of the recall process must be recorded. This documentation of the incident could be used for continuous improvement to prevent such an incident from happening again or for managing litigation cases. The documentation also helps to remember the causes and actions put in place to be better prepared in the future. It also enables to transmit information to competent authorities on the management of the crisis and to learn from this experience.

Despite the challenges that can entail a recall, the scientific literature in this area is not very abundant (Magno, 2012). Existing studies are often on the impacts (financial, brand image, etc.) that a recall can have and management organization for this type of crisis. To successfully complete the various actions recommended by the different approaches, companies must have reliable and detailed knowledge of the failure causing the recall and the severity and extent of the incident. This knowledge is best acquired through a comprehensive and reliable traceability system. The fast and reliable determination of root causes of the incident and the identification of the other potentially affected products allows for a targeted recall and limits the direct and indirect impacts of the recall. The analysis of root causes requires a lot of time and resources according to the stage of the life cycle from which the defect comes (design, manufacturing, storage, distribution, use). When the fault is likely to endanger users and if the root cause is not easily identifiable, the recall is made first, and the search for the root cause is subsequently carried out (Fig. 14.1).

The recall done this way most often concerns the entire production lot. For batch production and in the case of contamination, recalling the entire batch is justified. But in the case of other types of production and when the fault does not concern the whole production lot, recalling the whole lot is unjustified and entails additional costs that are avoidable. In the latter case, one should look for the root causes first and only recall the products identified as potentially faulty (Fig. 14.2).

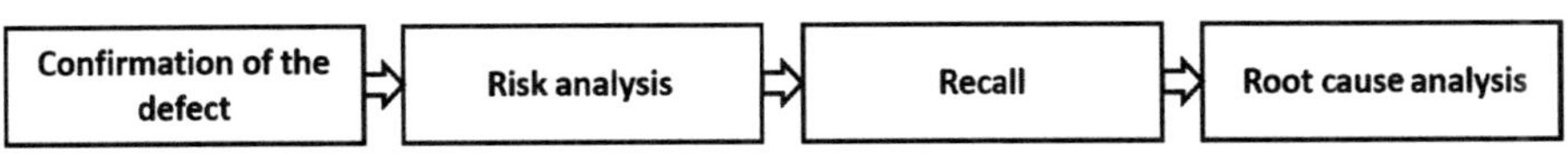

Figure 14.1 The sequence of recall tasks when the analysis of root causes is performed after the recall itself.

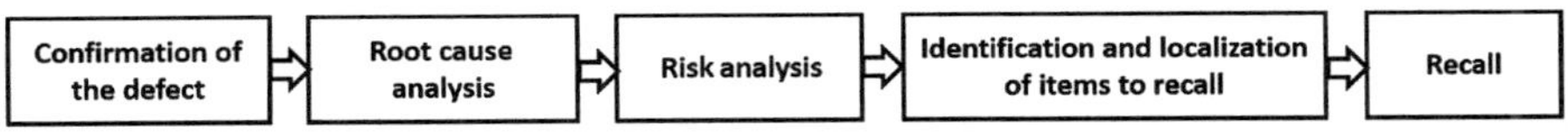

Figure 14.2 Recall procedure with the analysis of root causes before performing the recall.

Berman (1999) provides some examples of direct and indirect costs of a product recall. Investigation costs, transportation and storage costs, reimbursement, and fine costs are examples of expenses faced by a company performing a recall. Knowing the life cycle of the product from the origin of raw materials and its path is essential to identify the products to be recalled and set an effective procedure for the recall. That is why traceability plays a key role in the product recall process (Kumar 2014; Wynn et al., 2011; Kumar and Budin, 2006). Given that the causes of a recall can come from materials or from the process itself, traceability must cover both the materials and processes. The EU requires, for certain products (food and drugs), the possibility to trace all the items from suppliers to final consumers (Kumar, 2014; Wynn et al., 2011; Dabbene et al., 2014). Through good traceability, we can determine, locate, and date with certainty the events that occurred during the product life cycle. It will therefore be possible to search for the root causes and to identify the products likely to present the same failure. The ability to trace the product to be recalled is a common need for all recall procedures previously mentioned (Wynn et al., 2011). The determination of the causes of noncompliance where these are not obvious will depend, in large part, on the quality of the existing traceability system. In the following section, we will examine the relationship that exists between the product recall procedure and the traceability system.

3. Traceability and Product Recall

If traceability is imposed in some cases by the regulation, it is primarily to allow the identification of all potentially defective items in the event of product recalls (Storøy et al., 2013; Thakur, 2010). According to the ISO 22005 standard, traceability systems contribute to the search for the cause of nonconformity and enable, if necessary, withdraw and/or recall products (ISO, 2007).

Within the framework of the product recall, two types of traceability are necessary: tracking or forward traceability and tracing or backward traceability (Wynn et al., 2011; Storøy et al., 2013).

The tracking or forward traceability makes it possible to determine the finished products containing a particular ingredient or having undergone a particular process. Tracing or backward traceability, in turn, is useful in identifying suppliers and processes that have contributed to the production of a particular product.

Traceability data may be minimum and may contain only the date of manufacture, expiry date, and batch number, for example. It can likewise be richer, including process data serialized at item level. The finer the traceability data, the most targeted the recall (Moe, 1998; Berman, 1999). Targeted recall borderline the recall to only defective items. It thus reduces the direct and indirect impacts of the recall. Kumar and Budin (2006) give examples of recalls in which a massive recall was conducted beyond the noncompliant products because that traceability was lacking.

The recall also requires internal and external traceability. External traceability serves to determine the location of items to be recalled and to coordinate recall actions. The internal traceability contributes to the achievement of root cause analysis to determine the causes of noncompliance and to perform a targeted recall.

4. Minimizing the Size of the Recall Through the Traceability System by the Reduction of the Dispersion and Root Cause Analysis

The usual strategy, which consists of recalling entire lots without knowing the status (compliant or not) of recalled products, is expensive and does not foster continuous improvement (Diallo et al., 2014). This strategy induces sometimes unjustified and avoidable costs. These are direct costs related to the recall activities and indirect costs, especially the impact on the brand image. These massive recalls are generally carried out without knowing the status of all the recalled products. There are several examples where companies do very large recalls because they cannot identify really defective items (Hora et al., 2011). In this section, we discuss the different two possibilities to reduce the number of recalled products: reduction of the dispersion and determination of root causes to optimize the size of the recalled lots.

Some work has addressed the issue of reducing the number of recalled products. In Kumar (2014), the FMEA method and fault tree were used to determine the causes of noncompliance and to assess the reliability of the recall supply chain. In Kumar and Budin (2006), the HACCP method is used to prevent a recall or to better manage the crisis caused by the recall. In order to estimate the cost borne by each actor of the supply chain during a recall, the authors of Piramuthu et al. (2013) drew on a probabilistic model of the place of contamination. They determined the level of the supply chain at which the recall should be done. The considered supply chain consists of three levels: producer, distributor, and retailer. Conze and Kruger (2013) define the recall strategy to adopt based on a probabilistic risk analysis. The work published by Chen and Schweickert (2004) propose to determine the conditional probability of a product recall knowing that the products just before or after are recalled. They calculate the probability of recalling products adjacent to a recalled product.

Among the approaches determining root causes to optimize the size of the recalled lots, we can distinguish those using deterministic reliability engineering tools such as FMECA, HACCP, cause-effect diagram, and fault tree from those taking a probabilistic approach. In the industrial context with complicated and difficult model processes and involving uncertainty, the use of deterministic methods with categorical decisions is not always justifiable. It is therefore necessary to define a probabilistic causal analysis model, taking into account these uncertainties (Kumar, 2014). The other solution proposed to reduce the recall size is to reduce the dispersion.

To reduce the size of the recalled lots, other authors have proposed to reduce the batch dispersion by reducing the size and the mixing of batches using linear programming (Dupuy et al., 2005) or genetic algorithms and neural networks (Tamayo et al., 2009). Dupuy et al. define the notions of downward dispersion, upward dispersion, and batch dispersion. The downward dispersion of a raw material batch is "the number of finished product batches which contain parts of this raw material batch," the upward dispersion of a finished product batch is "the number of different raw material batches used to produce this batch," and the batch dispersion is "the sum of all raw material downward dispersion and all finished products upward dispersion" (Dupuy et al., 2005). The principle of the batch dispersion methodology is to identify and recall all the finished products containing the noncompliant raw material batch. Thus if the downward dispersion is large, the amount of recalled products will be too. In addition, if a noncompliant product is detected, all raw material batches used to produce it must be analyzed to investigate the causes of the nonconformity. Also in this case, low upward dispersion will make analysis of the raw material batches much quicker and easier to accomplish. In brief, a low batch dispersion will minimize recall size. However, before considering this solution of the reduction of the dispersion to minimize the quantity of recalled items, one should first assess how much leeway is possible in changing recipes and the costs that it involves. Indeed, in many industries, raw material batches from different suppliers for which some characteristic parameters are different are often mixed together (Memon et al., 2014). Furthermore, the definition of a batch depends on the type of production (continuous production or batch production) and the intended use. The input raw material is not always sufficient to characterize a batch. The series of manufacturing operations should also be considered. The definition of Dupuy et al. applies primarily to batch production systems.

For continuous or job production, the concept of a lot may be different from that of the batch production. In the case of continuous or job production, a lot-based recall may not be optimal because the articles of the same production lot may have different compositions or be produced in different conditions. Thereby, certain products of the recalled production batch may not have the noncompliance for which the recall is made.

Minimizing the size of the recall by reducing the batch dispersion as explained previously is applicable if the noncompliance is related to materials contained in the final product. Production faults, operating errors, and design mistakes are other potential causes of a recall to consider.

5. The Need of Internal Traceability in Addition to External Traceability

To successfully complete actions that contribute to achieve an effective recall, we need a reliable and thorough knowledge about the fault which caused the recall and about the severity and the extent of the problem. This knowledge is needed to quickly determine the root cause of the incident and the other potentially affected products. An effective collaboration between the different actors of the supply chain is another critical success factor. A good traceability system is necessary to achieve an effective recall. In terms of visibility and management policy, there are two levels of traceability: internal and external traceability (AISBL, 2013). Internal traceability concerns private data such as product design data and production parameters, origin of raw materials, and quality records. External traceability is related to public data that the different partners in a supply chain exchange between them to ensure end-to-end traceability. The types of data, the means for data collection, and storage for the two levels of traceability are different. Both level of traceability are useful for the recall process. To find the root causes of noncompliance, one mostly needs the internal traceability data (Diallo et al., 2014). The external traceability is necessary to determine the locations of items to recall, the supplier of a given raw material, and to manage the return of the noncompliant products.

6. Use of Batch Traceability and Unitary Traceability in Recall Procedure

Depending on the level of detail of the traceability process, a distinction could be drawn between batch traceability and unitary traceability. For batch traceability, the TRU (Traceable Resource Unit) is a batch of products. Product items belonging to the same batch are considered to be homogenous and share a common batch number. Under batch traceability procedure, for each tracked parameter, only one value is retained for all the items in the batch. The unitary traceability enables a serialized unique identification at the item level. Each item is uniquely identified using a serial number. Each item has a value for each of the tracked parameters. This requires a more complex tracking system and generates a very large amount of data. The unitary traceability is useful to achieve a targeted recall by the fact it offers extensive knowledge of the production processes and the supply chain. However, for some types of batch production systems, including the food industry, it is not necessary to implement a unitary traceability. The batch traceability is enough for these systems because incidents derive from materials and production recipes most of the time.

7. Conclusion and Perspectives

The product recall is a curative action that involves withdrawing from the market all products likely to present a defect that could affect the health or safety of consumers. To achieve an efficient recall, a reliable and adapted traceability system must be established. The actions to be implemented in case of recall should also be anticipated in order to avoid any type of impromptu action. In the case of complex industrial systems with many stakeholders, the search for root causes of a defect may be difficult and time-consuming. When the variability of the process is high, with complicated and difficult to model processes and involving uncertainty, the use of deterministic methods such as FMEA, HACCP, and fault tree with categorical decisions for causal analysis is not possible, hence the need to develop more sophisticated analysis tools. Probabilistic approaches such as Bayesian network offer interesting perspectives to address these challenges through the exploitation of traceability data to complement the expert knowledge.

References

AISBL, G., 2013. GS1 Traceability Standard. [Online]. Available: http://www.gs1.org/gsmp/kc/traceability (accessed 16.02.03.).

Berman, B., 1999. Planning for the inevitable product recall. Business Horizons 42, 69–78.

Branch, P.S.H.R., Commission, A.C.C., 2014. Product Safety Recalls Australia. [Online]. Available: http://www.recalls.gov.au/content/index.phtml/itemId/952401 (accessed 07.04.14.).

Chen, S., Schweickert, R., 2004. Predicting immediate ordered recall of lists with products of probabilities. Journal of Mathematical Psychology 48, 135–154.

Commission, D.-G.F.H.A.C.O.T.E., 2013a. 2012 Annual Report on the Operation of the Rapid Alert System for Non-Food Dangerous Products RAPEX. UE.

Commission, H.A.C.D.-G.O.T.E., 2013b. The Rapid Alert System for Food and Feed 2012 Annual Report. UE.

Conze, D.B., Kruger, C., 2013. Guiding product recall strategy with probabilistic modeling. Food Technology 67, 21.

CPSC, U.S.C.P.S.C., 2014. U.S. Consumer Product Safety Commission. [Online]. Available: http://www.cpsc.gov/en/ (accessed 09.04.14.).

Dabbene, F., Gay, P., Tortia, C., 2014. Traceability issues in food supply chain management: a review. Biosystems Engineering 120, 65–80.

Diallo, T.M.L., Henry, S., Ouzrout, Y., 2014. Using unitary traceability for an optimal product recall. In: Grabot, B., Vallespir, B., Gomes, S., Bouras, A., Kiritsis, D. (Eds.), Advances in Production Management Systems. Innovative and Knowledge-based Production Management in a Global-local World. Springer, Berlin, Heidelberg.

Dupuy, C., Botta-Genoulaz, V., Guinet, A., 2005. Batch dispersion model to optimise traceability in food industry. Journal of Food Engineering 70, 333–339.

EU, 2001. Directive 2001/95/Ec of the European Parliament and of the Council of 3 December 2001 on General Product Safety.

FDA, F.A.D.A., 2014. Guidance for Industry: Product Recalls, Including Removals and Corrections. [Online]. Available: http://www.fda.gov/safety/recalls/industryguidance/ucm129259.htm (accessed 07.04.14.).

GS1, 2012. Product Recall in Multiple Recall Jurisdictions Implementation Guideline.

Hora, M., Bapuji, H., Roth, A.V., 2011. Safety hazard and time to recall: the role of recall strategy, product defect type, and supply chain player in the U.S. toy industry. Journal of Operations Management 29, 766–777.

ISO, 2007. NF EN ISO 22005: Traceability in the Feed and Food Chain – General Principles and Basic Requirements for System Design and Implementation.

Kramer, M.N., Coto, D., Weidner, J.D., 2005. The science of recalls. Meat Science 71, 158–163.

Kumar, S., 2014. A knowledge based reliability engineering approach to manage product safety and recalls. Expert Systems with Applications 41, 5323–5339.

Kumar, S., Budin, E.M., 2006. Prevention and management of product recalls in the processed food industry: a case study based on an exporter's perspective. Technovation 26, 739–750.

Magno, F., 2012. Managing product recalls: the effects of time, responsible vs. opportunistic recall management and blame on consumers' attitudes. Procedia – Social and Behavioral Sciences 58, 1309–1315.

Malickson, D.L., 1983. Are you ready for a product recall? Business Horizons 26, 31–35.

Memon, M., Lee, Y., Mari, S., 2014. A model for optimizing traceability of product in a supply chain based on batch dispersion. In: Golinska, P. (Ed.), Logistics Operations, Supply Chain Management and Sustainability. Springer International Publishing.

Moe, T., 1998. Perspectives on traceability in food manufacture. Trends in Food Science & Technology 9, 211–214.

Piramuthu, S., Farahani, P., Grunow, M., 2013. RFID-generated traceability for contaminated product recall in perishable food supply networks. European Journal of Operational Research 225, 253–262.

Potter, A., Murray, J., Lawson, B., Graham, S., 2012. Trends in product recalls within the agri-food industry: empirical evidence from the USA, UK and the Republic of Ireland. Trends in Food Science & Technology 28, 77–86.

PROSAFE, 2011. Guidelines for Businesses to Manage Product Recalls & Other Corrective Actions.

Riggs Duffy, K., Hamory, B.H., 1987. Product recalls—what to do. American Journal of Infection Control 15, 131–134.

Services, E.S.O.P.A., 2014. Safety of Products and Services.

Storøy, J., Thakur, M., Olsen, P., 2013. The TraceFood Framework – Principles and guidelines for implementing traceability in food value chains. Journal of Food Engineering 115, 41–48.

Tamayo, S., Monteiro, T., Sauer, N., 2009. Deliveries optimization by exploiting production traceability information. Engineering Applications of Artificial Intelligence 22, 557–568.

Thakur, M., 2010. Operational Techniques for Implementing Traceability in Bulk Product Supply Chains. Iowa State University.

Wynn, M.T., Ouyang, C., Ter Hofstede, A.H.M., Fidge, C.J., 2011. Data and process requirements for product recall coordination. Computers in Industry 62, 776–786.

Analytical Approaches for the Evaluation of Food Protected Designation of Origin

15

S. Armenta, M. de la Guardia
University of Valencia, Burjassot, Valencia, Spain

1. Social and Law Frame of Protected Designation of Origin Labels

Agriculture in particular and food sectors in general have become global industries, transcending national boundaries and supplying a worldwide market place. This globalization of food markets, facilitated by the fast transport of goods around the world, caused a growing public distrust as a consequence of the increasing number of incidents related to food safety and mislabeling worldwide. Consumers around the world increasingly demand safe and high quality products, understanding quality as the sum of features, characteristics, and properties of a product, which bear on its ability to satisfy stated or implied needs (Will and Guenther, 2007). Quality is a flexible term in which intrinsic attributes such as taste, texture, and shelf life and extrinsic ones such as the use of pesticides, type of packaging material, processing technology, or the use of genetically modified organisms are important features. In summary, consumers demand agricultural products or foodstuffs with identifiable origins and specific characteristics. In particular, those linked to the geographical origin of food and their production conditions are of important concern.

The World Intellectual Property Organization (WIPO) defines a geographical indication (GI) as "a sign used on goods that have a specific geographical origin and possess qualities, a reputation, or characteristics that are essentially attributable to that origin" (WIPO, 2015). Appellations of origin (AO) are a special kind of GI. The term was used in the Paris Convention in 1883 (WIPO, 1883) and was defined in the Lisbon Agreement in 1958 (WIPO, 1958). Article 2 of the Lisbon Agreement defines AO as "… the geographical denomination of a country, region, or locality, which serves to designate a product originating therein, the quality or characteristics of which are due exclusively or essentially to the geographical environment, including natural and human factors". At present, no "worldwide" GI right exists. As explained by the WIPO (WIPO, 2015), "intellectual property rights are governed by the 'territoriality principle'. The effects of a right obtained in a particular jurisdiction are limited to the territory of that jurisdiction.

Advances in Food Traceability Techniques and Technologies. http://dx.doi.org/10.1016/B978-0-08-100310-7.00015-6

Thus, where a right over a GI is obtained in one jurisdiction, it is protected there but not abroad. In other jurisdictions, the GI would face the risks usually associated with lack of protection". There are three main ways of protecting a GI abroad: (1) by obtaining protection directly in the concerned jurisdiction, (2) by taking advantage of bilateral/plurilateral agreements concluded between countries, and (3) through WIPO Lisbon System for the International Registration of Appellations of Origin.

At the end of the 20th century, the European Union (EU) recognized and supported the potential of differentiating quality products on a regional basis. The first regulation on geographical indications was adopted in the EU in 1992 (EU Council, 1992), to harmonize diverse protection instruments existing in some Member States and to create a system of registration and protection of names compatible with the single common market. Since then, EU law lays down stringent requirements guaranteeing the standards of all European products. The EU has created three labels regarding Protected Designation of Origin (PDO), Protected Geographical Indication (PGI), and Traditional Specialty Guarantee (TSG) (see Fig. 15.1 for details regarding the European labels) which must follow quality schemes to identify products and foodstuffs farmed and produced under exact specifications. Those schemes encourage diverse agricultural production, protect product names from misuse and imitation, and help consumers by giving them information concerning the specific character of the labeled products.

PDO covers agricultural products and foodstuffs which are produced, processed, and prepared in a given geographical area (a specific place, region, or, in exceptional cases, a country) using recognized know-how, whose quality or characteristics are essentially or exclusively due to a particular geographical environment with its inherent natural and human factors and the production steps which all take place in the defined geographical area.

PGI protects agricultural products and foodstuffs closely linked to the geographical area (a specific place, region, or country), whose given quality, reputation, or other characteristics are essentially attributable to its geographical origin, and at least one of the stages of production, processing, or preparation takes place in the defined geographical area.

TSG indicates the traditional character of food, either in the composition or means of production. It describes a specific product or foodstuff that results from a mode of production, processing, or composition corresponding to traditional practice for that product or foodstuff or is produced from raw materials or ingredients than are those traditionally used.

Since 1992, more than 900 names have been registered for agricultural products and foodstuffs and systems for wine and spirits ensure protection for 1800 and 323 names, respectively. Thus it is clear that food products protected by PDO, PGI, or TSG provide big business in the EU, as the worldwide sales value of GI products registered in the EU were estimated at €54.3 billion in 2010.

In the EU, the list of registered names by country of origin is led by Italy by 33% regarding cheese, meat, and olive oil; Germany by 25% (beer, pastry, and meat); and France by 17% (dairy products and fresh and processed meat), followed by the United Kingdom (beef, lamb, and salmon), Spain (fruit and vegetables, meat and cheese), and Greece (dairy sector and olive oil). However, the incorporation of PDO

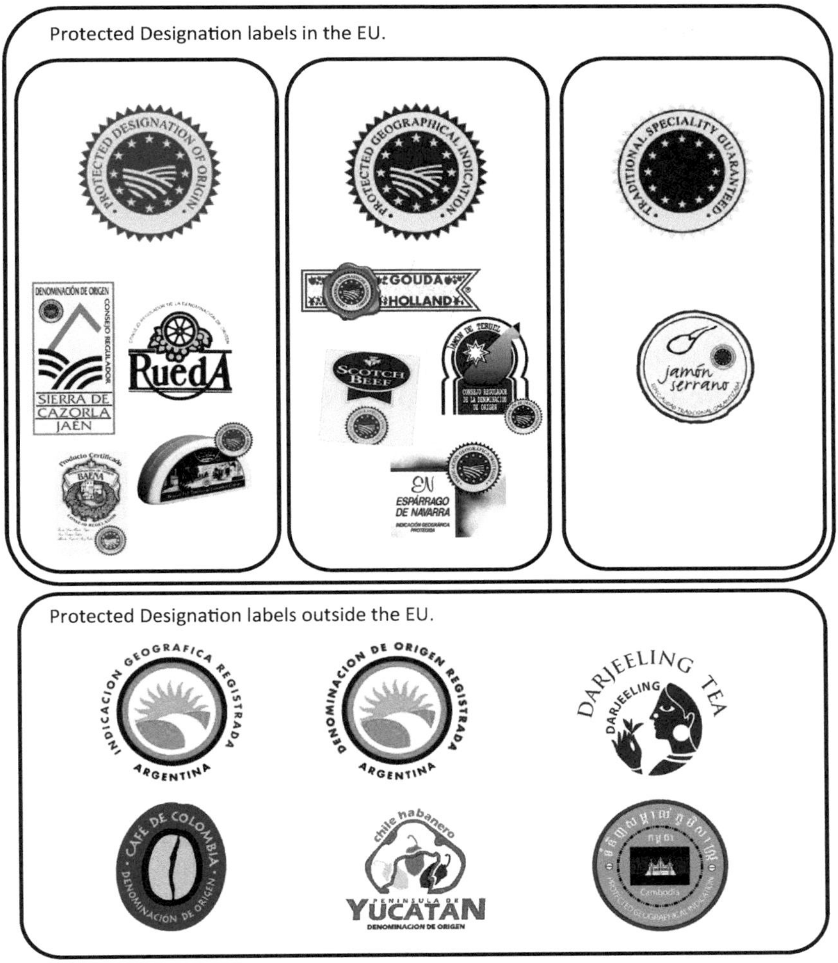

Figure 15.1 Protected designation labels inside and outside the EU.

food is not only a European history. Mexico was the first non-European country to establish a system of GI protection in 1974. Brazil and Peru passed legislation on GIs in 1996, followed by South Korea and India in 1999, Colombia in 2000, and Chile in 2005 (Bowen and Valenzuela-Zapata, 2009). GI food protected by collective or certification marks in their country of origin includes bananas from Costa Rica, Jaffa oranges from Israel, Idaho potatoes (United States), Puer tea (China), Pinggu peaches (China), tequila, coffee, mango fruit, vanilla, rice, and chilies from Mexico, Lima beans, white corn, Crookneck squash, and fresh and dried Maca from Peru, coffee from Colombia, and wine from Algeria, to name just a few.

In summary, GI work as product differentiators on the market by enabling consumers to distinguish between products with geographical origin-based characteristics and others without those characteristics. Usually, the price of such GI products is higher than those without GI, as it is absolutely necessary to have accurate methods for distinguishing them because financial incentives continue to drive retailers/resellers to misidentify the geographic origin of commodities and food products. In other words, determination of the GI is important for enforcement options for the food industry and protection of the consumer and manufacturers, and that is a great value to guarantee regional development and an appropriate distribution of the population in countryside areas (de la Guardia and Gonzalvez, 2013). Eastern EU countries also will increase the number of PDO foods, especially in the present situation, in which agricultural products from emerging economy countries offer a cheaper alternative than those produced in the frame of the EU.

2. Analytical Methods for Food Protected Designation Authentication

The use of analytical techniques to determine the GI of food products is the best way to avoid adulterations and mislabeling after the incorporation of food to the market (de la Guardia and Gonzalvez, 2013). Strategies employed to detect adulterated or mislabeled products have relied on instrumental techniques mainly because of the sophistication of fraudulent procedures. Different instrumental techniques based on the determination of organic compounds such as Liquid Chromatography (LC) (Cserháti et al., 2005), Gas Chromatography (GC) (Heenan and van Ruth, 2013), UV–Visible (UV–Vis) (Karoui and Dufour, 2008), and Infrared (IR) spectroscopy (Abbas et al., 2012) have been proposed for food authenticity in order to control the presence of the main components of the sample or some organic compounds that can be characteristic of a PDO. However, the normal range of organic compounds in food varies with fertilization, climatic conditions in the year of cultivation, history of fields, and variety or species, as well as geographical location and soil characteristics, as it is sometimes difficult to be definitive about the authenticity of a foodstuff. So, strategies for labeled food authentication could be classified in two main categories based on (1) trace element composition and stable isotope ratio and (2) organic compound and index-based analysis (see Fig. 15.2).

Additional developed analytical procedures for determining the authenticity of foodstuffs rely on genetic and DNA-based techniques (Pereira et al., 2013).

2.1 *Trace Element Composition and Stable-Isotope Ratio for Discrimination of Food With Protected Designation of Origin*

It is well-known that the content of selected minerals and trace elements clearly reflects the soil type of a cultivation area and the environmental growing conditions for food productions, and because of that, evaluation of trace element content has been proposed to assure the geographical origin of food samples (Drivelos and Georgiou, 2012; Gonzalvez et al., 2009a).

The techniques of choice to obtain the elemental fingerprint of food are mainly those with multielement detection capability such as ICP-based ones: Inductively

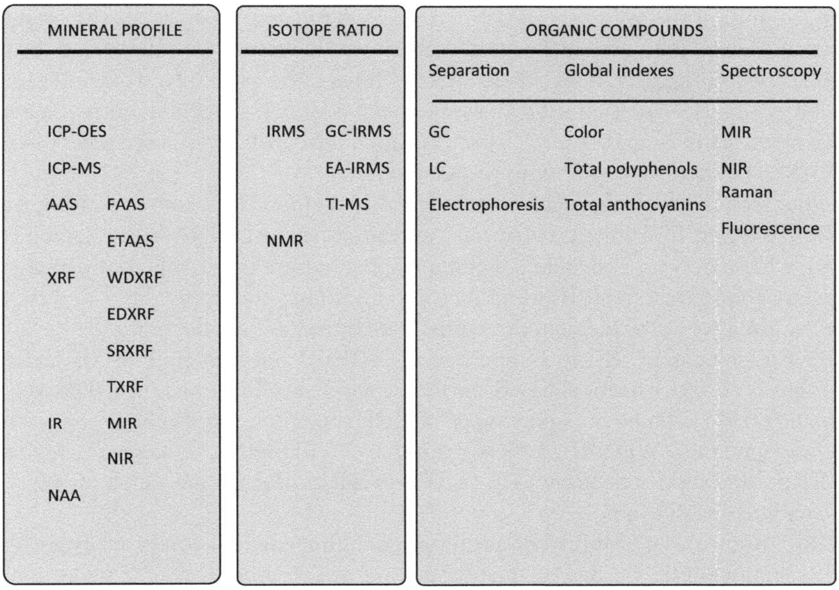

Figure 15.2 Analytical techniques employed for food origin authentication.

Coupled Plasma Optical Emission Spectroscopy (ICP-OES) and Inductively Coupled Plasma Mass Spectrometry (ICP-MS) (Iglesias et al., 2007). ICP-MS typically provides limits of detection (LOD) for 70 elements in the range of parts per billion and parts per trillion. On the other hand, ICP-OES is a fast multielement technique with an extended dynamic linear range and moderate-low LODs in the range of few parts per million. Up to 60 elements can be screened per sample run in less than one minute and trace element composition can be determined in a variety of aqueous or organic matrices. However, the main drawbacks of ICP-MS and ICP-OES are the expensive instrumentation and operation costs, the requirement for trained operators, and in most cases, the need of sample pretreatment steps, which frequently includes the complete mineralization of samples.

As alternative analytical techniques for determining the geographical origin of food, classical Atomic Absorption Spectrometry (AAS) based on both, flame (FAAS), or electrothermal atomization systems (ETAAS) can be used. AAS provides a sensitive and highly selective spectrometric tool suitable for the determination of many elements at trace and ultra-trace levels. The main advantages of FAAS are low operational costs and good analytical performance. On the other hand, the direct analysis of solid samples with a reduced pretreatment is possible using the ETAAS technique, also called Graphite Furnace Atomic Absorption Spectrometry (GFAAS), offering an enhanced sensitivity with LOD of ppb and appropriate selectivity at a reasonable cost with the possibility of determining low levels of nonmetals (eg, Cl, F, P, or S), which are not so accessible using plasma instrumentation due to interferences.

Emission spectrometry has also been used as an analytical technique to guarantee the geographical origin of food (Frias et al., 2003). The main drawback of flame atomic emission spectrometry (FAES) is the effect of small changes of temperature on the level of excitation and consequently on the production of free atoms and the presence of spectral interferences due to spectral lines overlapping. However, the limited capability of AAS techniques for the simultaneous or fast sequential determination of many elements in a single sample strongly limits the capability of classical FAAS and ETAAS to be used for food origin authentication.

Other less common approaches include Radiochemical Neutron-Activation Analysis (RNAA) and Instrumental Neutron-Activation Analysis (INAA) (Pellerano et al., 2008), which are very convenient techniques that require no sample pretreatment or a reduced sample treatment. However, the use of atomic energy facilities and the long time required for some element determinations limit their applicability.

To a lesser extent, X-ray Fluorescence (XRF) is another appropriate technique for PDO food classification based on the mineral profile of samples (Haswell and Walmsley, 1998). There are many types of XRF spectrometers available today: wavelength-dispersive (WDXRF), energy-dispersive (EDXRF), synchrotron radiation (SRXRF), and total-reflection XRF (TXRF), which offer a wide range of sensitivity and analytical conditions.

XRF offers the possibility of performing multielemental direct analysis (qualitative, semiquantitative, and/or quantitative) of solid samples over a wide dynamic range. Additionally, TXRF is a micromethod that needs only a few mL of solution and offers high-intensity synchrotron sources (four or five orders of magnitude greater than WDXRF or EDXRF). The main disadvantage of XRF instrumentation has been its limited sensitivity for some elements (eg, Cd and Pb) due to the high background intensities attributable to the high degree of scattering of the X-ray source and the cost of sophisticated X-ray-based alternatives. The background can be reduced significantly by using TXRF or SRXRF spectrometers, but one of the problems of applying TXRF is the preparation of clean sample reflectors. Even using ultra-pure reagents, impurities can appear, so careful measurement of blanks is mandatory. So, in fact, food authentication based on the mineral profile of samples from their X-ray spectra is quite limited (Bizjak Bat et al., 2012).

In some cases, the mineral composition of foods can be discriminated based on the intrinsic and extrinsic information provided by infrared spectra in both Near Infrared Range (NIR) and Mid Infrared Range (MIR), and there are some studies which use the aforementioned techniques for food authentication (Garrigues and de la Guardia, 2015). However, those approaches still remain at the bench academic level far from real applications to discriminate food produced on different areas.

Determination of Stable Isotope Ratio (SIR) has demonstrated to be a good descriptor for the characterization of food samples with PDOs. The Isotope-Ratio Mass Spectrometry (IRMS) is the technique usually employed to measure the relative abundance of natural isotopes in a given sample. Several different interfaces are used to introduce samples into the IRMS, the most common being Elemental Analyzers (EA-IRMS) and Gas Chromatographs (GC-IRMS). EA-IRMS involves sample pretreatment and is appropriate for the analysis of nonvolatile substances, even though it provides only an average isotope-ratio value for the entire sample. However, GC-IRMS can provide isotopic analysis of a complex mixture, including additional information and a high discriminatory power.

IRMS is often used for 13C/12C, 15N/14N, and 18O/16O, measuring their atomic mass according to magnetic deviation values. However, for stable-isotope measurement of Sr or Pb, it is more advisable to use Thermal-Ionization MS (TIMS) techniques that are based on thermal ionization of the solid sample loaded into the source of the mass spectrometer. This technique is less expensive than IRMS and is suitable for the characterization of species with low ionization potential.

Nuclear Magnetic Resonance (NMR) is commonly used in combination with IRMS for stable-isotope measurements. NMR allows the measurement of ^{1}H and ^{13}C and other nuclei. NMR has been demonstrated to be a useful tool in the analysis of different classes of foodstuffs and beverages for classification and discrimination (geographical and botanical origin as well as validating checking the brand authenticity) (Monakhova et al., 2013).

2.2 Organic Compound Analysis for Discrimination of Foods With Protected Designation of Origin

The search for the GI authenticity of foodstuffs has been the object of numerous studies using different physical and chemical determinations (Kilcast, 2013; Picó, 2012). Those analyses, when they are based on the determination of organic compounds or global indexes, can be classified in three main categories. The first category includes methodologies in which the components of the sample are separated prior to their detection, such as LC, GC, and electrophoresis. For instance, separation and detection of fatty acids and triacylglycerols (Aranda et al., 2004), sterols (Cunha et al., 2006), and aroma (Delgado et al., 2011) have been used for the authentication of food. The second category is based on those methods in which the sample is treated and a global index is obtained as, for example, color intensity, tonality, total polyphenols, total anthocyanins, etc. A third category is based on food characterization from direct spectroscopy measurements of samples without any previous sample treatment: Fourier Transform Infrared Spectroscopy (FTIR) in the middle (MIR) (Karoui et al., 2007) and near (NIR) ranges (de la Haba et al., 2014). Raman and fluorescence spectroscopy (Guimet et al., 2005) have been proposed for the discrimination of food origin.

Regarding the use of separation techniques for food characterization, LC has been extensively used in food analysis for measuring numerous compounds, eg, carbohydrates, vitamins, additives, mycotoxins, amino acids, proteins, triglycerides, lipids, chiral compounds, and pigments (Di Stefano et al., 2012). LC is a robust, relatively low cost and reproducible technique which can be coupled to different detectors depending on the studied analytes, and this has been extensively used in food analysis and authentication.

On the other hand, GC is one of the most universal separation techniques used in food analysis, mainly for volatile and semivolatile composition studies, aromas, and pesticides (Hajslova and Lehotay, 2002). In the same way, different detectors can be used depending on the analytes and the required selectivity: Flame Ionization

Detector (FID), Thermal Conductivity Detector (TCD), Electron Capture Detector (ECD), Photo-Ionization Detector (PID), and Mass Spectrometry (MS) detector, among others.

Global parameters have also been proposed in the literature for the evaluation of the origin and quality of foods. UV–VIS spectroscopy was mainly used to examine global parameters or indexes, such as the anisidine value in the quality of edible oils, being defined as the absorbance of a solution of a fat sample containing aldehydes which have reacted with *p*-anisidine, the process of fat deterioration by the peroxide value, or the general color, determined by the saturation of chlorophyll or carotenoid pigments, among others (Nawrocka and Lamorska, 2013).

Sample spectra could provide a good characterization of samples, and because of that, spectroscopy techniques, associated to chemometric methods, present a real alternative to the chromatography fingerprinting of foods in order to evaluate their origin (Ashurst and Dennis, 1996). In this case, IR spectroscopy provides a rapid, relatively inexpensive, noninvasive, and nondestructive technique (Karoui et al., 2004). The combination of NIR spectroscopy and chemometrics provides a powerful tool for classification of a wide variety of foodstuffs as a function of their origin (Blanco and Pages, 2002). Additionally, NIR spectroscopy provides real-time measurements of raw or finished foodstuffs at all stages of production, thus offering a fast, nondestructive, and cost-effective method of food analysis (Woodcock et al., 2008).

MIR spectroscopy is among the most preferred method for milk and dairy product analysis. This is due to the fact that almost every chemical substance (apart from some salts and very simple chemical compounds) has its own distinctive spectrum. Only substances occurring in very low concentrations can be difficult to determine, as the noise level of the method might be encountered in such a case. Moreover, the development of the Attenuated Total Reflectance (ATR) device avoids the sampling problems encountered when collecting spectra from water, opaque, and/or viscous samples. So vibrational spectroscopy has become an important tool for food origin characterization independently on the presence of high contents of water in food (Fernandez-Gonzalez et al., 2014).

On the other hand, molecular fluorescence spectroscopy can be used to identify and analyze fluorescent compounds at very low concentration levels (in the parts per billion range), and thus it could be a good technique for the authentication of foods with aromatic compounds. The main advantage of fluorescence is the high sensitivity of the measurements. Furthermore, fluorescence spectroscopy is a simple, noninvasive, nondestructive, and relatively inexpensive technique that allows rapid analysis, providing information on proteins, peptides, and free amino acids which include tryptophan, tyrosine, and phenylalanine residues (Karoui et al., 2005).

Another less commonly used technique for spectroscopy fingerprinting authentication of food is Raman spectroscopy. This technique has been applied to study water, carbohydrates, protein, and fat structures in food samples and to determine the level of adulteration of virgin olive oil by some vegetable oils such as soybean, corn, and raw olive residue oils (Baeten et al., 1996).

2.3 Genetic Techniques for Discrimination of Foods With Protected Designation of Origin

DNA-based food authentication is, at present, a vast area that intersects population genetics, forensic genetics, phylogenetics, phylogeography, and general molecular technologies. In this section, we will provide a short description of the most common genetic-based methods to food authentication in the context of the protected designation of origin (PDO).

DNA has several characteristics that make it a useful target for the analysis of food products: a high chemical and thermal stability and presence in the majority of cells and the fact that virtually identical information can be obtained from a specific individual, independently of the tissue considered (Lockley and Bardsley, 2000).

DNA-based methods consist mainly in three different steps: (1) DNA isolation or extraction, (2) polymerase chain reaction (PCR) amplification of DNA fragments containing the appropriate genetic markers, and (3) detection of the DNA profiles defined by those markers. At present, the most commonly used methods target DNA using PCR primers, followed by electrophoretic separation and comparison of the resulting profiles or sequences with adequate controls or references.

Genetic studies related to the analysis of foods with PDO are mainly based on the species identification in order to assess food authenticity. It is usually performed to detect fraudulent substitution of a given species by a similar low-value ingredient. DNA-based methods have demonstrated to be an appropriate way to unambiguously identify the species, breed, or variety present in food products when the examination of morphological characteristics is neither possible nor reliable because of processing or semiprocessing. The selection of reliable genetic markers that allow for the discrimination between species, populations, breeds, or varieties, as a function of their PDO, is essential. In theory, any genomic region that is polymorphic, including those that are known not to have a biological function, can be used as a marker for the purpose of genetic discrimination.

3. Multivariate Statistics Applied to the Identification of Food Origin

According to the most common definition, chemometrics is a chemical discipline that uses mathematics, statistics, and other methods based on formal logic for the construction or selection of optimum measurement methods and experimental designs and also to extract the most important information in the analysis of experimental data (Otto, 2007). The basic idea of most of the chemometric methods is based on the use of as much information as possible about the samples, followed by a reduction of the original variables, by projecting the multidimensional data onto a few new directions, called latent variables or principal components. Thus, based on spectroscopy or chromatography signals, chemical composition of samples, and multivariate statistical analysis, food products can be characterized and classified according to their origin, quality, variety, type, or other features (Benincasa et al., 2007).

The most common procedures used to discriminate foods from their origin can be grouped into two main classes: (1) unsupervised and (2) supervised methods. In unsupervised methods the classification of samples is performed without a prior knowledge of their origin. Ordinary analysis of variance (ANOVA), where samples are classified based on one of their attributes, or a multiple ANOVA (MANOVA), where the classification depends on several variables, have been used to compare foods produced from different regions (Alda-Garcilope et al., 2012).

Cluster Analysis (CA) has been also used to establish an unsupervised classification that involves measurement of either the distance or the similarity between the samples to be clustered. CA provides a way to discover potential relationships between objects grouped in the same cluster or class in terms of their similarity (Gonzalvez et al., 2009b). Hierarchical Cluster Analysis (HCA) is used to obtain relatively homogeneous clusters of samples based on measured characteristics. It starts considering each sample as a separate cluster and then combines the clusters sequentially until only a single cluster remains. HCA is most appropriate for the treatment of a small number of samples. On the other hand, when the number of samples is large, K-Means Cluster is the most appropriate methodology.

Principal Component Analysis (PCA) identifies, in the hyperspace of the variables studied, the directions on which most of the information is retained, thus reducing the dimensionality of the system (Li-Chan et al., 2010). By projecting the objects of the data set inside the space of the first few components (one or two), it is possible to evidence differences between the various objects, determining at the same time which variables are principally involved (Fig. 15.3 shows, as an example, the graphics obtained for the classification of PDO products by using HCA and PCA).

However, in front of the aforementioned unsupervised methods, supervised classification (discriminant) techniques are used to establish if an unknown is similar or not to a series of authentic materials. Linear Discriminant Analysis (LDA) hypothesizes that the distribution is multivariate normal, and that the covariance matrix of each category (dispersion of the category) is not significantly different from one case to another (Gonzalvez et al., 2009b). The Mahalanobis distances of each object from the centroids of the established categories are computed, and the resulting object is assigned to the category with lowest distance, the limit between two categories being a linear function.

Partial Least Squares (PLS) is an asymmetric method permitting us to predict the origin of one set of samples from another reference set and treating both separately. This modeling technique establishes the relationship between two sets of predictor and response variables. It establishes a correlation analysis that estimates the values of one variable from a set of independent variables.

Artificial Neural Networks (ANN) comprise a set of interconnected processing algorithms working in parallel to mimic the process of the human brain. ANN is a supervised method that needs a minimum of known data to train the system correctly (Marini, 2009).

As an indication of the potentiality of the supervised methods of sample classification, Fig. 15.4 provides the graphics found by using LDA, PLS, and ANN for the authentication of different kinds of food.

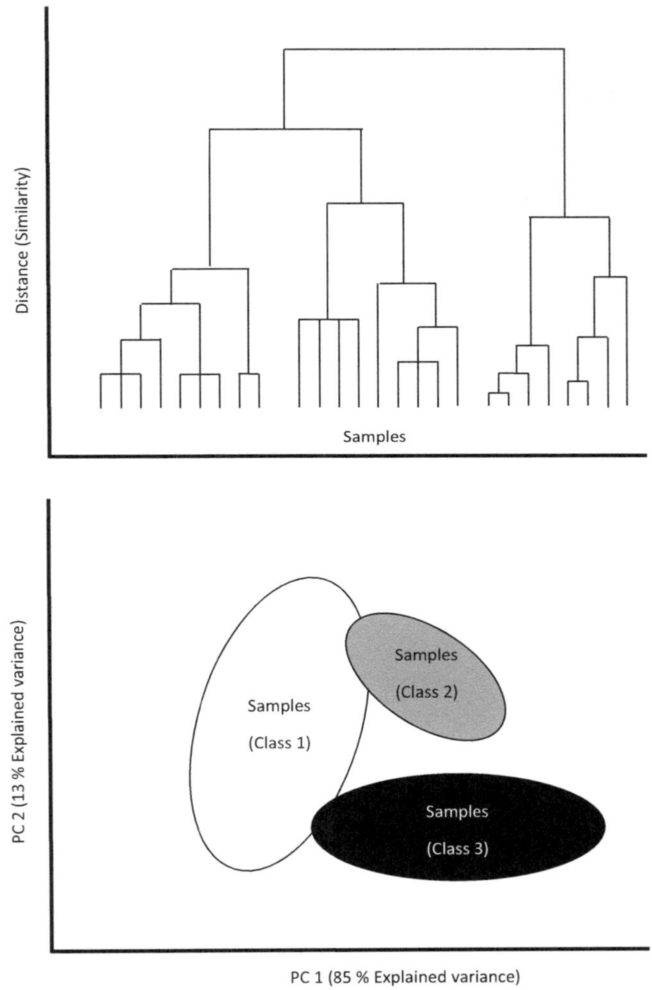

Figure 15.3 Images obtained for the classification of samples using HCA and PCA tools.

4. Main Applications

We have selected some examples of recently published papers about the characterization of food origins in order to provide evidence about the potentiality of the aforementioned analytical and statistical tools to correctly authenticate labeled products.

4.1 Wine, Spirits, and Other Beverages

As it has been previously mentioned, the worldwide sales concerning GI products registered in the EU involved a billion euros in 2010 at the wholesale stage in the region

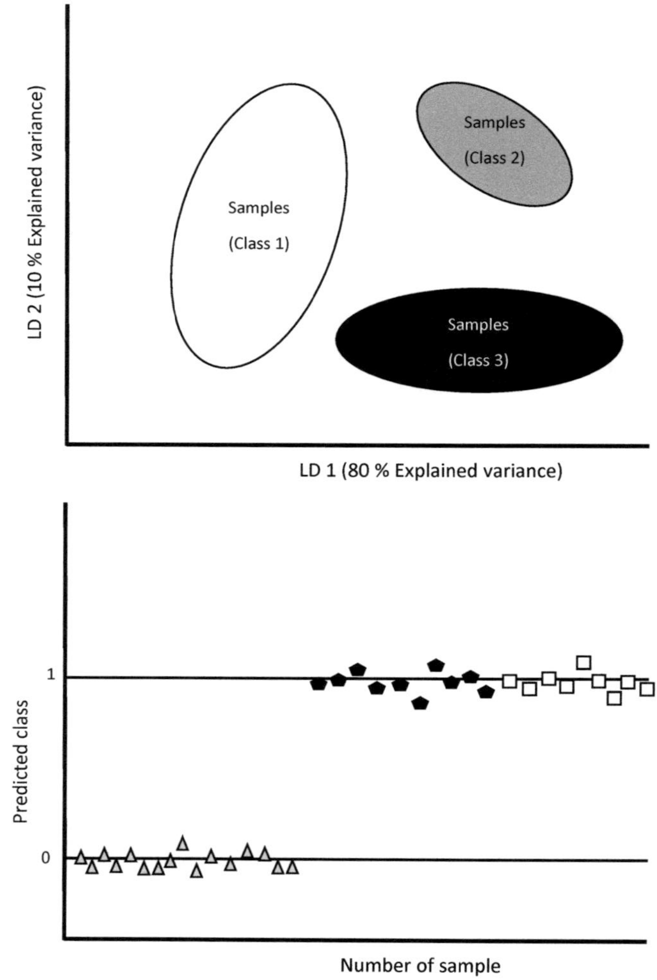

Figure 15.4 Graphics obtained for the authentication of PDO products using supervised chemometric tools.

of production. Wines accounted for 56% of total sales (€30.4 billion), agricultural products and foodstuffs for 29% (€15.8 billion), spirit drinks for 15% (€8.1 billion), and aromatized wines for 0.1% (€31.3 million).

The legal recognition of the geographical origin of wines, by certifications such as the AO or the PGI, is of great interest to both consumers and producers, since it ensures a certain quality level and helps to maintain the individuality of a product within a global market (Fabani et al., 2010). To achieve these certifications, particular grape varieties must be cultivated in well-defined geographical areas and the wines produced subjected to regulations and controls carried out by regulatory councils (Pérez-Magariño and González-San José, 2001).

The great number of parameters affecting wine quality was in the origin of the development of different protocols for their analysis, and thus wine constituents are strictly regulated by international organizations or government agencies to avoid fraud and health risks.

Typical examples of recently published methodologies (from 2010 to 2015) for the GI authentication of wines includes:

Lambrusco, a Protected Designation of Origin (PDO) red wine of Modena (Italy) for which the Sr-87/Sr-86 isotopic ratio, traditionally used as origin marker, has been studied, and samples coming from four different vintages (2009, 2010, 2011, and 2012) were analyzed. The results indicated no significant variability among the different vintages of wines and a perfect agreement between the isotopic range of the soils and wines. Nevertheless, the Sr-87/Sr-86 isotopic ratio was not powerful enough to discriminate between similar products (Durante et al., 2015).

The use of H-1 NMR spectroscopy has been used to improve previous results. Application of PLS-DA revealed a good varietal discrimination by analyzing the low-frequency spectral region. This model explains 68.8% of the variance for the Y vector or classification factor regarding the varietal source. In particular, the signals of 2,3-butanediol, lactic, succinic and malic acids, and threonine were found to be the most statistically significant variables in the model to characterize Lambrusco wines (Papotti et al., 2013).

The mineral composition of Turkish wines was used to classify native and nonnative varieties. Wines from four vintages (2006–2009) were analyzed by ICP-OES and ICP-MS followed by multivariate analysis. The effective variables for discrimination analysis were natural minerals (Sr, Li, Al, Ba, and B) and minerals originating from agricultural activities, processing, or pollution (Ca, Cu, Mg, Co, Pb, and Ni). Characteristics of Turkish wines from native and nonnative grape varieties, such as Cabernet Sauvignon, Merlot, Syrah, and Chardonnay were defined in terms of their mineral content (Sen and Tokatli, 2014).

On the other hand, free amino acids (AA) and biogenic amines (BA) were quantified in Cannonau and Vermentino wines, two "Controlled Designation of Origin" wines from Sardinia (Italy). The analytical method used, in this case, for the simultaneous determination of AA and BA was based on the selective derivatization with dansyl chloride followed by LC with fluorescence detection. γ-aminobutyric acid, 4-hydroxyproline, glycine, leucine + isoleucine, and putrescine proved to be useful for differentiating Cannonau wines from Vermentino wines (Tuberoso et al., 2015).

A similar procedure to the previous one, based on LC with *o*-phthaldialdehyde precolumn derivatization and fluorescence detection, was used to determine biogenic amines in wines from two wine-producing areas (Jumilla and Bullas PDOs) situated in the Murcia region (Spain). Histamine and putrescine were the most prevalent amines in Jumilla wines (49% and 48%, respectively), whereas tryptamine and putrescine were the most prevalent ones in Bullas wines (19% and 38%, respectively) (Villalba-Rodriguez et al., 2011), evidencing that very low levels of biogenic amines could offer a valuable tool for wine authentication.

Other analytical methods used to classify wines according to their origin were based on spectroscopy measurements. ANN treatment of combined signals obtained in the UV–Vis and NIR spectroscopy ranges was evaluated as a rapid method to classify wines belonging to Designation of Origin (DO) Rias Baixas, Ribeira Sacra, Monterrei, Ribeiro, and Valdeorras (NW Spain). Results showed the feasibility of applying ANN to UV–Vis–NIR analysis for the authentication of DO wines (Martelo-Vidal and Vazquez, 2014).

Information on genetic polymorphism given by simple sequence repeats (SSR; microsatellite markers) proved helpful when applied to discriminate "Barbera" and "Dolcetto" from "Nebbiolo" musts, using capillary microelectrophoresis (lab-on-chip technology) (Recupero et al., 2013).

The number of PDOs and PGIs related to beers in the EU increased from 17 in 2005 to 23 in 2010. The total sales value of beer with PDOs and PGIs remains stable in the period of 2005–2010 around 2300 million €, with Germany (12), the Czech Republic (9), and the United Kingdom (2) as the countries with registered PGIs.

Beers from Spain were analyzed in order to use trace elements as chemical descriptors to differentiate them by their geographical origin. The content of Zn, P, B, Mn, Fe, Mg, Al, Sr, Ca, Ba, Na, and K determined by ICP-OES in combination of several pattern-recognition techniques (eg, PCA and LDA) was successfully used with the finding that the elements with the greatest discriminatory power were P, Mg, K, Mn, Ca, and Na (Alcazar et al., 2003).

Other registered beverages protected by origin-certified labels include cider, with the United Kingdom, Ireland, France, and Spain as the major producers, and tea, with China and India as the major producers.

To evaluate the authenticity of ciders, Garcia-Ruiz et al. combined isotopic ratio and mineral profile by using MC-ICP-MS and ICP-OES to determine a total of 37 elements. Concentrations of Na, Mg, Al, K, Ca, Ti, V, Mn, Zn, As, Rb, Sr, Mo, Ba, and the 87Sr/86Sr ratio were processed by PCA, ANOVA, and LDA. The use of 87Sr/86Sr as a unique variable was not enough to distinguish between samples coming from regions close to each other. However, the combination with trace element concentration allowed the cider samples to be separated (Garcia-Ruiz et al., 2007) and opened the way for a correct authentication of ciders.

The polyphenolic profiles of apple juice of different varieties and regions from China were used to characterize and classify apple juices from PDO districts. Polyphenols were determined by LC and MS combined with PCA and LDA, providing respectively 98.3% and 91.2% success rate in terms of prediction ability (Guo et al., 2013).

Moreda-Pineiro et al. determined trace elements in tea from various Asian and African countries by using ICP-OES and ICP-MS. The concentrations of Co, Cr, Cs, Cu, Ni, Pb, Rb, Ti, and V measured by ICP-MS and Al, Ba, Ca, Fe, Mg, Mn, Sr, and Zn were employed as variables in the treatment of data using different pattern-recognition techniques (eg, PCA, CA, LDA, and SIMCA) (Moreda-Pineiro et al., 2003). Data obtained correctly distinguished samples from China, India, and Sri Lanka using LDA.

Discrimination of brandies and wine distillates has been performed by front face fluorescence spectroscopy and multivariate data analysis. Total luminescence and synchronous scanning fluorescence spectra were recorded followed by a classification of samples using

PCA and HCA. Both, carried out on the emission spectra (360–650nm) recorded at an excitation wavelength 350nm and synchronous fluorescence spectra (200–700nm) collected at wavelength interval of 90nm provided very good differentiation between the two spirit classes. The obtained results indicate that the front face fluorescence spectroscopy offers a promising approach for the authentication of brandies (Sadecka et al., 2009).

4.2 Olive Oil

The importance of vegetable origin fats for healthy nutrition justifies the increasing interest in the production, protection, and authentication of vegetable oils, especially olive oil, and the need for a clear discrimination of that from other oils and olive oil produced under different conditions and in different places.

Spain is the leading olive oil producing country in the world, with an average annual production of 700,000–800,000t, even reaching 1,400,000t in recent years. Because the price, in Euros per ton of olive oil, depends on its quality and origin (eg, refined, virgin, or extra virgin olive oil and PDO certified origin or not), it seems evident the need to determine precisely the origin of olive oil through the use of advanced scientific methods. This is particularly important for the EU, given the rising demand for high quality and healthy foods among consumers.

Many physicochemical characteristics, such as acidity, peroxide value, K indexes, phenolic, volatile carotenoids, and chlorophyll composition, are highly dependent on the quality of the olives before harvest, the extraction process to obtain oils, and their preservation, and in this sense, it has been suggested that fatty acid composition could provide the most reliable results for the characterization of PDO olive oils (Ollivier et al., 2003). However, many different procedures are available in the scientific literature for olive oil authentication.

Volatile organic compound concentration measured by PTR-MS sample headspace analysis has been used as a fingerprint to classify olive oils from five different European countries (Araghipour et al., 2008). Developed PLS-DA models were fitted to the data to classify samples into "country," "region," and "district" of origin, respectively, with correct classification rates.

Tunisian virgin olive oil (VOO) varieties have been characterized by GC and NIR spectroscopy. GC of Fatty Acid Methyl Esters (FAME) and NIR spectra of oils, associated to chemometric treatment, allowed the study of the intervarietal variability and the verification of the variety origins of some Tunisian commercial VOO. The specificity of Tunisian VOOs was evaluated by comparing the samples with those produced in Algeria, Morocco, and France PDO VOO. Classification in varietal origins by SIMCA used the FAME compositions and NIR spectra of the most commonly used varieties and showed a high potential to authenticate the varietal origin of Tunisian VOO (Laroussi-Mezghani et al., 2015). Furthermore, the authentication of olive oils achieved by MIR-NIR spectroscopy combined with chemometrics has been successfully developed for the traceability of extra VOO from the PDO Sabina (Italy) (Bevilacqua et al., 2012).

Moreover, the prediction of the geographical origin of monovarietal VOO samples, from three regions of southern Greece collected in five harvesting years (2001–2006), was performed by means of ^{1}H and ^{31}P NMR spectroscopy and their content in fatty acids, phenolics, diacylglycerols, total free sterols, free acidity, and iodine number (Petrakis et al., 2008). DA showed that the geographical prediction at the level of three regions is very high (87%). Additionally, the use of classification and binary trees made it possible to build a geographical prediction algorithm for unknown samples in a self-improvement way, which can be readily extended to other varieties and areas.

Multielement composition of VOOs from different Italian regions, measured by ICP-MS, was used to assess the traceability of the origin of oils (Benincasa et al., 2007). Data were processed by LDA and unknown samples correctly classified by comparison with those of known origin.

Electronic nose technology and UV–Vis spectroscopy have been used to verify the geographical origin of extra VOO from three different areas of Liguria (Casale et al., 2007). The application of LDA, after feature selection, was enough to differentiate the three geographical denominations of Liguria ("Riviera dei Fiori," "Riviera del Ponente Savonese," and "Riviera di Levante"), obtaining 100% success in classification and close to 100% correct prediction.

The applicability of SSR markers in High Resolution Melting (HRM) analysis for the identification of the olive cultivars used in the "Terra di Bari" PDO extra virgin olive oil was demonstrated (Montemurro et al., 2015). A panel of nine cultivars was tested with 17 SSR primer pairs, and the PCR products were at first analyzed with a Genetic Analyzer automatic sequencer. An identification key was obtained for the nine cultivars, which showed the unambiguous discrimination among the varieties constituting the "Terra di Bari" PDO extra virgin olive oil: Cimadi Bitonto, Coratina, and Ogliarola. Subsequently, an SSR-based method was set up with the DCA18 marker, coupled with HRM analysis for the distinction of the Terra di Bari olive oil from non-Terra di Bari olive oil using different mixtures. Thus this analysis enabled the distinction and identification of the PDO mixtures.

4.3 Cheese

The number of PDO and PGI related to dairy products has been increased, especially concerning cheese products, for which registered products in the EU increased continuously from 155 in 2005 to 176 in 2010. The total sales value of cheese products with PDO and PGI was more than 6300 million € in 2010, with France (46), Italy (36), Spain (23), and Greece (20) being the countries with the highest number of registered PGIs.

Trace element content of different cheese samples from PDOs located in Northern Spain was analyzed to study the influence of processing conditions on the final concentration of cheese (Moreno-Rojas et al., 2012). The use of sheep milk, animal rennet, salting by immersion in a brine bath, and ripening in cheeses reported, in general, the highest levels of trace elements. Moreover, pattern-recognition analysis of multielemental data, based on mineral and trace element content, was used to determine geographical traceability and to detect possible imitations with a correct classification in 98.5% of cases.

Mineral elements such as Ba, Ca, Cr, Cu, Hg, K, Mg, Mn, Mo, Na, Ni, and V were determined by FAAS and ETAAS in 278 cheese samples from different European countries (Suhaj and Korenovska, 2008). The use of PCA, CA, FA, and CDA as multivariate statistic tools demonstrated that the concentration of selected trace elements can be used as markers to authenticate cheese, with CDA being the most suitable chemometric tool for this purpose.

Different semihard and hard cheeses from the Marche region (Italy), made from either raw or pasteurized ewes, goats, and cows or blended milk, were analyzed for pH, water activity, gross composition, bacterial ecology, and volatile compound profile (Aquilanti et al., 2013). PCA and PLS-DA of the outcomes allowed the cheeses to be grouped according to type of employed milk and cheese production process treatment. However, no clear separation between PDO, niche, and specialty cheeses from similar productions or industrial competitors was attainable, thus indicating the need of using complementary techniques for an accurate PDO authentication.

A purge/cryotrap/thaw/static headspace Solid Phase Microextraction (SPME) was developed as a means for preconcentrating Volatile Organic Compounds (VOC) in cheese (Tompa et al., 2013). Monoterpenes were evaluated as potential markers for the respective geographic origin of five traditional Slovenian cheeses featuring PDO. Data of nine monoterpenes (alpha-pinene, camphene, alpha-phellandrene, beta-pinene, 3-carene, 2-carene, limonene, tricyclene, and gamma-terpinene) in cheese samples showed clustering related to the area of production.

Ultra-performance LC-electrospray ionization-tandem MS/MS methodology based on Multiple Reaction Monitoring (MRM) has been developed for the sensitive and selective detection of buffalo mozzarella adulteration (Russo et al., 2012). The targeted quantitative analysis was performed by monitoring specific transitions of the phosphorylated beta-casein f33-48 peptide, identified as a novel species-specific proteotypic marker. The high sensitivity of MRM-based MS and the wide dynamic range of triple quadrupole spectrometers have proved to be a valuable tool for the analysis of food matrices such as dairy products, thus offering new opportunities for monitoring food quality and adulterations.

NIR spectroscopy has proven to be an useful tool to both accurately estimate several chemical properties and to classify samples according to their origin. Latent variable models applied to NIR spectral data were developed and used to estimate several chemical properties and to classify the available dataset according to the location and management of the cheese making factory (lowland and alpine), the ripening age (6, 12, 18, and 36 months), the altitude of milk production (low, medium, medium–high, and high), and the period of the year of the cheese production (May, July, and September) in Italian cheeses (Ottavian et al., 2012). In a similar way, ATR-FTIR combined with chemometric analysis was used to classify grated Parmigiano-Reggiano cheese from other grana-type cheeses from Italy and Central and Northern Europe (Gori et al., 2012). LDA and PCA were used to analyze spectral data. Both methods successfully classified the four classes of grated cheese samples, and LDA was found to be the best chemometric approach.

A genetic traceability of the geographical origin of Italian Protected Designation of Origin (PDO) water buffalo Mozzarella from different producers was made based on a molecular technique (Bonizzi et al., 2007).

4.4 Fish and Seafood Products

Fish and fishery products represent a valuable source of protein and essential micronutrients for balanced nutrition and good health. The consumer demand for fish and seafood is growing with the contribution of fish to the average annual diet, reaching a record of 18.8 kg per person per year in 2011 (FAO, 2012). It is partly due to the increase in aquaculture and the increased diversity of fish species and seafood products available on the global market. While species identification can usually be made from morphological characteristics from fish or shellfish in their whole form, seafood is often processed before reaching the consumer. Additionally, many consumers are increasingly aware of nutritional and environmental issues regarding fisheries, size of acceptable species, catch location, and methods (Potts et al., 2011).

In 2010, in the EU, there was 19 PDO labels related to fish, mollusks, and crustaceans with a total value of 443,303 K€ (less than 3% of the total value of foods with PDO labels, 15,789,569 K€). Some examples are Glückstädter Matjes (Germany), Carnikavas nēģi (Letonia), Cozza di Scardovari (Italy), Fal Oyster (UK), Isle of Man Queenies (UK), Moules de Bouchot de la Baie du Mont-Saint-Michel (France), and Mejillón de Galicia; Mexillón de Galicia (Spain).

Chemical composition, concerning macro and trace elements, fatty acid profiles, and stable isotopes have been used as traceability tools to assess geographic origin and seasonality of croaker (Micropogonias furnieri) (Chaguri et al., 2015). Nonetheless, further investigation is still required with larger samples of croaker to enable the implementation of fatty acids, elements, or stable isotope as authenticity tools by food control agencies.

Stable isotope analysis has demonstrated to be a promising approach to identify the authenticity and traceability of seafood and aquaculture products. Carbon and nitrogen stable isotope ratios of three commercial fish, viz, Mackerel, Yellow Croaker, and Pollock, originating from various countries, has been used as discriminative criteria (Kim et al., 2015). Apart from the species-dependent variation in the isotopic values, marked differences in the delta C-13 and delta N-15 ratios were also observed with respect to the country of origin. This suggests that C and N isotopic signatures could be reliable tools to identify and trace the origin of commercial fish.

The production method (wild or farmed), geographical origin, and biological species have been differentiated in shrimps using stable isotope ratio analysis of carbon and nitrogen and/or multielement composition (Ortea and Gallardo, 2015). The correct classification of 100% of samples according to their actual origin and method of production, and 93.5% according to biological species, was obtained.

ICP-MS in combination with different supervised chemometric approaches has been successfully used to classify cultivated mussels in Galicia (Northwest Spain) under the European PDO label (Costas-Rodríguez et al., 2010).

Hyperspectral imaging is a rapid, reagentless, and nondestructive analytical technique that integrates traditional spectroscopic and imaging techniques into a system to attain both spectral and spatial information from an object that cannot be achieved with either digital imaging or conventional spectroscopic techniques. This technique has been used for detecting the adulteration, contamination, and authenticity of meat, poultry, and fish (Kamruzzaman et al., 2015). However, molecular and genetic techniques have been more commonly used to detect mislabeling and frauds related to fish species rather than to assess the PDO of fish or seafood. The genetic identification of the main commercialized species of fish maws has been carried out, based on the amplification of a fragment of the mitochondrial 16S rRNA gene and subsequent basic local alignment search tool (BLAST) analysis (Wen et al., 2015). Results indicated that this method was applicable to all of them and demonstrated that 53.3% of the products were incorrectly labeled and 58.3% of "croaker" products were substituted with catfish or perch species.

DNA barcoding can be used to accurately identify a seafood product to species based on its genetic signature and so provides a means to test the authenticity and accuracy of seafood labeling (Lamendin et al., 2015). This can be especially useful for products such as fillets which have few or no unambiguous identifying characters and can easily be mislabeled.

Cross et al. (2006) developed a multiplex PCR based on 5S rDNA gene sequences for identification of four oyster species (*Ostrea edulis*, *Ostrea stentina*, *Crassostrea angulata*, and *Crassostrea gigas*). Those species are directly related to PDO, but generally the PDO regulation implies other aspects that cannot be identified by molecular methods. For example, Fal Oyster is a UK PDO given to oysters caught in a designated area using only traditional sailing and rowing vessels between the period of October 1st and March 31st.

A clear example of this is the paper of Lapegue et al. (2006), in which *Ostrea edulis*, the European flat oyster, and *Ostreola stentina*, the Provence oyster or dwarf oyster, both present along the European and African, Atlantic, and Mediterranean coasts, were differentiated. Data supported a high degree of differentiation between *O. stentina* and *O. edulis* and a close relationship between *O. stentina* and both *Ostrea aupouria* (from New Zealand) and *Ostreola equestris* (from Mexico Gulf/Atlantic).

Molecular-based methods are only useful in the case of species that can be directly related to a PDO. As an example, the fish roe of Messolongi is a famous product from Greece with PDO (Klossa-Kilia et al., 2002). This processed sea food product originates from the whole ovaries of the fish *M. cephalus*, which is captured in Messolongi lagoon in Greece. A PCR–RFLP-based method has been developed for the authentication of fish roe of Messolongi, enabling a clear discrimination of the fish roe of Messolongi from the fish roe originating from the other Mugilidae species coexisting in the same area.

4.5 Other Foods

Other products covered by the PDO/PGI regulations due to their high value and quality are fresh and cured meat, fruits and vegetables, and fresh fish, among others.

Recent examples of the assessment of PDO of fruits, vegetables, and legumes includes "Fava Santorinis," a PDO yellow split pea cultivated in Santorini that has been classified by means of their concentration of rare earth elements measured by ICP-MS (Drivelos et al., 2014). Tropea PGI onions (Italy) have been also distinguished from non-Tropea PGI onions using the mineral profile measured by ICP-MS (Furia et al., 2011). Cherries from Alicante's Mountain PGI (Matos-Reyes et al., 2013) and Arròs de Valencia PDO (Gonzalvez et al., 2011) were differentiated using the mineral content measured by ICP-OES; potatoes from Galicia PGI (Spain) using the mineral profile measured by FAAS and FAES (Herrero Latorre et al., 2013); Italian potatoes from Sicily, Apulia, and Tuscany were also classified using the headspace solid-phase microextraction GC–MS and IRMS data (Longobardi et al., 2011); tomatoes with Italian PDOs have been discriminated from non-Italian ones using the mineral profile obtained by dynamic reaction cell (DRC) ICP-MS with CH_4 as reaction gas (Lo Feudo et al., 2010); and almonds from Spain and the United States were discriminated by ATR-FTIR thermal analysis techniques, differential scanning calorimetry (DSC), and thermogravimetric analysis (TGA) (Valdes Garcia et al., 2013), among others.

Mislabeling of commercial tomato products has been detected by means of molecular genetic analysis (Scarano et al., 2015). SSR were tested on 20 tomato varieties, validated on foodstuff, and then employed to establish the genetic identity of 25 commercial tins labeled as "San Marzano."

Moreover, the headspace composition measured by proton transfer reaction time of the flight mass spectrometry (PTR-ToF-MS) technique was used for the rapid characterization of dry cured hams produced according to four of the most important PDOs: an Iberian one (Dehesa de Extremadura) and three Italian ones (Prosciutto di San Daniele, Prosciutto di Parma, and Prosciutto Toscano) (Sanchez del Pulgar et al., 2011). Fatty acid profile measured by GC-FID has been used as an effective tool to differentiate two Portuguese PDO bovine breeds (Mirandesa and Barrosã) and two production subsystems (traditional and organic farming), allowing a 100% correct classification (Dias et al., 2008).

The use of single nucleotide polymorphisms in combination with Bayesian statistics for the traceability of four European PGI beef products has been successfully evaluated using a dataset comprising 24 breeds from Italy, France, Spain, Denmark, the Netherlands, Switzerland, and UK genotyped with 90 polymorphic markers (Negrini et al., 2008).

Other products, such as honey, have been classified as a function of their geographical origin (Miel de Corse (Corsica) PDO) by H-1 NMR and multivariate analysis techniques (Donarski et al., 2008). Another study (Schellenberg et al., 2010) reported that the stable isotope ratios of carbon, nitrogen, hydrogen, and sulfur in honey proteins can be applied to verify the origin of honey. The geographical origin of three Slovenian unifloral honeys was investigated by the elemental content using TXRF and the stable carbon and nitrogen isotope ratios using IRMS (Kropf et al., 2010).

5. Conclusions and Future Trends

Food safety is an important matter in our society and to guarantee the quality and safety of human foods, the traceability of the foods from the farm to the fork has been revealed as the best tool. Based on that, governments and agencies have tried to protect foods by using guarantee labels. However, when foods are in the market, it is important to develop analytical processes to make food authentication, including the origin of commercialized products, and to do it, there are a series of proposals in the scientific literature.

In order to obtain a fast and accurate authentication of the characteristics and origin of foods, it is mandatory to have a detailed description of food qualities and composition and, additionally, the spectroscopy, chromatography, and/or DNA-based typical recordings for every labeled protected products must be included to assure a good comparability between original products and suspicious ones.

On considering the methodologies suggested in the scientific literature to verify the authenticity of the origin of protected food, it seems clear that future efforts will be focused on the simultaneous use of several parameters obtained from different methods. Thus incorrect assignments due to the variability of crops as a function of the climate conditions of each year will be avoided, reducing false negative results.

On the other hand, a clear confirmation of the origin of products must be done as fast as possible in order to avoid food spoilage, and to do it, easy and cheap screening methods are required to support the work of inspection authorities with the highest guarantees to the consumers and producers.

The use of the mineral profile of foods together with the natural isotopic ratios have been evidenced as the most powerful methodologies. However, we hope that in the near future, the combination of spectroscopy, chromatography, and DNA-based data obtained directly from the samples in the presence of well-selected internal standards could provide clear evidence about the authenticity of products. However, to move the aforementioned ideas from the bench to the real world, a strong cooperation between policy makers, regulatory councils, and research teams will be required. In this sense, by using clear descriptions of the labeled products and well-validated variabilities as a function of the harvest year, the development of useful tools for a fast authentication of market products will be much more useful.

An additional aspect to be taken into consideration in the near future is the development and application of chemometrics to improve the quality of information obtained from market foods, and in this sense, both the available software and the new ones must be included in the modern instrumentation and methodologies proposed for food authentication.

So we must be optimistic about the future solutions of the complex questions related to the food production conditions and the geographical origin of market foods.

References

Abbas, O., Dardenne, P., Baeten, V., 2012. Near-infrared, mid-infrared and Raman spectroscopy. In: Picó, Y. (Ed.), Chemical Analysis of Food: Techniques and Applications. Academic Press, Elsevier, Oxford.

Alcazar, A., Martin, M.J., De Pablos, F., Gonzalez, A.G., 2003. Diferenciación del origen geográfico de cervezas mediante técnicas de reconocimiento de patrones según su contenido mineral. Cerveza Y Malta 40, 21–25.

Alda-Garcilope, C., Gallego-Picó, A., Bravo-Yagüe, J.C., Garcinuño-Martínez, R.M., Fernández-Hernando, P., 2012. Characterization of Spanish honeys with protected designation of origin "Miel de Granada" according to their mineral content. Food Chemistry 135, 1785–1788.

Aquilanti, L., Santarelli, S., Babini, V., Osimani, A., Clementi, F., 2013. Quality evaluation and discrimination of semi-hard and hard cheeses from the Marche region (Central Italy) using chemometric tools. International Dairy Journal 29, 42–52.

Araghipour, N., Colineau, J., Koot, A., Akkermans, W., Rojas, J.M.M., Beauchamp, J., Wisthaler, A., Mark, T.D., Downey, G., Guillou, C., Mannina, L., van Ruth, S., 2008. Geographical origin classification of olive oil by PTR-MS. Food Chemistry 108, 374–383.

Aranda, F., Gómez-Alonso, S., Rivera del Álamo, R.M., Salvador, M.D., Fregapane, G., 2004. Triglyceride, total and 2-position fatty acid composition of Cornicabra virgin olive oil: comparison with other Spanish cultivars. Food Chemistry 86, 485–492.

Ashurst, P.R., Dennis, M.J., 1996. Food Authentication. Blackie Academic and Professional, London.

Baeten, V., Meurens, M., Morales, M.T., Aparicio, R., 1996. Detection of virgin olive oil adulteration by Fourier transform Raman Spectroscopy. Journal of Agricultural and Food Chemistry 44, 2225–2230.

Benincasa, C., Lewis, J., Perri, E., Sindona, G., Tagarelli, A., 2007. Determination of trace element in Italian virgin olive oils and their characterization according to geographical origin by statistical analysis. Analytica Chimca Acta 585, 366–370.

Bevilacqua, M., Bucci, R., Magrì, A.D., Magrì, A.L., Marini, F., 2012. Tracing the origin of extra virgin olive oils by infrared spectroscopy and chemometrics: a case study. Analytica Chimca Acta 717, 39–51.

Bizjak Bat, K., Vidrih, R., Necemer, M., Mozetic Vodopivec, B., Mulic, I., Kump, P., Ogrinc, N., 2012. Characterization of Slovenian apples with respect to their botanical and geographical origin and agricultural production practice. Food Technology and Biotechnology 50, 107–116.

Blanco, M., Pages, J., 2002. Classification and quantification of finishing oils by near infrared spectroscopy. Analytica Chimca Acta 463, 295–303.

Bonizzi, I., Feligini, M., Aleandri, R., Enne, G., 2007. Genetic traceability of the geographical origin of typical Italian water buffalo Mozzarella cheese: a preliminary approach. Journal of Applied Microbiology 102, 667–673.

Bowen, S., Valenzuela-Zapata, A., 2009. Geographical indications, terroir, and socioeconomic and ecological sustainability: the case of tequila. Journal of Rural Studies 25, 108–119.

Casale, M., Armanino, C., Casolino, C., Forina, M., 2007. Combining information from headspace mass spectrometry and visible spectroscopy in the classification of the Ligurian olive oils. Analytica Chimca Acta 589, 89–95.

Chaguri, M.P., Maulvault, A.L., Nunes, M.L., Santiago, D.A., Denadai, J.C., Fogaca, F.H., Sant'Ana, L.S., Ducatti, C., Bandarra, N., Carvalho, M.L., Marques, A., 2015. Different tools to trace geographic origin and seasonality of croaker (*Micropogonias furnieri*). LWT-Food Science and Technology 61, 194–200.

Costas-Rodríguez, M., Lavilla, I., Bendicho, C., 2010. Classification of cultivated mussels from Galicia (Northwest Spain) with European Protected Designation of Origin using trace element fingerprint and chemometric analysis. Analytica Chimca Acta 664, 121–128.

Cross, I., Rebordinos, L., Diaz, E., 2006. Species identification of Crassostrea and Ostrea oysters by polymerase chain reaction amplification of the 5S rRNA gene. Journal of AOAC International 89, 144–148.

Cserháti, T., Forgács, E., Deyl, Z., Miksik, I., 2005. Chromatography in authenticity and traceability tests of vegetable oils and dairy products: a review. Biomedical Chromatography 19, 183–190.

Cunha, S.S., Fernandes, J.O., Oliveira, M.B., 2006. Quantification of free and esterified sterols in Portuguese olive oils by solid-phase extraction and gas chromatography-mass spectrometry. Journal of Chromatography A 1128, 220–227.

Delgado, F.J., González-Crespo, J., Cava, R., Ramírez, R., 2011. Formation of the aroma of a raw goat milk cheese during maturation analysed by SPME-GC-MS. Food Chemistry 129, 1156–1163.

Dias, L.G., Correia, D.M., Sá-Morais, J., Sousa, F., Pires, J.M., Peres, A.M., 2008. Raw bovine meat fatty acids profile as an origin discriminator. Food Chemistry 109, 840–847.

Donarski, J.A., Jones, S.A., Charlton, A.J., 2008. Application of cryoprobe H-1 nuclear magnetic resonance spectroscopy and multivariate analysis for the verification of Corsican honey. Journal of Agricultural and Food Chemistry 56, 5451–5456.

Drivelos, S.A., Georgiou, C.A., 2012. Multi-element and multi-isotope-ratio analysis to determine the geographical origin of foods in the European Union. TrAC-Trends in Analytical Chemistry 40, 38–51.

Drivelos, S.A., Higgins, K., Kalivas, J.H., Haroutounian, S.A., Georgiou, C.A., 2014. Data fusion for food authentication. Combining rare earth elements and trace metals to discriminate "Fava Santorinis" from other yellow split peas using chemometric tools. Food Chemistry 165, 316–322.

Durante, C., Baschieri, C., Bertacchini, L., Bertelli, D., Cocchi, M., Marchetti, A., Manzini, D., Papotti, G., Sighinolfi, S., 2015. An analytical approach to Sr isotope ratio determination in Lambrusco wines for geographical traceability purposes. Food Chemistry 173, 557–563.

EU Council, 1992. Council Regulation (EEC) No 2081/92 "Protection of geographical indications and designations of origin for agricultural products and foodstuffs". Official Journal of the European Union. 208, 0001–0008. http://eur-lex.europa.eu/legal-content/ES/TXT/?uri=CELEX:31992R2081.

Fabani, M.P., Arrúa, R.C., Vázquez, F., Dáaz, M.P., Baroni, M.V., Wunderlin, D.A., 2010. Evaluation of elemental profile coupled to chemometrics to assess the geographical origin of Argentinean wines. Food Chemistry 119, 372–379.

FAO, 2012. The State of the World's Fisheries and Aquaculture. Food and Agriculture Organization of the United Nations, Rome.

Fernandez-Gonzalez, A., Montejo-Bernardo, J.M., Rodríguez-Prieto, H., Castaño-Monllor, C., Badía-Laíño, R., Díaz-García, M.E., 2014. Easy-to-use analytical approach based on ATR–FTIR and chemometrics to identify apple varieties under Protected Designation of Origin (PDO). Computers and Electronics in Agriculture 108, 166–172.

Frias, S., Conde, J.E., Rodriguez-Bencomo, J.J., Garcia-Montelongo, F., Perez-Trujillo, J.P., 2003. Classification of commercial wines from the Canary Islands (Spain) by chemometric techniques using metallic contents. Talanta 59, 335–344.

Furia, E., Naccarato, A., Sindona, G., Stabile, G., Tagarelli, A., 2011. Multielement fingerprinting as a tool in origin authentication of PGI food products: Tropea red onion. Journal of Agricultural and Food Chemistry 59, 8450–8457.

Lo Feudo, G., Naccarato, A., Sindona, G., Tagarelli, A., 2010. Investigating the origin of tomatoes and triple concentrated tomato pastes through multielement determination by inductively coupled plasma mass spectrometry and statistical analysis. Journal of Agricultural and Food Chemistry 58, 3801–3807.

Garcia-Ruiz, S., Moldovan, M., Fortunato, G., Wunderli, G., Garcia Alonso, J.I., 2007. Evaluation of strontium isotope abundance ratios in combination with multi-elemental analysis as a possible tool to study the geographical origin of ciders. Analytica Chimca Acta 590, 55–66.

Garrigues, S., de la Guardia, M., 2015. Handbook of Mineral Elements in Food. Wiley-Blackwell, Oxford.

Gonzalvez, A., Armenta, S., de la Guardia, M., 2009a. Trace-element composition and stable-isotope ratio for discrimination of foods with Protected Designation of Origin. TrAC-Trends in Analytical Chemistry 28, 1295–1311.

Gonzalvez, A., Llorens, A., Cervera, M.L., Armenta, S., de la Guardia, M., 2009b. Elemental fingerprint of wines from the protected designation of origin Valencia. Food Chemistry 112, 26–34.

Gonzalvez, A., Armenta, S., de la Guardia, M., 2011. Geographical traceability of "Arros de Valencia" rice grain based on mineral element composition. Food Chemistry 126, 1254–1260.

Gori, A., Maggio, R.M., Cerretani, L., Nocetti, M., Caboni, M.F., 2012. Discrimination of grated cheeses by Fourier transform infrared spectroscopy coupled with chemometric techniques. International Dairy Journal 23, 115–120.

de la Guardia, M., Gonzalvez, A., 2013. Food Protected Designation of Origin: Methodologies and Applications. Comprehensive Analytical Chemistry 60, Elsevier, Amsterdam.

Guimet, F., Boqué, R., Ferré, J., 2005. Study of oils from the protected denomination of origin "Siurana" using excitation-emission fluorescence spectroscopy and three-way methods of analysis. Grasas Y Aceites 56, 292–297.

Guo, J., Yue, T.L., Yuan, Y.H., Wang, Y.T., 2013. Chemometric classification of apple juices according to variety and geographical origin based on polyphenolic profiles. Journal of Agricultural and Food Chemistry 61, 6949–6963.

de la Haba, M.J., Arias, M., Ramírez, P., López, M.I., Sánchez, M.T., 2014. Characterizing and authenticating Montilla-Moriles PDO vinegars using near infrared reflectance spectroscopy (NIRS) technology. Sensors (Basel) 14, 3528–3542.

Hajslova, J., Lehotay, S.J., 2002. Application of gas chromatography in food analysis. TrAC-Trends in Analytical Chemistry 21, 686–697.

Haswell, S.J., Walmsley, A.D., 1998. Multivariate data visualisation methods based on multi-elemental analysis of wines and coffees using total reflection X-ray fluorescence analysis. Journal of Analytical Atomic Spectrometry 13, 131–134.

Heenan, S.P., van Ruth, S.M., 2013. Emerging flavor analysis methods for food authentication. In: Kilcast, D. (Ed.), Instrumental Assessment of Food Sensory Quality: A Practical Guide. Woodhead Publishing, Elsevier, Cambridge.

Herrero Latorre, C., Barciela Garcia, J., Garcia Martin, S., Pena Crecente, R.M., 2013. Chemometric classification of potatoes with protected designation of origin according to their producing area and variety. Journal of Agricultural and Food Chemistry 61, 8444–8451.

Iglesias, M., Besalu, E., Antico, E., 2007. Internal standardization-atomic spectrometry and geographical pattern recognition techniques for the multielement analysis and classification of Catalonian red wines. Journal of Agricultural and Food Chemistry 55, 219–225.

Kamruzzaman, M., Makino, Y., Oshita, S., 2015. Non-invasive analytical technology for the detection of contamination, adulteration, and authenticity of meat, poultry, and fish: a review. Analytical Chimca Acta 853, 19–29.

Karoui, R., Dufour, E., Pillonel, L., Picque, D., Cattenoz, T., Bosset, J.O., 2004. Determining the geographic origin of Emmental cheeses produced during winter and summer using a technique based on the concatenation of MIR and fluorescence spectroscopic data. European Food Research and Technology 219, 184–189.

Karoui, R., Martin, B., Dufour, E., 2005. Potentiality of front-face fluorescence spectroscopy to determine the geographic origin of milks from the Haute-Loire department (France). Lait 85, 223–236.

Karoui, R., Mazerolles, G., Bosset, J.O., De Baerdemaeker, J., Dufour, E., 2007. Utilisation of mid-infrared spectroscopy for determination of the geographic origin of Gruyere PDO and L'Etivaz PDO Swiss cheeses. Food Chemistry 105, 847–854.

Karoui, R., Dufour, E., 2008. Spectroscopic technique: fluorescence and ultraviolet-visible (UV-Vis) spectroscopy. In: Sun, D.W. (Ed.), Modern Techniques for Food Authentication. Academic Press, Elsevier, Burlington.

Kilcast, D., 2013. Instrumental Assessment of Food Sensory Quality: A Practical Guide. Woodhead Publishing, Elsevier, Cambridge.

Kim, H., Kumar, K.S., Shin, K.H., 2015. Applicability of stable C and N isotope analysis in inferring the geographical origin and authentication of commercial fish (Mackerel, Yellow Croaker and Pollock). Food Chemistry 172, 523–527.

Klossa-Kilia, E., Papasotiropoulos, V., Kilias, G., Alahiotis, S., 2002. Authentication of Messolongi (Greece) fish roe using PCR–RFLP analysis of 16s rRNA mtDNA segment. Food Control 13, 169–172.

Kropf, U., Korosec, M., Bertoncelj, J., Ogrinc, N., Necemer, M., Kump, P., Golob, T., 2010. Determination of the geographical origin of Slovenian black locust, lime and chestnut honey. Food Chemistry 121, 839–846.

Lamendin, R., Miller, K., Ward, R.D., 2015. Labelling accuracy in Tasmanian seafood: an investigation using DNA barcoding. Food Control 47, 436–443.

Lapègue, S., Ben Salah, I., Batista, F.M., Heurtebise, S., Neifar, L., Boudry, P., 2006. Phylogeographic study of the dwarf oyster, *Ostreola stentina*, from Morocco, Portugal and Tunisia: evidence of a geographic disjunction with the closely related taxa, *Ostrea aupouria* and *Ostreola equestris*. Marine Biology 150, 103–110.

Laroussi-Mezghani, S., Vanloot, P., Molinet, J., Dupuy, N., Hammami, M., Grati-Kamoun, N., Artaud, J., 2015. Authentication of Tunisian virgin olive oils by chemometric analysis of fatty acid compositions and NIR spectra. Comparison with Maghrebian and French virgin olive oils. Food Chemistry 173, 122–132.

Li-Chan, E., Chalmers, J., Griffiths, P., 2010. Applications of Vibrational Spectroscopy in Food Science: Instrumental and Fundamentals Applications. Wiley, Chichester.

Lockley, A.K., Bardsley, R.G., 2000. DNA-based methods for food authentication. Trends in Food Science and Technology 11, 67–77.

Longobardi, F., Casiello, G., Sacco, D., Tedone, L., Sacco, A., 2011. Characterisation of the geographical origin of Italian potatoes, based on stable isotope and volatile compound analyses. Food Chemistry 124, 1708–1713.

Marini, F., 2009. Artificial neural networks in foodstuff analyses: trends and perspectives A review. Analytical Chimca Acta 635, 121–131.

Martelo-Vidal, M.J., Vazquez, M., 2014. Rapid authentication of white wines. Part 1: classification by designation of origin. Agro Food Industry Hi-Tech Journal 25, 44–48.

Matos-Reyes, M.N., Simonot, J., Lopez-Salazar, O., Cervera, M.L., de la Guardia, M., 2013. Authentication of Alicante's Mountain cherries protected designation of origin by their mineral profile. Food Chemistry 141, 2191–2197.

Monakhova, Y.B., Kuballa, T., Lachenmeier, D.W., 2013. Chemometric methods in NMR spectroscopic analysis of food products. Journal of Analytical Chemistry 68, 755–766.

Montemurro, C., Miazzi, M.M., Pasqualone, A., Fanelli, V., Sabetta, W., di Rienzo, V., 2015. Traceability of PDO olive oil "Terra di Bari" using high resolution melting. Journal of Chemistry 496986. http://dx.doi.org/10.1155/2015/496986.

Moreda-Pineiro, A., Fisher, A., Hill, S.J., 2003. The classification of tea according to region of origin using pattern recognition techniques and trace metal data. Journal of Food Composition and Analysis 16, 195–211.

Moreno-Rojas, R., Camara-Martos, F., Sanchez-Segarra, P.J., Amaro-Lopez, M.A., 2012. Influence of manufacturing conditions and discrimination of Northern Spanish cheeses using multi-element analysis. International Journal of Dairy Technology 65, 594–602.

Nawrocka, A., Lamorska, J., 2013. Determination of food quality by using spectroscopic methods. In: Grundas, S., Stepniewski, A. (Eds.), Advances in Agrophysical Research. InTech.

Negrini, R., Nicoloso, L., Crepaldi, P., Milanesi, E., Marino, R., Perini, D., Pariset, L., Dunner, S., Leveziel, H., Williams, J.L., Ajmone Marsan, P., 2008. Traceability of four European protected geographic indication (PGI) beef products using single nucleotide polymorphisms (SNP) and bayesian statistics. Meat Science 80, 1212–1217.

Ollivier, D., Artaud, J., Pinatel, C., Durbec, J.P., Guérère, M., 2003. Triacylglycerol and fatty acid compositions of French virgin olive oils. Characterization by chemometrics. Journal of Agricultural and Food Chemistry 51, 5723–5731.

Ortea, I., Gallardo, J.M., 2015. Investigation of production method, geographical origin and species authentication in commercially relevant shrimps using stable isotope ratio and/or multi-element analyses combined with chemometrics: an exploratory analysis. Food Chemistry 170, 145–153.

Ottavian, M., Facco, P., Barolo, M., Berzaghi, P., Segato, S., Novelli, E., Balzan, S., 2012. Near-infrared spectroscopy to assist authentication and labeling of Asiago d'allevo cheese. Journal of Food Engineering 113, 289–298.

Otto, M., 2007. Chemometrics. Wiley-VCH, Weinheim.

Papotti, G., Bertelli, D., Graziosi, R., Sivestri, M., Bertacchini, L., Durante, C., Plessi, M., 2013. Application of one- and two-Dimensional NMR spectroscopy for the characterization of protected designation of origin Lambrusco wines of Modena. Journal of Agricultural and Food Chemistry 61, 1741–1746.

Pellerano, R.G., Mazza, S.S., Marigliano, R.A., Marchevsky, E.J., 2008. Multielement analysis of Argentinean lemon juices by instrumental neutronic activation analysis and their classification according to geographical origin. Journal of Agricultural and Food Chemistry 56, 5222–5225.

Pereira, F., Amorim, A., van Asch, B., 2013. Genetic and DNA-based techniques. In: de la Guardia, M., Gonzalvez, A. (Eds.), Food Protected Designation of Origin. Elsevier, Amsterdam.

Pérez-Magariño, S., González-San José, M.L., 2001. Differentiation parameters of Ribera del Duero wines from other Spanish denominations of origin. Food Science and Technology International 7, 237–244.

Petrakis, P.V., Agiomyrgianaki, A., Christophoridou, S., Spyros, A., Dais, P., 2008. Geographical characterization of Greek virgin olive oils (Cv. Koroneiki) using ^{1}H and ^{31}P NMR fingerprinting with canonical discriminant analysis and classification binary trees. Journal of Agricultural and Food Chemistry 56, 3200–3207.

Picó, Y., 2012. Chemical Analysis of Food: Techniques and Applications. Academic Press, Elsevier, Oxford.

Potts, T., Brennan, R., Pita, C., Lowrie, G., 2011. Sustainable Seafood and Ecolabelling: The Marine Stewardship Council, UK Consumers, and Fishing Industries Perspectives SAMS Report: 270–211. Scottish Association for Marine Science, Oban.

Recupero, M., Garino, C., De Paolis, A., Cereti, E., Coisson, J.D., Travaglia, F., Arlorio, M., 2013. A method to check and discover adulteration of nebbiolo-based monovarietal musts: detection of barbera and dolcetto cv via SSR analysis coupled with lab-on-chip (R) microcapillary electrophoresis. Food Analytical Methods 6, 952–962.

Russo, R., Severino, V., Mendez, A., Lliberia, J., Parente, A., Chambery, A., 2012. Detection of buffalo mozzarella adulteration by an ultra-high performance liquid chromatography tandem mass spectrometry methodology. Journal of Mass Spectrometry 47, 1407–1414.

Di Stefano, V., Avellone, G., Bongiorno, D., Cunsolo, V., Muccilli, V., Sforza, S., Dossena, A., Drahos, L., Vékey, K., 2012. Applications of liquid chromatography-mass spectrometry for food analysis. Journal of Chromatography A 1259, 74–85.

Sadecka, J., Tothova, J., Majek, P., 2009. Classification of brandies and wine distillates using front face fluorescence spectroscopy. Food Chemistry 117, 491–498.

Sanchez del Pulgar, J., Soukoulis, C., Biasioli, F., Cappellin, L., Garcia, C., Gasperi, F., Granitto, P., Maerk, T.D., Piasentier, E., Schuhfried, E., 2011. Rapid characterization of dry cured ham produced following different PDOs by proton transfer reaction time of flight mass spectrometry (PTR-ToF-MS). Talanta 85, 386–393.

Scarano, D., Rao, R., Masi, P., Corrado, G., 2015. SSR fingerprint reveals mislabeling in commercial processed tomato products. Food Control 51, 397–401.

Schellenberg, A., Chmielus, S., Schlicht, C., Camin, F., Perini, M., Bontempo, L., Heinrich, K., Kelly, S.D., Rossmann, A., Thomas, F., Jamin, E., Horacek, M., 2010. Multielement stable isotope ratios (H, C, N, S) of honey from different European regions. Food Chemistry 121, 770–777.

Sen, I., Tokatli, F., 2014. Characterization and classification of Turkish wines based on elemental composition. American Journal of Enology and Viticulture 65, 134–142.

Suhaj, M., Korenovska, M., 2008. Study of some European cheeses geographical traceability by pattern recognition analysis of multielemental data. European Food Research and Technology 227, 1419–1427.

Tompa, G., Susic, R., Rogelj, I., Pompe, M., 2013. Cryotrap/SPME/GC/MS method for profiling of monoterpenes in cheese and their clustering according to geographic origin. Acta Chimca Slovenica 60, 595–603.

Tuberoso, C.I.G., Congiu, F., Serreli, G., Mameli, S., 2015. Determination of dansylated amino acids and biogenic amines in Cannonau and Vermentino wines by HPLC-FLD. Food Chemistry 175, 29–35.

Valdes Garcia, A., Beltran Sanahuja, A., Garrigos Selva, M.C., 2013. Characterization and classification of Almond cultivars by using spectroscopic and thermal techniques. Journal of Food Science 78, C138–C144.

Villalba-Rodriguez, A., Fernandez-Fernandez, J.I., Martinez-Cutillas, A., Gil-Munoz, R., 2011. Study of different biogenic amines in wines from two Protected Designations of Origin (PDOS) in Murcia (Spain): multivariate classification. Journal International des Sciences de la Vigne et du Vin 45, 111–119.

Wen, J., Zeng, L., Sun, Y.L., Chen, D.H., Xu, Y.H., Luo, P., Zhao, Z., Yu, Z.H., Fan, S.G., 2015. Authentication and traceability of fish maw products from the market using DNA sequencing. Food Control 55, 185–189.

Will, M., Guenther, D., 2007. Food Quality and Safety Standards. Deutsche Gesellschaft fuer Technische Zusammenarbeit (GTZ) GmbH, GTZ Division 45, Eschborn.

WIPO, 1883. World Intellectual Property Organization. Switzerland, Geneva. http://www.wipo.int/treaties/en/ip/paris/summary_paris.html.

WIPO, 1958. World Intellectual Property Organization. Switzerland, Geneva. http://www.wipo.int/lisbon/en/legal_texts/lisbon_agreement.html.

WIPO, 2015. World Intellectual Property Organization. Switzerland, Geneva. http://www.wipo.int/geo_indications/en/faq_geographicalindications.html.

Woodcock, T., Downey, G., O'Donnell, C.P., 2008. Better quality food and beverages: the role of near infrared spectroscopy. Journal of Near Infrared Spectroscopy 16, 1–29.

Effective Use of Food Traceability in Cereal Grain Food Supply Chains

16

H. Dong, D. Luo, H. Luo
Guangzhou Quality Supervision and Testing Institute, Guangzhou, Guangdong, China

1. Introduction

With the continuous improvement of living standards and the increase of the globalization of food markets, the enthusiasm of consumers pursuing high-quality food with a clear geographical origin identity has always grown rapidly. However, mislabeling and adulteration are still two of the most serious problems in food markets; these issues, on one hand, threaten the livelihood of those honest traders, and on the other, seriously infringe the consumers' rights (Federica et al., 2004; Ritz et al., 2005; Sacco et al., 2009). That is why food safety and food authenticity are becoming of primary importance for both consumers and food industries, in all of the production processes, from raw materials purchased to finished products distributed to all over the world, which, of course, makes the geographical origin identification of food products more and more important.

Cereal grain, which mainly contains rice, wheat, soybean, corn, and so on, is one of the most important staple food crops and plays a vital role as the main source of energy, protein, and fat intake for almost all of the world's population. For most food products, the authenticity and geographical origin can be distinguished by botanical or cultivar type, as well as the absence of adulterants, but as for cereal grain, there are more complex problems since they can depend on both origin and botanical or cultivar type (Wu et al., 2015). As a matter of fact, cultivar and cultivation area are very important factors in the market prices of cereal grain. Unfortunately, whenever price differences exist, there is also a potential for those dishonest traders to attempt to make money by adding inferior cereal grain to premium ones or even passing off the cheaper cereal grain as the more expensive one, resulting in extremely widespread adulteration or mislabeling of cereal grain (Yaeko et al., 2008). Under these kinds of circumstances, appropriate analytical methods to ensure a fair competition among producers and to protect consumers against fraud are required.

Currently, several different analytical techniques, for example, Isotope Ratio Mass Spectrometry (IRMS), Inductively Coupled Plasma-Mass Spectrometry (ICP-MS), Liquid Chromatography (LC), Near Infrared Spectroscopy (NIR), and so on, have been well utilized in the authenticity identification of cereal grain, with the ultimate

Advances in Food Traceability Techniques and Technologies. http://dx.doi.org/10.1016/B978-0-08-100310-7.00016-8

aim for the determination of the geographical origin of cereal grain. Most importantly, the application of multivariate statistical analysis of data obtained by the analytical instruments is also essentially helpful and allows effective discrimination of cereal grain according to their geographical origins.

The main aim of this chapter is to provide an overview concerning the current and the most effective uses, especially IRMS, Infrared spectroscopy (IR), multielement analysis, and some other analytical techniques that have been developed for determining the geographical origin in order to discriminate different origins of cereal grain, confirming their authenticity.

2. Effective Use in Cereal Grain Traceability

2.1 Isotope Ratio Mass Spectrometry (IRMS)

In general, the isotopic compositions of food materials are governed by the geoclimatic conditions under which a plant grows (Longobardi et al., 2015). Specifically, the isotopic compositions of plant materials reflect various factors such as isotopic compositions of source materials (eg, CO_2, H_2O and NO_2) and their assimilation processes, as well as growth environments, especially the climate and altitude. For example, the C isotopic composition strongly depends on the carbon fixation process such as the C-3, C-4, or Crassulacean Acid Metabolism (CAM) cycle (Smith and Epstein, 1971), while the N isotopic composition mainly depends on soil nutrition (Kohl et al., 1973; Meints et al., 1975; Shearer and Legg, 1975). Therefore the differences in the isotopic compositions make IRMS a suitable method for the geographical origin traceability of cereal grain. So far, several studies have been reported on tracing the geographical origin of different kinds of cereal grain by isotopic compositions. Different kinds of isotope ratio analysis and various samples were demonstrated to be amenable to cereal grain traceability by isotopic analysis, and according to which several successful results were achieved (Oda et al., 2002; Nietner et al., 2014; Asfaha et al., 2011). Relevant literature is summarized in Table 16.1.

The determination and investigation of the stable isotope ratios have been used primarily in the geological sciences of age determination. For a very long time, IRMS technique has been widely used to trace the geographical origin of organic materials in the biochemistry field, archaeology field, petroleum chemistry field, geochemistry field, and so on (Ambrose and DeNiro, 1986; Hayes et al., 1990; Rozanski et al., 1992). It has also become increasingly important in the geographical origin traceability of food products. Studies concerning the geographical origin of meat (Schmidt et al., 2005), dairy products (Rossmann et al., 2000; Ritz et al., 2005), honey (Luo et al., 2015b), and cereal grain (Simon et al., 2002), and some other kinds of food products, using the IRMS method, have been published successively by scientists all over the world.

As IRMS technique of differential analyses of samples and standards permits very small differences in the isotopic composition of samples to be determined reliably and accurately, and isotopic abundance of a sample relative to a reference is normally expressed by the symbol "δ", that is to say, the δ notation is used to describe the

Table 16.1 **Summary of the Recent Literature on the Use of Stable Isotope Analysis (or Together with Other Relevant Technique Analysis) in Cereal Grain Traceability**

Cereal Grain Type	Isotope Ratios Measured	Instrument	Samples Sources	References
Wheat	$^{13}C/^{12}C$, $^{15}N/^{14}N$, and $^{18}O/^{16}O$	EA-IRMS[a]	Canada, Australia, Turkey, and Italy	Brescia et al. (2002a,b)
Wheat	$^{13}C/^{12}C$ and $^{18}O/^{16}O$	EA-IRMS	Southern Italy	Brescia et al. (2002a,b)
Wheat	$^{13}C/^{12}C$ and $^{15}N/^{14}N$	EA-IRMS and ICP-MS	North America, Canada, France, and Germany	Branch et al. (2002)
Wheat	$^{13}C/^{12}C$, $^{15}N/^{14}N$, $^{18}O/^{16}O$, $^{34}S/^{32}S$, and $^{87}Sr/^{86}Sr$	EA-IRMS	Europe	Asfaha et al. (2011)
Wheat	$^{13}C/^{12}C$ and $^{15}N/^{14}N$	EA-IRMS	Australia, USA, Canada, and China	Luo et al., (2015a,b)
Wheat kernel	$^{13}C/^{12}C$, $^{15}N/^{14}N$, and $^{2}H/^{1}H$,	IRMS	China	Liu et al. (2015)
Wheat, rice, corn, soybean, millet	$^{13}C/^{12}C$ and $^{15}N/^{14}N$	EA-IRMS	China	Wu et al. (2015)
Rice	$^{13}C/^{12}C$ and $^{18}O/^{16}O$	EA-IRMS and ICP-MS	USA, Europe, and Basmati	Simon et al. (2002)
Rice	$^{87}Sr/^{86}Sr$	EA-IRMS	China and Vietnam, Australia	Kawasaki et al. (2002)
Rice	$^{87}Sr/^{86}Sr$ and $^{11}B/^{10}B$	IRMS	Japan, USA, China, Australia, Korea, and Vietnam	Oda et al. (2002)
Rice	$^{13}C/^{12}C$, $^{15}N/^{14}N$, and $^{18}O/^{16}O$	EA-IRMS	Australia, Japan, and USA	Yaeko et al. (2008)
Corn, rice, wheat	$^{2}H/^{1}H$, $^{13}C/^{12}C$, $^{15}N/^{14}N$, $^{18}O/^{16}O$, and $^{34}S/^{32}S$	IRMS	China, USA, Canada, and European Union	Nietner et al. (2014)

[a]EA-IRMS is short for elemental analyzer-isotope ratio mass spectrometry.

isotopic difference between the sample and an international standard, which can be defined as the following equation (marked as Eq. [16.1]),

$$\delta\,(‰) = (R_{\text{sample}} - R_{\text{standard}})\ /\ R_{\text{standard}} \times 1000 \qquad [16.1]$$

in which, R_{sample} is the isotope ratio (eg, $^{13}C/^{12}C$, $^{15}N/^{14}N$, $^{18}O/^{16}O$, and so on) of the sample and R_{standard} is that of reference material.

As for the determination of $^{2}H/^{1}H$, $^{13}C/^{12}C$, $^{15}N/^{14}N$, $^{18}O/^{16}O$, and $^{34}S/^{32}S$ isotope ratios, the analyte needs to be first converted into a gas phase, such as H_2, N_2, CO, CO_2, and SO_2, just before they are ionized. For this purpose, CO_2 is usually generated from organic compounds by oxidation with copper oxide and cryogenically trapped in liquid nitrogen. Hydrogen is obtained from the reductive conversion of water using an appropriate metal (eg, zinc). For the determination of the ^{18}O content, an equilibrium technique is employed using CO_2, which serves as the analyte. The common interface for introducing samples into the IRMS cannot be anything but the elemental analyzer (EA) (Drivelos and Georgiou, 2012).

It has been demonstrated by Kawasaki et al. (2002) that the geographical origin of rice can be identified by single $\delta^{87}Sr$ ($^{87}Sr/^{86}Sr$) value. In this study, the $^{87}Sr/^{86}Sr$ ratio of rice from China, Vietnam, and Australia were determined, and the results found that the $^{87}Sr/^{86}Sr$ ratio of rice of these regions ranged from 0.710 to 0.711 and 0.715 to 0.717, respectively. It is obviously indicated that strontium stable isotope ratio could be a potential indicator for estimating the rice provenance. In an extending work researched by Oda et al. (2002), the combination of strontium ($^{87}Sr/^{86}Sr$) and boron ($^{11}B/^{10}B$) isotope ratios was used to differentiate the geographical origin of rice samples obtained from Japan, the United States, China, Australia, Korea, and Vietnam. The results indicated that rice samples from these countries could be distinguished from each other by a single strontium isotope ratio alone or together with the boron isotope ratio. The stable carbon and nitrogen isotopic compositions ($\delta^{13}C$ and $\delta^{15}N$ values) of different cereal grains from different regions were determined in the research of Wu et al. (2015). The results demonstrated that the $\delta^{13}C$ values of rice, soybean, millet, wheat, and corn were significantly different ($P<0.05$) within different origins (Heilongjiang, Shandong, and Jiangsu province of China), while $\delta^{15}N$ values were not. Further study showed that the $\delta^{13}C$ values of wheat from Australia, the United States, Canada, and the Jiangsu and Shandong provinces of China were also significantly different ($P<0.01$). Thus the comparison of $\delta^{13}C$ values would be potentially useful for rapid and routine discrimination of the geographical origin of cereal grains. In a relative study conducted by Luo et al. (2015a,b), $\delta^{13}C$ and $\delta^{15}N$ values of 35 wheat samples originated from different regions were determined by EA-IRMS. The results indicated that wheat from Australia, the United States, Canada, and China could be potentially discriminated by using analyte $\delta^{13}C$ and $\delta^{15}N$ values. The $\delta^{13}C$ values of wheat from Australia, the United States, Canada, and the Jiangsu and Shandong provinces of China ranged from −22.63‰ to −22.33‰, −24.425‰ to −24.20‰, −25.47‰ to −25.37‰, −25.22‰ to −25.12‰, and −25.65‰ to −25.53‰, respectively. Wheat samples originated from these regions can be clearly differentiated by $\delta^{13}C$ values. As for nitrogen isotope ratio, $\delta^{15}N$ values of 35 wheat samples

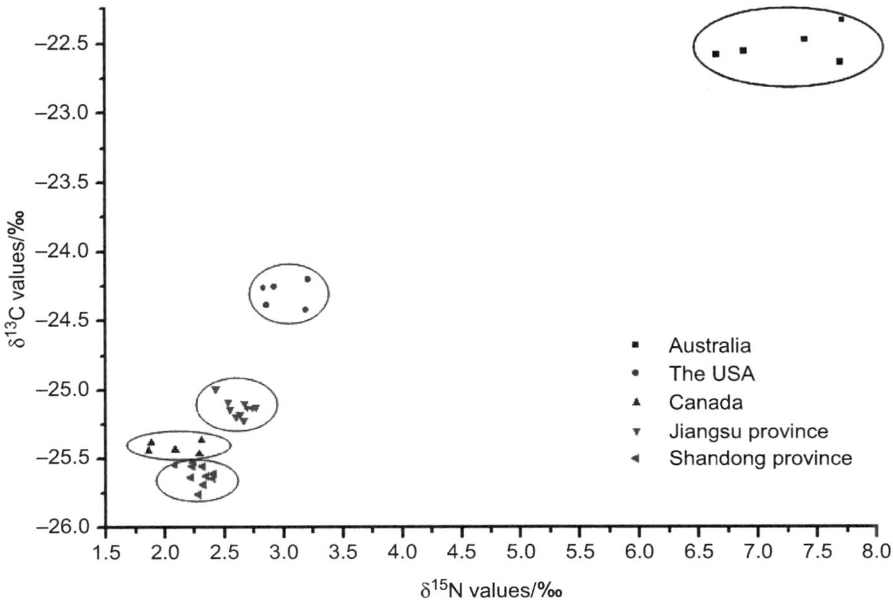

Figure 16.1 The distribution of C and N isotopic compositions of wheat originated from different regions (Luo et al., 2015a,b).

were calculated between 1.859‰ and 7.712‰, and the results illustrated regional distributions of $\delta^{15}N$ values of wheat as Australia > the United States > Jiangsu province of China > Shandong province of China > Canada. Two-dimensional distribution diagrams of carbon and nitrogen isotopic composition of wheat samples from different regions were presented in Fig. 16.1, and it obviously showed that there is a regional concentration in the wheat samples from these regions. In addition, researchers conducted a preliminary investigation using stable isotope analysis of $\delta^{13}C$ and $\delta^{15}N$ combined with ICP-MS analysis of Cd, Pb, Se, and Sr to identify the geographical origin of wheat samples cultivated in North America ($n=9$), Canada ($n=7$), France ($n=3$), and Germany ($n=1$) (Branch et al., 2002). Results indicated that discriminant analysis could correctly predict the geographical origin of all the wheat samples. However, the concentration of Se and the $\delta^{13}C$ values were consistently higher in America and Canadian samples than in France and Germany samples, and most importantly, the use of $\delta^{13}C$ values alone was found to be sufficient to determine the geographical origin of all 20 wheat samples in this work.

Not only stable carbon and nitrogen isotopic compositions can be used in identifying the geographical origin of cereal grain, but oxygen isotopic composition can also be applied for the cereal grain traceability. The use of oxygen isotopic composition has been reported by Japanese scientists (Yaeko et al., 2008) for the rice samples from Australia ($n=1$), Japan ($n=12$), and the United States ($n=1$). The C and N contents and stable C, N, and O isotopic compositions ($\delta^{13}C$, $\delta^{15}N$, $\delta^{18}O$) of polished rice (a single cultivar, Koshihikari rice, was chosen) were determined in order to discriminate the geographical origin of rice. The results implied that C and N contents and the isotopic compositions are consistent with those of general plant materials, being 37.2% to 40.0% (C content), 0.8 to 1.4% (N content), −27.1 to −25.4% ($\delta^{13}C$), +0.4 to +9.0% ($\delta^{15}N$), and +18.8 to +22.9% ($\delta^{18}O$), respectively. However, the cultivated area is clearly distinguished by a pentagonal radar plot based on the elements combined with isotopic compositions. Thus they concluded that the comparison of C and N contents and $\delta^{13}C$, $\delta^{15}N$, and $\delta^{18}O$ values would potentially be useful for routine discrimination of the geographical origin of the polished rice.

With regard to sulfur isotope ratio, the situation is quite different. Sulfur occurs in nature in multiple oxidation forms and isotopic characteristics. In addition, sulfur assimilation in most plants is a kind of unidirectional process. In this process, the total originally ingested element remains completely in the plant, independent of (partial) biochemical conversions with isotope fractionation and binding forms. The bulk sulfur isotope ratio thus preserves and reflects the isotopic property of the primary material in question and hence provides an ideal biomarker for the geographical origin of the sample. Nevertheless, isotope fractionations on defined reactions in plants and food chains lead to organic compounds with S from the same source but different $\delta^{34}S$ values (Krivachy et al., 2015). That is why $\delta^{34}S$ values can also be used in the geographical origin identification of cereal grain. In a paper published by Asfaha et al. (2011), more than 500 cereal samples collected over two years from 17 sampling sites across Europe were obtained and not only the $\delta^{13}C$, $\delta^{15}N$, and $\delta^{18}O$ values, but also $\delta^{34}S$ values were determined. The results showed that these parameters were the most discriminatory indicators in identifying the geographical origin of cereal grain.

2.2 Infrared Spectroscopy

In general, IR spectroscopy, which can measure the vibrations of molecules, is the measurement of the wavelength and intensity of the absorption of infrared light by samples. In fact, each structural characteristic or functional group of a molecule has the unique vibrational frequency that can be used to determine what functional groups are present in samples. When the effects of all the different functional groups are taken together, the result is a unique molecular "fingerprint" which can be subsequently used to confirm the identity of a sample (Putzig et al., 1994; Luykx and Ruth, 2008).

It is widely known that infrared radiation is the region of the electromagnetic spectrum between the visible and the microwave wavelengths. Actually, the infrared portion of the electromagnetic spectrum can be divided into three regions: the Near-Infrared (NIR), Mid-Infrared (MIR), and Far-Infrared (FIR), which are named for their relation to the visible spectrum. The nominal range of wavelengths for NIR is between 750 and 2500 nm (13,400 cm^{-1} to 4000 cm^{-1}), while for the MIR, the spectral range is from 2500 to 25,000 nm (4000 cm^{-1} to 400 cm^{-1}). This means that NIR can provide more complex structural information than MIR. The FIR (approx. 400 cm^{-1} to 10 cm^{-1}), lying adjacent to the microwave region, has low energy and may be used for rotational spectroscopy. Different samples, not only solid, liquid, but also gaseous, can absorb some of the incoming infrared radiation at specific wavelengths, resulting in a "fingerprint" or spectrum (Cozzolino, 2009, 2014; Smyth and Cozzolino, 2011, 2013).

Spectral "signatures" in the MIR result from the fundamental stretching, bending, and rotating vibrations of the sample molecules, while NIR spectra result from complex overtone and high-frequency combinations at the shorter wavelengths.

Several researches using IR technique to identify the geographical origin of cereal grain have been reported. A total of 240 wheat samples from two growing years (2007/2008, 2008/2009) from four major wheat producing regions (Hebei province, Henan province, Shandong province, and Shaanxi province) in China were analyzed by Zhao et al. (2013a,b,c). In this study, after spectral data pretreatment to wheat kernel and whole wheat flour, Linear Discriminant Analysis (LDA) based on principal component scores and Discriminant Partial Least Squares (DPLS) were applied to classify geographical origins of samples excluding water spectrum. The results indicated that there were significant differences among wheat NIR spectra (presented in Figs. 16.2 and 16.3) from different regions. Wheat samples could be classified based on geographical origins with chemometrics. In conclusion, NIR combined with chemometrics is a promising method for identifying wheat according to geographical origin, but the effectiveness should be verified further.

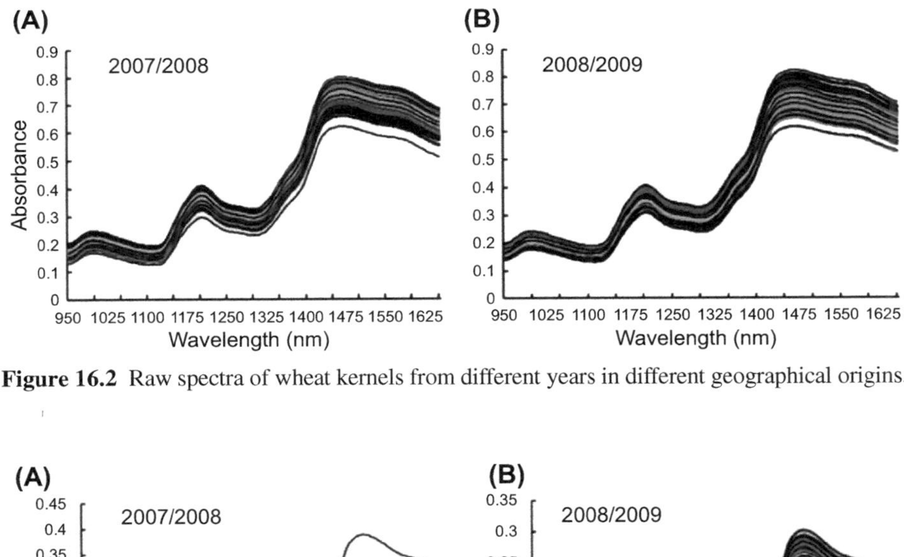

Figure 16.2 Raw spectra of wheat kernels from different years in different geographical origins.

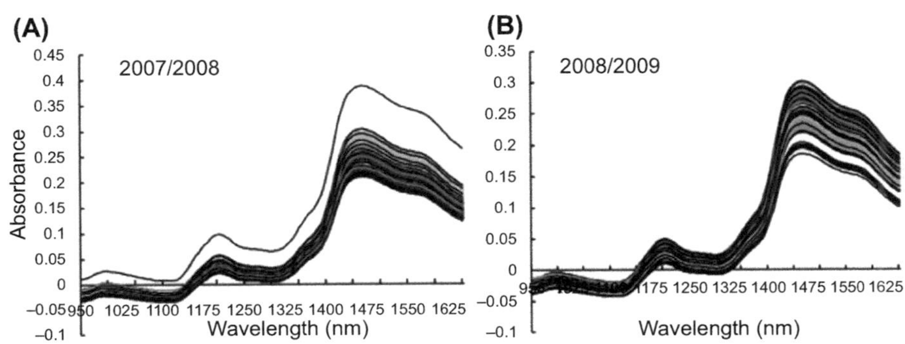

Figure 16.3 Raw spectra of whole wheat flours from different years in different geographical origins.

In an extending work conducted by Zhao et al. (2013b), the effects of not only geographical origin, but also genotype, harvest year, and their interactions on wheat NIR spectra were studied to find the reasons for differences in NIR fingerprints of wheat from different geographical origins and the stability of NIR fingerprints among different years. In this work, 10 varieties of wheat kernel samples (180 samples in total) grown in three regions of China for 2 years were analyzed by NIR. Likewise, the spectra after pretreatment were analyzed by Principal Component Analysis (PCA), multiway analysis of variance, and discriminant partial least-squares analysis. The results showed that geographical origin, genotype, year, and their interactions all had significant effects on wheat kernel NIR fingerprints. Most importantly, the second overtones of N-H and C-H stretching vibrations and a combination of stretch and deformation of the C-H group in wheat samples were mainly influenced by the geographical origin. The wavelength ranges 975–990 nm, 1200 nm, and 1355–1380 nm contained plenty of information to build robust discriminant models of wheat geographical origin. In another study, NIR has been used in combination with chemometric methods to predict the geographical origin of both wheat grain and flour samples produced in different regions of Chile (González-Martín et al., 2014). In this work, the spectral data obtained with NIR were analyzed using a supervised pattern recognition method, DPLS. As a result, the method correctly classified 76% of the wheat grain samples and between 90% and 96% of the flour samples according to their geographical origin. They concluded that NIR, together with chemometric methods, could provide a rapid tool for the classification of wheat grain and flour samples according to their geographical origin. NIR spectroscopy, combined with chemometrics methods, was also used to discriminate wheat varieties (Miralbés, 2008). A total of 249 samples of different wheat varieties from the 2003/2004 harvest year were used to develop the best discriminant equation by applying various scatters and mathematical treatments in the range of 400–2500 nm. The results obtained showed that the routine application of discriminant equations through the use of NIR spectroscopy is a rapid method of varietal identification that does not require qualified personnel in milling industry wheat delivery. In some cases, when individual quality parameters (eg, protein, wet gluten, and deformation energy) between wheat varieties are very similar the discriminant equation would act as a preliminary evaluation before applying quantitative calibration equations. This study has worked well for wheat varieties from France and Spain. However, further studies have to be done for mixed varieties or classes of wheat samples that include a large number of wheat varieties. A study investigated applying Attenuated Total Reflection Fourier Transform Mid-Infrared spectroscopy (ATR-FT-MIR) to the oil fraction extracted from samples in situ in order to identify the geographical origin of dried distiller grains with soluble (DDGS) was published in 2014 in *Food Chemistry* (Vermeulen et al., 2014). In this work, the use of spectroscopic and chemometric tools enabled the botanical and geographical origins of DDGS, as well as the industrial process used to produce them, to be identified. The models developed during the study provided a classification higher than 95% using an external validation set. The ATR-FT-MIR developed in this work is a rapid and nondestructive method that needs a small sample and enables only the oil fraction and fatty acid composition to be analyzed. This method could be particularly helpful when labeling is missing or

Table 16.2 **Summary of the Recent Literature on the Use of IR Spectroscopy in the Cereal Grain Traceability**

Cereal Grain Type	IR Spectroscopy Used	Samples Sources	References
Rice	NIR	Korea	Kim et al. (2003)
Rice	Raman spectroscopy	Korea	Hwang et al. (2012)
Wheat	NIR	Canada	Choudhary et al. (2009)
Wheat	NIR	China	Zhao et al. (2013a,b,c)
Wheat kernel	NIR	China	Zhao et al. (2013b)
Wheat	NIR	Spain and France	Miralbés (2008)
Wheat grain	NIR	Chile	González-Martín et al. (2014)
Wheat	NIR	Italian regions	Foca et al. (2009)
Wheat and corn	ATR-FT-MIR	USA, Canada, China, and Europe	Vermeulen et al. (2014)
Maize	ATR-FTIR	Brazil	Kuhnen et al. (2010)

during a crisis situation due to contamination, when geographical origin identification could be essential in terms of feed safety.

Another technique, Raman spectroscopy, has also been effectively utilized for the discrimination of rice samples according to geographical origin (Hwang et al., 2012). Ultimately, relevant literature are summarized in Table 16.2.

2.3 Multielement Analysis

As it is well-known that organic components of a food-crop production depend on various conditions, for example, fertilization, history of the field, climatic conditions in the year of cultivation, geographical location, and soil composition, and so on, it is not always possible to determine the geographical origin of a product by analyzing the organic components. However, the mineral content of crops is influenced by the soil composition and local environmental factors (eg, rainfall, temperature, sunshine, etc.). Certain regions have specific "fingerprints" of elements, and no two regions are likely to have identical soil maps. Consequently, element composition can effectively characterize geographical origin of cereal grain. Furthermore, minerals are significantly more stable in cereal grain than organic compounds, so they may be useful markers in terms of geographical classification (Zhao et al., 2011).

The techniques used for the elemental fingerprinting of food are mostly those with multielement detection capability, and the most common techniques mainly include the following: Inductively Coupled Plasma-Mass Spectrometry (ICP-MS), Inductively Coupled Plasma-Atomic Emission Spectroscopy (ICP-AES), and Atomic Absorption Spectroscopy (AAS). ICP-MS is a powerful tool for multielement analysis, which can provide limits of detection (LODs) for more than 70 elements at low concentrations, typically parts per billion or trillion. However, conventional ICP-MS systems cannot provide the lowest LODs, because of the polyatomic interferences that can enhance the background signal or overlap the signals of the most abundant isotopes. Actually, a DRC used in the analysis can remove unwanted interferences by creating specific chemical reactions with a supplementary gas, which improves the selectivity and the sensitivity effectively and dramatically. Moreover, sensitivity and selectivity can also be improved by replacing the quadrupole with a sector field analyzer. ICP-AES provides high reproducibility and a quantitative linear range greater than conventional methods with arc or spark, allowing the simultaneous determination of metals and nonmetals. This technique uses a very high-temperature excitation source and molecular interferences can be greatly reduced, which makes it a fast, multielement technique with an extended dynamic linear range and moderate-to-low LODs. As a matter of fact, up to 60 elements can be screened per sample run in less than one minute. ICP-AES monitors wavelengths down to 165 nm that provide high sensitivity for phosphorous and sulfur determinations. However, the analysis is more expensive, and additionally, all samples have to be introduced in a dissolved form. In conclusion, the main drawbacks of ICP techniques, not only ICP-MS, but also ICP-AES mentioned above, are their cost and the requirement for trained operators. AAS is also a technique that can be used for food authentication. AAS instruments can analyze only one or a few elements simultaneously. AAS requires that the analyte atoms are in the gas phase, so ions or atoms in a sample must undergo desolvation and vaporization in a high-temperature source (eg, flame or electrically heated graphite furnace). The main advantages of flame AAS (FAAS) with atomization in a common air-acetylene flame are low operational costs and good analytical performance. The use of FAAS for food sample analysis is usually performed to avoid possible spectral or polyatomic interferences, which can appear in plasma techniques. The direct analysis of solid samples with a reduced pretreatment is possible using the electrothermal AAS (ET-AAS) technique that offers high sensitivity and selectivity at a reasonable cost with the possibility of determining low levels of nonmetals, which are not so accessible in plasma instrumentation due to various interferences. The most important drawback of AAS techniques is apparently the lack of multielement analysis. The other techniques applying for the element analysis of food also include Inductively Coupled Plasma-Optical Emission Spectrometry (ICP-OES), Neutron-Activation Analysis (NAA), X-ray Fluorescence (XRF) spectroscopy, Particle-Induced X-ray Emission or Proton-Induced X-ray Emission (PIXE) spectroscopy, and gamma ray spectroscopy.

In the geographical origin identification of cereal grain by multielement analysis, several studies have been conducted, and many gratifying results are obtained. The distinction of the geographical origin of Japanese brown rice samples by their multielement concentrations was studied by Yasui and Shindoh (2000) using ICP-AES and ICP-MS. In the study, 34 kinds of unhulled rice samples were collected from 27 different locations. Nine elements (Mn, Zn, Fe, Cu, Rb, Mo, Ba, Sr, Ni) were used in a PCA that permitted distinction of rice samples cultivated in the Tohoku/Kanto area from those in Hokuriku, and those in Tohoku from those in the Kanto regions. In addition, Simon et al. (2002) reported the use of IRMS and ICP-MS to determine the geographical origin of rice samples cultivated in the United States, Europe, and Basmati regions (India and

Pakistan). Nine key parameters (boron, holmium, gadolinium, magnesium, rubidium, selenium and tungsten, and stable carbon and oxygen isotope ratios) were identified by stepwise canonical discriminant analysis and permitted 100% discrimination between rice samples from these production origins. The authors observed high levels of boron (>2.5 μg/g) associated with rice samples from America and notably high levels of holmium were found in rice samples from the state of Arkansas. European rice samples generally contained relatively high levels of magnesium and Indian/Pakistani samples were characterized by relatively low ^{18}O abundance, characteristic of the relatively high altitude at which Basmati rice crops were cultivated and irrigation with Himalayan water. In another study published by Cheajesadagul et al. (2013), the concentrations of 21 elements were analyzed and evaluated by High Resolution-Inductively Coupled Plasma-Mass Spectrometry (HR-ICP-MS), then these elements were used as chemical indicators to discriminate the geographical origin of rice samples. The results found that the classification of rice samples was carried out based on elemental composition by a radar plot and multivariate data analysis, including PCA and Discriminant Analysis (DA). Thai jasmine rice can be differentiated from foreign rice samples by radar plots and multivariate data analysis. Furthermore, the DA can differentiate Thai jasmine rice samples according to each region of origin (northern, northeastern, or central regions of Thailand). Therefore they concluded that multielement fingerprinting combined with the use of multivariate statistical techniques could be considered as a powerful tool for rice geographical origin traceability. Moreover, in a paper published in *Food Chemistry* by Gonzálvez et al. (2011), trace element determination of rice samples by ICP-OES has been addressed to achieve a complete geographical origin classification. The training set was formed by 107 samples from Spain, including Valencia, Tarragona, Murcia and Extremadura, Japan, Brazil, and India, whereas the validation set was formed by 46 samples of the aforementioned geographical origins. Data were processed by LDA, using the concentration of 32 elements as an analytical variable, enabling a 91.30% of correct classification for the validation set (Table 16.3).

As for wheat, Zhao et al. (2011) published a research work in *Journal of Agriculture and Food Chemistry*, which was about determining the geographical origin of wheat using multielement analysis and multivariate statistics. In this article, the element contents of wheat from four major wheat producing regions of China were analyzed and used in multivariate statistical analysis to classify wheat samples according to the geographical origin. The concentrations of 15 elements in 240 samples from the 2007/2008 and 2008/2009 harvest years were determined by ICP-MS (Table 16.3). The analysis of variance and LDA were applied to classify wheat geographical origin, and the effects of region, variety, and harvest year on the element contents were also analyzed in this study. The results indicated that the multielement analysis is a promising method to provide reliable origin information for wheat, although the element profiles and discriminant models were affected by wheat varieties, harvest years, and agricultural practices. In their extending research published in 2013 (Zhao et al., 2013a,b,c), the concentrations of 22 elements in wheat grains and their provenance soils from Hebei and Henan provinces in China were analyzed by HR-ICP-MS and XRF (Table 16.3). The relationships of concentrations of 22 elements between wheat grain and soil were also studied. The elements associated with parent soil were used to

Table 16.3 Summary of the Recent Literature on the Use of Multielements of Cereal Grains Traceability

Cereal Grain Type	Elements Measured	Instrumental Techniques	Samples Sources	References
Rice	B, Ho, Gd, Mg, Rb, Se, and W	ICP-MS	USA, Europe, India, and Pakistan	Simon et al. (2002)
Rice	C and N	EA-IRMS	Australia, Japan, and USA	Yaeko et al. (2008)
Rice	Mn, Zn, Fe, Cu, Rb, Mo, Ba, Sr, and Ni	ICP-AES and ICP-MS	Japan	Yasui et al. (2000)
Rice	B, Co, Sr, Mo, Cd, Cs, Ba, Pb, Ti, V, As, Se, Mn, Cu, Rb, Mg, Al, Cr, Fe, Ni, and Zn	HR-ICP-MS	Thailand, France, Japan, Italy, India, and Pakistan	Cheajesadagul et al. (2013)
Rice	Al, As, Ba, Bi, Cd, Ca, Cr, Co, Cu, Fe, Pb, Li, Mg, Mn, Mo, Ni, K, Se, Na, Sr, Tl, Ti, Zn, La, Ce, Pr, Nd, Sm, Eu, Ho, Er, and Yb	ICP-OES	Spain	Gonzálvez et al. (2011)
Wheat	Be, Na, Mg, Al, K, Ca, V, Mn, Fe, Cu, Zn, Mo, Cd, Ba, and Th	ICP-MS	China	Zhao et al. (2011)
Wheat	Be, Na, Mg, Ca, Sc, Ti, V, Cr, Mn, Fe, Co, Cu, Zn, Ga, Se, Rb, Sr, Y, Zr, Cd, Cs, and Pb	ICP-MS and XRF	China	Zhao et al. (2013a,b,c)
Wheat	Cd, Pb, Se, and Sr	ICP-MS	North America, Canada, France, and Germany	Branch et al. (2002)
Wheat and corn	Na, K, Ca, Cu, and Rb	ICP-MS	Europe	Asfaha et al. (2011)

discriminate wheat origin with PCA and LDA. The results confirmed that significant correlations existed between the elements Cr, Mn, Ga, Rb, Sr, Zr, and Cd in wheat grain and its provenance soil. The models built with these seven elements obtained 95% of total correct classification for all of the samples. The reflection of geochemical provinces in certain element concentrations may serve as a classifier for determining wheat geographical origin. Relevant literature is summarized in Table 16.3.

2.4 The Other Techniques

Several other techniques, such as Nuclear Magnetic Resonance (NMR) spectroscopy, Liquid Chromatography (LC), even molecular biological techniques, and so on, were also gradually adopted in the traceability of cereal grain. For example, NMR spectroscopy provides the possibility of obtaining information on the presence of many compounds in a single spectrum. In a paper published by Lamanna et al. (2011), ^{1}H-NMR was used to detect metabolic profiles of wheat flour samples of different geographical and botanical origin. In this study, 100 durum wheat samples originated from the Altamura, Minervino, and Poggiorsini locations (Apulia region, Southern Italy) were considered and analyzed. In the statistical procedure, the analysis of variance (ANOVA) was used in combination with the LDA to maximize the ability of LDA in predicting unknown samples. Fig. 16.4 shows the first two linear discriminant functions for the

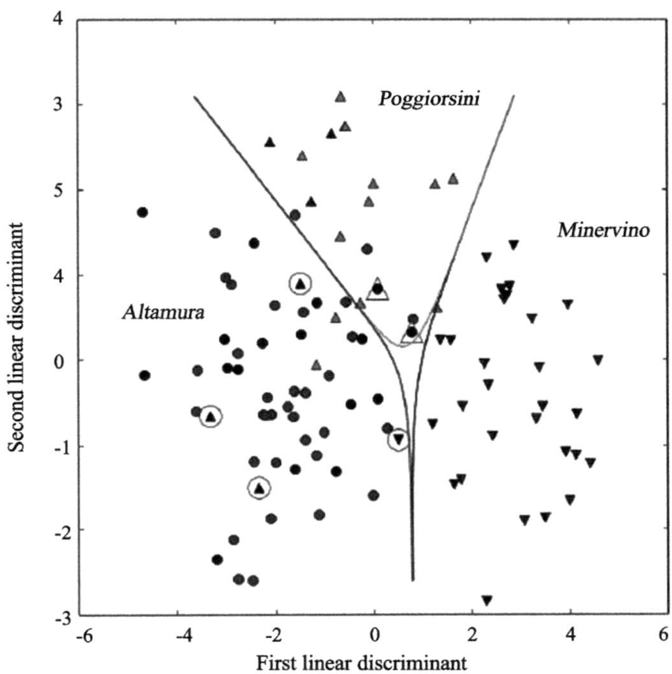

Figure 16.4 Linear discriminant analysis of a reduced number of NMR variables for the geographical origin classification (Lamanna et al., 2011).

geographical origin of wheat, and it is clear from Fig. 16.4 that the NMR spectra contain sufficient information to assess the geographical origin of wheat samples.

The red circle, blue down triangles, and yellow up triangles represent, respectively, "Altamura," "Minervino," and "Poggiorsini" geographical classes in the training set. The same symbols in black represent the class assigned by the LDA model to the validation samples. Misassigned samples are surrounded by larger symbols representing the real geographical origin of the corresponding validation sample.

3. Conclusions and Future Prospects

In this chapter, several effective techniques, such as IRMS, IR, multielement analysis, and so on, in application of geographical origin traceability of cereal gain, were discussed and summarized. For IRMS, the most commonly used isotope ratios were carbon, nitrogen, and oxygen isotope ratios ($^{13}C/^{12}C$, $^{15}N/^{14}N$, $^{18}O/^{16}O$), in addition, the $^{87}Sr/^{86}Sr$ and $^{11}B/^{10}B$, $^{2}H/^{1}H$, and $^{34}S/^{32}S$ analysis were also adopted in the cereal grain traceability. As it is discussed above, stable isotope analysis is a powerful tool for tracing the geographical origin and identifying the provenance of agro-products, even cereal grain. However, there are some limitations for this method. Concerning the geographical origin of cereal grain, the probable influence of the environment factors such as climate, altitude, and distance from oceans should be considered when using the stable isotope composition analysis. NIR can provide more complex structural information than MIR, making NIR a more frequently used method in the origin identification of cereal grain. As for multielement analysis, so far, more than 50 elements (B, K, Bi, W, C, N, V, Ho, Gd, Rb, Be, Al, Na, Cu, Rb, Mg, Ca, Sc, Ti, Cr, Fe, Co, Ga, Se, Rb, Zr, Mn, Ba, Sr, Mo, Cs, Zn, Th, As, Cd, Pb, Li, Mg, Ni, Sr, Tl, La, Ce, Pr, Nd, Sm, Eu, Er, Yb, and so on) were applied in the geographical origin analysis of cereal grain and obtained good results. Comparing multi-isotope-ratio analysis and multielement analysis, the latter appears to have the upper hand when combined with chemometric techniques, but the best possible achievable results come from combining the two techniques. During the literature summary, it has also been found that stable isotope analysis in combination with multielement analysis or NMR analysis is a promising way for the geographical origin of cereal grain. Many studies employed this combination with impressive results. However, implementation of this strategy is quite difficult in routine analysis. In addition, the abundant data collected can be efficiently processed by a statistic analysis method, such as PCA, LDA, and so on.

Cereal grain authentication and traceability will continue to be important in the near future, with new technological developments and analytical breakthroughs. In the longer-term, it is hoped that a deeper understanding of how geochemical and meteorological signatures are transferred into the cereal grain system may obviate the need for expensive comparative databases of authentic cereal grain, even the other food. Ultimately, this would allow generations of isotopic and multielement maps or IR "fingerprints" for different kinds of cereal grain from different geographical locations, which could be incorporated into traceability systems. Although for the moment, there is a need for comparative databases to be established for cereal grain, overcoming the need for comparative databases in cereal grain provenance determinations will be an extremely challenging work.

References

Ambrose, S.H., DeNiro, M.J., 1986. Reconstruction of African human diet using bone collagen carbon and nitrogen isotope ratios. Nature 319, 59–130.

Asfaha, D.G., et al., 2011. Combining isotopic signatures of $n(^{87}Sr)/n(^{86}Sr)$ and light stable elements (C, N, O, S) with multi-elemental profiling for the authentication of provenance of European cereal samples. Journal of Cereal Science 53, 170–177.

Branch, S., Burke, S., Evans, P., Fairman, B., Briche, C.S.J.W., 2002. A preliminary study in determining the geographical origin of wheat using isotope ratio inductively coupled plasma mass spectrometry with ^{13}C, ^{15}N mass spectrometry. Journal of Analytical Atomic Spectrometry 18, 17–22.

Brescia, M.A., Di Martino, G., Guillou, C., Reniero, F., Sacco, A., 2002a. Differentiation of the geographical origin of durum wheat Semolina samples on the basis of isotopic composition. Rapid Communication in Mass Spectrometry 16, 2286–2290.

Brescia, M.A., et al., 2002b. Characterization of Italian durum wheat Semolina by means of analytical and spectroscopic determinations. Cereal Chemistry 79, 238–242.

Cheajesadagul, P., Arnaudguilhem, C., Shiowatana, J., Siripinyanond, A., Szpunar, J., 2013. Discrimination of geographical origin of rice based on multi-element fingerprinting by high resolution inductively coupled plasma mass spectrometry. Food Chemistry 141, 3504–3509.

Choudhary, R., Mahesh, S., Paliwal, J., Jayas, D.S., 2009. Identification of wheat classes using wavelet features from near infrared hyperspectral images of bulk samples. Biosystems Engineering 102, 115–127.

Cozzolino, D., 2009. Near infrared spectroscopy in natural products analysis. Planta Medica 75, 746–757.

Cozzolino, D., 2014. An overview of the use of infrared spectroscopy and chemometrics in authenticity and traceability of cereals. Food Research International 60, 262–265.

Drivelos, S.A., Georgiou, C.A., 2012. Multi-element and multi-isotope-ratio analysis to determine the geographical origin of foods in the European Union. Trends in Analytical Chemistry 40, 38–51.

Federica, C., Karine, W., Anaisabel, B.C., Georg, H., Michéle, L., Giuseppe, V., 2004. Application of multielement stable isotope ratio analysis to the characterization of French, Italian, and Spanish Cheeses. Journal of Agricultural and Food Chemistry 52, 6592–6601.

Foca, G., Cocchi, M., Vigni, M.L., Caramanico, R., Corbellini, M., Ulrici, A., December 15, 2009. Different feature selection strategies in the wavelet domain applied to NIR-based quality classification models of bread wheat flours. Chemometrics and Intelligent Laboratory Systems 99, 91–100.

González-Martín, M.I., Moncada, G.W., González-Pérez, C., Martín, N.Z.S., López-González, F., Ortega, I.L., Hernández-Hierro, J., 2014. Chilean flour and wheat grain: tracing their origin using near infrared spectroscopy and chemometrics. Food Chemistry 145, 802–806.

Gonzálvez, A., Armenta, S., Guardia, M., 2011. Geographical traceability of "Arròs de Valencia" rice grain based on mineral element composition. Food Chemistry 126, 1254–1260.

Hayes, J.M., Freeman, K.H., Popp, B.N., Hoham, C.H., 1990. Compound-specific isotopic analyses: a novel tool for reconstruction of ancient biogeochemical processes. Organic Geochemistry 16, 1115–1128.

Hwang, J., Kang, S., Lee, K., Chung, H., 2012. Enhanced Raman spectroscopic discrimination of the geographical origins of rice samples via transmission spectral collection through packed grains. Talanta 101, 488–494.

Kawasaki, A., Oda, H., Hirata, T., 2002. Determination of strontium isotope ratio of brown rice for estimating its provenance. Soil Science and Plant Nutrition 48, 635–640.

Kim, S.S., Mee-Ra, R., Kim, M.J., Lee, S.H., 2003. Authentication of rice using near-infrared reflectance spectroscopy. Cereal Chemistry 80, 346–349.

Kohl, D.H., Shearer, G.B., Commones, B., 1973. Variation of ^{15}N in corn and soil following applications of fertilizer nitrogen. Soil Science Society of America Journal 37, 888–892.

Krivachy, N., Rossmann, A., Schmidt, H., 2015. Potentials and caveats with oxygen and sulfur stable isotope analyses in authenticity and origin checks of food and food commodities. Food Control 48, 143–150.

Kuhnen, S., Ogliari, J., Dias, B., Fernando, P., 2010. ATR-FTIR spectroscopy and chemometric analysis applied to discrimination of landrace maize flours produced in southern Brazil. International Journal of Food Science and Technology 45, 1673–1681.

Lamanna, R., Cattivelli, L., Miglietta, M.L., Troccoli, A., 2011. Geographical origin of durum wheat studied by ^{1}H-NMR profiling. Magnetic Resonance in Chemistry 49, 1–5. http://dx.doi.org/10.1002/mrc.2695.

Liu, H., Guo, B., Wei, Y., Wei, S., Ma, Y., Zhang, W., 2015. Effects of region, genotype, harvest year and their interactions on $\delta^{13}C$, $\delta^{15}N$ and δD in wheat kernels. Food Chemistry 171, 56–61.

Longobardi, F., Sacco, D., Casiello, G., Ventrella, A., Sacco, A., 2015. Characterization of the geographical and varietal origin of wheat and bread by means of nuclear magnetic resonance (NMR), isotope ratio mass spectrometry (IRMS) methods and chemometrics: a review. Agricultural Sciences 6, 126–136.

Luo, D., Dong, H., Luo, H., Xian, Y., Wan, J., Guo, X., Wu, Y., 2015a. The application of stable isotope ratio analysis to determine the geographical origin of wheat. Food Chemistry 174, 197–201.

Luo, D., Luo, H., Dong, H., Xian, Y., Guo, X., Wu, Y., 2015b. Hydrogen ($^{2}H/^{1}H$) combined with carbon ($^{13}C/^{12}C$) isotope ratios analysis to determine the adulteration of commercial honey. Food Analytical Methods 9, 255–262.

Luykx, D.M.A.M., Saskia Ruth, M., 2008. An overview of analytical methods for determining the geographical origin of food products. Food Chemistry 107, 897–911.

Meints, V.W., Shearer, G., Kohl, D.H., Kurtz, L.T., 1975. A comparison of unenriched vs $\delta^{15}N$ enriched fertilizer as a tracer for N fertilizer uptake. Soil Science 119, 421–425.

Miralbés, C., 2008. Discrimination of European wheat varieties using near infrared reflectance spectroscopy. Food Chemistry 106, 386–389.

Nietner, T., Haughey, S.A., Ogle, N., Fauhl-Hassek, C., Elliott, C.T., 2014. Determination of geographical origin of distillers dried grains and solubles using isotope ratio mass spectrometry. Food Research International 60, 146–153.

Oda, H., Kawasaki, A., Hirata, T., 2002. Determining the rice provenance using binary isotope signatures along with cadmium content. In: Proceedings of the 17th World Congress of Soil Science, 14th to 21st August, Thailand, Symposium No. 59 2002, pp. 336–338.

Putzig, C.L., Leugers, M.A., McKelvy, M.L., Mitchell, G.E., Nyquist, R.A., Papenfuss, R.R., et al., 1994. Infrared spectroscopy. Analytical Chemistry 66, 26–66.

Ritz, P., Gachon, P., Garel, J.P., Bonnefoy, J.C., Coulon, J.B., Renou, J.P., 2005. Milk characterization: effect of the breed. Food Chemistry 91, 521–523.

Rossmann, A., Haberhauer, G., Holzl, S., Horn, P., Pichlmayer, F., Voerkelius, S., 2000. The potential of multielement stable isotope analysis for regional origin assignment of butter. European Food Research and Technology 211, 32–40.

Rozanski, K., Araguás-Araguás, L., Gonfiantini, R., 1992. Relation between long term trends of oxygen-18 isotope composition of precipitation and climate. Science 258, 981–985.

Sacco, D., Brescia, M.A., Sgaramella, A., Casiello, G., Buccolieri, A., Ogrinc, N., Sacco, A., 2009. Discrimination between Southern Italy and foreign milk samples using spectroscopic and analytical data. Food Chemistry 114, 1559–1563.

Schmidt, O., Quilter, J.M., Bahar, B., Moloney, A.P., Scrimgeour, C.M., Begley, I.S., 2005. Inferring the origin and dietary history of beef from C, N and S stable isotope ratio analysis. Food Chemistry 91, 545–549.

Shearer, G., Legg, J.O., 1975. Variation in the natural abundance of $\delta^{15}N$ of wheat plants in relation to fertilizer nitrogen applications. Soil Science Society of America Journal 39, 896–901.

Simon, K., Malcolm, B., Stephen, C., Christopher, R., John, D., Paul, B., 2002. The application of isotopic and elemental analysis to determine the geographical origin of premium long grain rice. European Food Research and Technology 214, 72–78.

Smith, B.N., Epstein, S., 1971. Influence of the diet on the distribution of carbon isotope fractionation. Plant Physiology 47, 380–384.

Smyth, H.E., Cozzolino, D., 2011. Applications of infrared spectroscopy for quantitative analysis of volatile and secondary metabolites in plant materials. Current Bioactive Compounds 7, 66–74.

Smyth, H.E., Cozzolino, D., 2013. Instrumental methods (spectroscopy, electronic nose and tongue) as tools to predict taste and aroma in beverages: advantages and limitations. Chemical Reviews 113, 1429–1440.

Vermeulen, P., Pierna, J.A.F., Abbas, O., Dardenne, P., Baeten, V., 2014. Origin identification of dried distillers grains with solubles using attenuated total reflection Fourier transform mid-infrared spectroscopy after in situ oil extraction. Food Chemistry. http://dx.doi.org/10.1016/j.foodchem.2014.09.103.

Wu, Y., Luo, D., Dong, H., Wan, J., Luo, H., Xian, Y., Guo, X., Qin, F., Han, W., Wang, L., Wang, B., 2015. Geographical origin of cereal grains based on element analyzer-stable isotope ratio mass spectrometry (EA-SIRMS). Food Chemistry 174, 553–557.

Yaeko, S., Yoshito, C., Nanako, O.O., Naohiko, O., Takashi, K., 2008. Geographical origin of polished rice based on multiple element and stable isotope analyses. Food Chemistry 109, 470–475.

Yasui, A., Shindoh, K., 2000. Determination of the geographic origin of brown-rice with trace-element composition. Bunseki Kagaku 49, 405–410.

Zhao, H., Guo, B., Wei, Y., Zhang, B., 2013a. Multi-element composition of wheat grain and provenance soil and their potentialities as fingerprints of geographical origin. Journal of Cereal Science 57, 391–397.

Zhao, H., Guo, B., Wei, Y., Zhang, B., 2013b. Effects of grown origin, genotype, harvest year, and their interactions of wheat kernels on near infrared spectral fingerprints for geographical traceability. Food Chemistry 152, 316–322.

Zhao, H., Guo, B., Wei, Y., Zhang, B., 2013c. Near infrared reflectance spectroscopy for determination of the geographical origin of wheat. Food Chemistry 138, 1902–1907.

Zhao, H., Guo, B., Wei, Y., Zhang, B., Sun, S., Zhang, L., Yan, J., 2011. Determining the geographic origin of wheat using multielement analysis and multivariate statistics. Journal of Agriculture and Food Chemistry 59, 4397–4402.

Effective Use of Food Traceability in Meat Supply Chains

17

J.E. Hobbs
University of Saskatchewan, Saskatoon, SK, Canada

1. Introduction

Traceability is the subject of a deepening body of analysis within social sciences. The concept of "traceability" has many facets and is viewed through many different lenses: it is alternately defined broadly or in fairly narrow terms, and it is associated with a broad range of potential functions and objectives, from food safety, to biosecurity, animal health, market access, supply chain management, product differentiation, and quality verification. It is the subject of significant public policy regulatory initiatives yet, in its other guises, is a solely private sector endeavor. The motivation for implementing or enhancing food traceability is credited alternately to consumers, the food industry, technology providers, and regulators, individually or in concert (Hobbs, 2006).

The meat (and livestock) sector has been at the forefront of policy and industry developments concerning traceability. In a large part, this is due to a series of early high profile food safety crises affecting the meat and livestock sector (eg, Bovine Spongiform Encephalopathy [BSE] in beef, dioxin contamination of animal feed in Europe, *Escherichia coli 0157:H7* contamination in meat products) which served to focus public attention on food safety but also stimulated demand for more information on the source of food products and the production and processing methods in use within food supply chains (Hobbs et al., 2002; Meuwissen et al., 2003; Hobbs, 2004). The topic of traceability within meat supply chains therefore represents fertile ground from which to explore the drivers, outcomes, and implications of food traceability. While the focus of the chapter is on meat traceability, it is impossible to discuss meat traceability without also considering the traceability of livestock as a component of meat traceability systems.

Drawing upon insights from food economics, policy, and agribusiness literature, this chapter examines the roles, outcomes, and implications of traceability in meat supply chains. The chapter begins by examining perspectives on traceability in the context of meat supply chains: what is traceability, what functions do meat and livestock traceability systems perform, and how do their objectives differ? Building upon these concepts, the chapter then examines public and private sector approaches to meat and livestock traceability: why and how have mandatory traceability systems emerged and how do voluntary private sector traceability systems differ in terms of motivations and outcomes? A growing literature has emerged evaluating consumer responses to food traceability assurances; key insights from this literature and the

Advances in Food Traceability Techniques and Technologies. http://dx.doi.org/10.1016/B978-0-08-100310-7.00017-X

implications for the effective use of meat traceability in supply chains are discussed. The chapter concludes with a discussion of future trends and suggestions for further research.

2. Perspectives on Traceability

2.1 Defining Traceability: Tracking Versus Tracing

A number of official definitions of traceability exist within national and international regulatory bodies, for example, the International Organization for Standardization (ISO), a worldwide federation of national standards bodies and the Codex Alimentarius Commission. Both define traceability as "…the ability to follow the movement of a feed or food through specified stage(s) of production, processing, and distribution" (ISO-22005:2007; CAC/GL 60–2006, 2006).

At the supranational or national level the 2002 European Union (EU) Common Food Law defines traceability as "…the ability to trace and follow a food, feed, food-producing animal or substance intended to be, or expected to be incorporated into a food or feed, through all states of production, processing and distribution" (EC Regulation 178/02, 2002, clause 1.5). According to the OECD Working Party on Agricultural Policies and Markets, the United States Food and Drug Administration (FDA) proposed that product tracing be defined as "…the ability to identify by means of paper or electronic records a food product and its producer, from where and when it came, and to where and when it was sent" (OECD, 2003, p. 28). While, in consultations regarding the establishment of a national livestock and poultry traceability system in Canada, the Canadian Food Inspection Agency (CFIA) defined traceability as "… the ability to follow an item or group of items—including animals, plants, food products and agricultural inputs such as feed, seed or ingredients—from one point in the supply chain to another" (CFIA, 2012).

In a broader treatment, Opara (2003, p. 102) defines "agricultural traceability" as the "…collection, documentation, maintenance, and application of information related to all processes in the supply chain in a manner that provides guarantee (sic) to the consumer and other stakeholders on the origin, location and life history of a product as assisting in crises (sic) management in the event of a safety and quality breach".

A common thread within most of these definitions is the twin concepts of tracing and tracking, where *tracing* implies the ability to trace the source of a product (or ingredients) backward or upstream in the supply chain, while *tracking* implies the ability to follow a product forward, downstream in the supply chain. In this vein, tracing is focused on determining the history of the product, for example, identifying the source of contamination in a foodborne disease outbreak. Tracking can be used to locate and recall products that may pose a risk to health, or may be a component of identity preservation as a mean to provide specific quality assurances to consumers (Meuwissen et al., 2003).

The last point is important: while tracking *may* be a component of a quality verification system, the two concepts are not synonymous. The ability to trace a food product (backward) or track (forward) within a supply chain does not necessarily provide additional information to consumers about how the product was produced or whether it meets their quality expectations, particularly with respect to credence attributes.[1] Two additional dimensions are therefore relevant: *transparency*, defined as the public availability of information on the procedures and practices used to produce the food product, and *quality assurance*, related to specific management and quality measurement practices (Liddell and Bailey, 2001; Hobbs, 2006). In this vein, Smith et al., 2005 draw a useful distinction between the concepts of "identification," "traceability," and "verification" in the context of livestock and meat traceability, wherein identification (eg, of animals) is relatively easy to accomplish (ear tags, Radio Frequency Identification [RFID] tags), traceability through multiple stages of the supply chain is more complex, and verifying identity and traceability claims about livestock and meat through a supply chain is particularly challenging.

2.2 Breadth, Depth, and Precision

Three characteristics that help distinguish between different types of traceability systems are breadth, depth, and precision (Golan et al., 2003). *Breadth* refers to the amount of information (the number of attributes) recorded by a traceability system. In the context of meat and livestock traceability, this may include the farm from which the animals originated, what the animal beef was fed, the age of the animal, use of subtherapeutic antibiotics, vaccination records, animal housing (welfare), where and how the animal was slaughtered, where the meat was processed or packaged, the temperature at which the meat was stored during transportation, etc.

Clearly there are an almost infinite number of attributes within any single food product, and a traceability system that recorded and tracked all of this information would be enormous and likely uneconomic. The breadth of a meat traceability system is determined by two primary factors: the technological feasibility and cost of measuring, recording, and communicating information about a specific attribute, and the market demand for this information. In private sector traceability systems, firms must balance the costs and benefits of providing attribute information to determine the optimal breadth of the system (Golan et al., 2003). Good public policy, in mandating traceability requirements, should also consider the relative costs and benefits of attribute information, although for the most part, meat traceability systems with more breadth tend to be private sector systems responding to market demand for information about specific attributes.

Depth refers to how far backward or forward in the supply chain the system tracks (Golan et al., 2003). One-step-up, one-step-down traceability, the ability for a firm to identify all inputs and to know its customers for a specific product batch, is relatively straightforward and is the type of traceability enshrined in the mandatory traceability requirements within the EU (discussed below). As traceability systems deepen, so

[1] Credence attributes are qualities that, in the absence of labeling, cannot be detected by consumers prior to purchase or even after consumption: examples include animal welfare, environmental sustainability, use of subtherapeutic antibiotics in animal production, and the presence of genetically modified organisms (GMOs).

does the complexity of providing accurate traceability information. In the context of meat traceability, depth could entail whether a meat product on a retail shelf can be traced back to the distributor, the importer (where relevant), the processor, the farm (or farms) on which the animal resided, or even the genetic history (parents) of the animal (Tonsor and Schroeder, 2006). Clearly depth and breadth are related. A meat traceability system with the objective of providing assurances on the origin of the animals from which the meat is derived, or the use of certain animal welfare practices, requires a more extensive degree of depth than a traceability system with assurances about where the meat was processed and packaged.

Recent events have served to focus public attention on the depth and complexity of meat supply chains and the challenges in tracing the history of a meat product. The foremost of these was the so-called "horsemeat scandal" in Europe in 2013, where a number of processed beef products were found to contain substituted unlabeled horsemeat. According to media reports at the time, one such case was that of a Findus beef lasagna product found to contain horsemeat (in some cases up to 100% horsemeat) (The Guardian, 2013).

The case neatly illustrates the complexity of meat supply chains: Findus had ordered the beef lasagna from Comigel, a company in northern France, which asked its subsidiary in Luxembourg to make the lasagna; the subsidiary ordered the meat from a meat processing company in southern France, which put in an order with a trader in Cyprus; the trader subcontracted the meat order to another trader in The Netherlands, who placed an order with an abattoir in Romania. Finally, the Romanian abattoir sent the meat (including horsemeat represented as beef) to the food manufacturing company in southern France, who provided the meat to Comigel's subsidiary in Luxembourg (The Guardian, 2013). In this case, while the mandatory EU traceability requirement (discussed in Section 3) may have assisted in tracking down the history of the meat substitution after it was discovered, it did not (could not) prevent the fraudulent substitution from occurring in the first place.

The third element that distinguishes different types of traceability systems is the degree of traceability *precision* that is required. Golan et al. (2003, p.20) characterize precision as "the degree of assurance with which the tracking system can pinpoint a particular food product's movement." As Tonsor and Schroeder (2006) suggest, in meat supply chains, this may include tracking a particular primal from a specific carcass to a precise kill time and slaughter chain slot, in other words, individual animal and product traceability versus lot traceability (tracking a primal to a specific batch of carcasses). Combined with greater depth in traceability, higher degrees of traceability precision might consist of tracking a meat product to a specific animal and identifying all of the properties on which that animal resided prior to slaughter. While adding breadth might additionally entail the ability to identify all medical treatments (when and where) administered to the animal during its lifetime.

As Golan et al. (2003) point out, in private sector traceability systems, the first decision a firms makes with respect to the degree of precision warranted in its traceability system is the acceptable error rate. In a shipment of ground beef, how many grams of ground lamb are acceptable? Error-rate specifications, which are driven by quality requirements, determine the strictness of the segregation system with which the traceability system is associated. Low (or zero) tolerances for ground lamb in a shipment of beef will require stricter segregation systems (eg, the plant only processing meat from one species). The second decision pertaining to precision concerns the unit of analysis: ie, a processing plant, a refrigerated truck load, a day of production at the processing plant, a specific batch, a specific animal. Traceability based on larger units (an entire plant), while less costly, also afford less accuracy in isolating problems or facilitating food recalls, while a smaller unit of analysis (a specific batch or meat from a specific animal) allow greater traceability precision but at a greater cost (Golan et al., 2003). The extent to which the benefits of precision outweigh the costs will depend on the objective of the traceability system, for example, to facilitate product recalls in the event of foodborne disease or to provide assurances about credence quality attributes.

Two points emerge from this review of the dimensions of traceability. First, full, complete, and absolute traceability of all food attributes, ingredients, and products is probably an unattainable goal. Even if technologically feasible, the economic costs of implementing full, complete, and absolute traceability are likely to outweigh the benefits (Hobbs, 2006; Golan et al., 2003). Second, it is apparent that traceability systems differ on a number of dimensions in terms of scope, scale, and objectives, and that "traceability" is a multidimensional concept. Within this context, it is pertinent to consider the motivations behind meat and livestock traceability systems: why have they emerged and what roles or functions do they perform?

2.3 Roles of Meat and Livestock Traceability Systems

Meat and livestock traceability systems perform a diverse set of roles, in some cases driven by private sector incentives to improve supply-side cost management, reduce exposure to product recall costs, or enhance product differentiation, while in other cases driven by public sector goals with respect to reducing the social costs of foodborne illness. In general, traceability systems have been introduced in response to information asymmetry[2] arising from food safety and food quality attributes. It is useful to think meat and livestock traceability systems serve three broad economic functions (Hobbs, 2004). The first is a reactive function, allowing the trace back of products or animals in the event of a food safety or animal health problem. The second is a liability function, and the third is an information or quality verification function. Each of these functions will be considered in turn.

The reactive function, also termed *ex post cost reduction*, has been a major driver of the development and adoption of meat and livestock traceability systems in a number of countries. Motivated in large part by animal health and food safety scares related to meat products (eg, BSE in beef, foot and mouth disease), industries, firms, and regulators moved to put in place traceability systems for the livestock (particularly cattle) sector. In the case of animal identification and traceability systems, a strong motivation is to facilitate emergency management in the event of animal health problems,

[2] Information asymmetry arises when one party to a transaction (eg, the seller) has more information (eg, about the true quality of the product) than the other party (eg, the buyer).

such as foot and mouth disease. In these situations, a credible animal identification and traceability system can limit the number of herds quarantined or destroyed, as well as facilitate the reestablishment of access to international markets.

For the downstream food sector, the traceability of meat products allows more efficient and effective food recalls in the event of a food safety problem, both reducing the direct costs of the meat recall to the affected firm, as well as reducing the collateral damage to other firms within the industry by limiting recalls and reputational damage to the affected brand. In 2012, the discovery of *E. coli* contaminated beef from the XL beef packing plant in Canada led to the largest beef recall in Canadian history. Poor record keeping at the plant, and the challenges of identifying which batches were contaminated, when, and how, escalated the size of the recall. An independent review of the XL Foods case pointed to a weak food safety culture at the processing plant, a slow response once a problem was identified, and inadequacies in the food safety inspection processes. The widespread recall resulted in 1800 products being removed from the Canadian and US markets.

The collateral damage to the broader beef industry in Canada was significant. The event is estimated to have cost the Canadian beef industry $16 to $27 million, exacerbated by the fact that the XL Foods plant accounted for 35 percent of Canadian beef slaughter capacity. The temporary closure of the plant forced cattle producers to seek alternative markets for their fed cattle or caused delays in their ability to market the cattle at optimal weights (Independent Review of XL Foods Inc. Beef Recall 2012, 2013). The costs to XL Foods were considerable, both in terms of the food recall and in terms of damage to the company's reputation. A meat traceability system that facilitates a speedy and effective response to a food safety problem can help mitigate the costs to the firm, the broader industry, and society at large.

The *ex post cost reduction* function also has broader societal benefits. The ability to trace potentially contaminated food products more effectively can reduce the number of people exposed to unsafe food, thereby reducing the societal costs of foodborne illness. It is estimated, for example, that the annual cost of foodborne illness in the United States (accounting for health-related costs, productivity losses, and quality of life effects) is in the region of US $77 billion (Scharff, 2012). While a meat traceability system may not prevent a foodborne disease, it can help mitigate the scope and scale of the resulting costs to society at large.

The second major function of a meat traceability system is to facilitate *allocation of liability* for food safety breaches. The ability to trace meat products back along a supply chain facilitates the establishment of both civil and statutory liability for food safety problems, thereby encouraging firms to "play by the rules" in terms of food safety practices. The existence of statutory or civil liability risks can also stimulate the development of traceability systems. The 1990 Food Safety Act in the United Kingdom was notable for extending legal liability for food safety throughout the food chain, including to food retailers (who previously could rely upon a manufacturer's warranty defense). The change in regulatory liability prompted retailers to focus far more attention on their supply chain relationships and highlighted the need for greater traceability within their supply chains as a means of monitoring upstream suppliers (Hobbs and Kerr, 1992).

Firms may also be subject to civil (contractual or tort) liability resulting from negligence in the production, preparation, handling, storage, or distribution of meat products or from misrepresentation of products. Legal proof of responsibility is necessary for liability (both statutory and civil) to be an effective "stick" in encouraging good food safety practices. Traceability can be part and parcel of this proof. In practice, proving liability can be challenging, given the difficulty of determining fault in complex, multilayered supply chains, and given the need to prove mens rea, or actions that are willful, negligent, or reckless. This weakens the liability incentive for strong food safety systems (Hobbs, 2006).

In some senses, the liability function of a meat and livestock traceability system is a double-edged sword. On the one hand, concern over liability may have inhibited the development and adoption of livestock traceability systems; indeed, this concern featured prominently in the debate over whether the United States should establish a national cattle traceability system (Souza-Monteiro and Caswell, 2004). On the other hand, it could be argued that for firms (farms) practicing due diligence in food safety and animal health, effective traceability systems can exonerate them in the event of a food safety problem, thereby reducing their liability risks and the disruption to business caused by untargeted product recalls.

The third broad function of a meat and livestock traceability system, the *ex ante information or quality verification function*, reflects a set of demand-side incentives for the provision of information to consumers on quality attributes. Meat traceability systems featuring ex ante quality verification assurances reduce information costs for consumers through labeling the presence of credence attributes, for example, animal welfare attributes, country of origin information, assurances about non-GMO animal feed, etc. In the context of the earlier discussion, meat traceability systems focused on ex ante quality verification involve tracking attributes forward in the supply chain and contribute to transparency within meat supply chains. For firms, ex ante meat traceability systems are usually a component of product differentiation and value-added marketing strategies. Combined tracking and tracing within a meat quality verification system, the ability to provide consumers with information on the source of products or ingredients, lends credibility to quality assurances and has been an important incentive for the adoption of private supply chain-based traceability systems, as discussed in the next section. Increased transparency through traceability can also improve the pricing efficiency within a meat supply chain by providing more accurate price signals to livestock producers on the product qualities valued by the end user (Smith et al., 2005).

In some cases, ex ante quality verification and traceability systems may not be aimed at providing detailed traceability information to consumers per se. In the livestock sector, some farm quality assurance programs featuring traceability primarily provide assurances to food retailers and serve to reduce transaction (monitoring) costs within the supply chain (Hobbs, 2006). Again, in this case, traceability serves to facilitate downstream tracking of attribute information as well as lending credibility to the quality claim.

It should be clear from this overview that meat and livestock traceability systems can be viewed from many perspectives and have multiple potential roles. Not surprisingly, therefore, a diverse array of traceability systems are in evidence in the meat and livestock sector, featuring both regulated mandatory traceability requirements and purely private sector traceability initiatives.

3. Public and Private Sector Approaches to Meat Traceability

3.1 Public Sector Traceability Initiatives

Public sector traceability systems are defined as regulatory initiatives imposing mandatory traceability requirements. Regulatory intervention typically occurs in the presence of market failure: in the case of food safety, this implies that, to a firm, the costs of producing unsafe food are less than the costs to society, and left to its own devices, the market will provide a suboptimal level of food safety. In the context of traceability a similar argument pertains. In addition, there may be economies of scale in the adoption of a single national (or supranational) traceability standard to facilitate food recalls or the management of livestock diseases, for example, to enhance data compatibility or allow integration of traceability databases across regions or species.

One of the earliest public sector meat and livestock traceability initiatives was the establishment of rules for cattle identification and beef labeling in Member States of the EU in 1997 (subsequently amended in 2000) (EC Regulation 1760/2000, 2000). The regulation ushered in mandatory cattle identification, along with a beef labeling requirement that would facilitate "farm to fork" traceability of beef products. Each EU Member State was responsible for introducing an animal identification system at the national level, while the beef labeling requirement took the form of a reference code allowing identification of the animal or group of animals from which the meat was derived, plus information on the countries in which the animal was born, reared, slaughtered, and processed.

The catalyst for this regulatory initiative was the BSE crisis in Europe in the 1990s, which resulted in a severe loss of consumer confidence in beef products and the closure of many export markets to beef produced in countries with cases of BSE. In the context of the earlier discussion, the EU cattle identification and beef labeling regulation is an example of an ex post or reactive traceability system: the primary objective being to more effectively manage animal disease problems and assist in meat recalls in the event of food safety events. A later regulatory initiative expanded the provenance meat labeling requirements to other species (*fresh, chilled, and frozen meat of swine, sheep, goats, and poultry*) (EC Regulation No. 1337/2013, 2013).

The notion that the EU regulations, because they encompass a meat labeling component, also provide consumers with ex ante quality verifications is somewhat of a red herring. On one level, one could argue that the provision of information on the countries in which the animal was born, reared, slaughtered, and processed provides valuable information to consumers who derive utility from knowing the source of their food. Beyond identifying origin, however, the regulation does little to reduce consumers' information asymmetry with respect to other credence attributes that may be relevant to consumers, such as animal welfare, animal feed ingredients, and other production-related attributes. Neither were these features the intent of the original EU cattle identification and beef labeling regulation, which was about addressing market failure and protecting livestock and public health. Instead, stronger ex ante quality verifications are typically the purview of private sector traceability systems.

The EU subsequently deepened its regulation of food traceability with the General Food Law, which introduced a legal framework for traceability across all food products (EC Regulation 178/2002, 2002). In essence, this regulation imposes a one-up, one-down traceability requirement for firms, requiring that food firms have the ability to identify the immediate source of their inputs (one-step-back) and the initial buyers of their outputs (one-step-forward). Again, the primary objective here is ex post cost and risk reduction in the event of a food safety problem. It is worth noting that the existence of this mandatory food traceability regulation within the EU did not prevent the "horsemeat scandal" of 2013, although arguably it may have assisted in eventually tracing back the history of the fraudulently substituted horsemeat.

The EU is not alone in mandating traceability requirements that affect the meat and livestock sector. In Canada, a National Agriculture and Food Traceability System (NAFTS), in the first instance focusing on lifecycle traceability in livestock and poultry, is under consideration. A primary motivation of the proposed livestock and poultry traceability system is emergency management in the event of animal health problems. If implemented, the mandatory traceability requirements for all livestock and poultry would build upon an existing traceability system for Canadian cattle (discussed in the next section).

The United States considered, but eventually abandoned, the development of a national animal identification system, as discussed in the next section. Instead, livestock tracking requirements for the movement of livestock across state borders only were introduced in March 2013 when the USDA Animal Disease Traceability Rule (ADTR) became operational. The mandatory animal identification system tracks animals only when they cross state borders, for the sole purpose of animal disease management. Cattle/bison, horse (equines), swine, sheep/goats, poultry, and captive cervidae that move between or among states are subject to the ADTR, unless specifically exempted. All livestock subject to the rule must be officially identified and accompanied by an interstate certificate of veterinary inspection or other documentation such as owner/shipper statements or brand certificates (USDA, 2013). Information pertaining to ranch or farm of origin, age, movement tracking, or diet is not recorded for animals that do not cross state boundaries under the ADTR (Schroeder and Tonsor, 2012).

While public sector meat and livestock traceability initiatives are primarily focused on ex post cost and risk reduction functions, private sector traceability systems encompass a broader range of objectives. Private sector meat and livestock initiatives include industry-wide traceability systems and firm or supply chain-level traceability systems.

3.2 Industry-Wide Traceability Initiatives

Industry-wide initiatives to enhance traceability have tended to focus on livestock traceability and involve strategic decisions by industry associations or producer groups to implement industry-wide animal identification and traceability. A good example is the decision by the Canadian Cattlemen's Association to implement a national identification and traceability system for cattle and bison in Canada through the formation of the Canadian Cattle Identification Agency (CCIA). Initially implemented in 2001 as a voluntary system, it became mandatory in 2002 with mandatory enforcement powers granted under the Canadian Health of Animals Act. For the purposes of this chapter, the mandatory cattle identification system is discussed under "industry initiatives" since in essence it was driven primarily by a recognition that the industry needed cattle traceability in the event of an animal disease problem as a risk reduction strategy, both to reduce the scope and costs of animal disease outbreaks and to maintain consumer confidence and market access in the event of animal health and food safety problems (Hobbs, 2004). The CCIA system began with animal identification, allowing the trace back of an animal from point of slaughter to farm of origin, and later added premises identification and animal movement tracking. It remains a preslaughter animal identification system and does not encompass postslaughter meat traceability.

Similar industry-wide livestock traceability initiatives appear in other countries, including Australia's National Livestock Identification System (NLIS), a whole-of-life individual animal identification system allowing an animal to be traced from its place of birth to the point of slaughter through the use of RFID tags which are read and recorded as the animal moves through the production system and at each transaction (Tonsor and Schroeder, 2006).

The United States explored the establishment of a National Animal Identification System (NAIS) as a voluntary, federal traceability system to be administered by the Animal and Plant Health Inspection Service (APHIS), using premise registration, animal identification, and animal movement tracking. NAIS could potentially also have monitored vaccination programs, animal health inspection, and certification programs and enabled regional documentation in disease outbreaks in order to maintain access to international markets. Faced with concerted resistance from some industry stakeholders and legislators, however, NAIS was abandoned in 2010, leaving the United States without a comprehensive industry-wide national livestock traceability system (Schroeder and Tonsor, 2012).

3.3 Individual Supply Chain Traceability Initiatives

While public sector and industry-wide traceability initiatives have primarily focused on livestock (and sometimes meat) tracking and traceability with an explicit ex post cost reduction function and an implicit liability enhancement function, private sector initiatives at the level of the individual supply chain tend to be driven by ex ante quality verification objectives. A private sector solution to the consumer's information problem with respect to credence attributes (animal welfare, use of subtherapeutic antibiotics, origin, and so on) is credible voluntary labeling by firms. A market premium for verification of credence attributes that requires a system to track these attributes through the supply chain provides the incentive for firms to develop proprietary meat traceability systems and to require traceability information from their suppliers. Traceability assurances may also be marketed as an attribute of interest per se.

Early examples of supply chain partnerships between retailers, processors, and beef producers offering various levels of traceability emerged in the UK beef industry following the BSE crisis (Fearne, 1998). Typically, these partnerships combined an ability to trace the history of individual meat cuts to the farm or animal of origin with assurances about on-farm production methods. Another early example of meat traceability is the so-called "Story Meats" in Japan. Clemens (2003) discusses how, in the wake of a widespread loss in consumer confidence following the BSE crisis in Japan, supermarkets implemented meat traceability systems that provided consumers with the story of how the meat they were considering for purchase was produced. Of varying degrees of complexity, in some cases the "story meats" enabled consumers to scan a barcode on a meat package for information on BSE testing of the animal, information on where the animal was reared and slaughtered, information on the production methods (such as feed) used, and on the producer who reared the animal (including a photograph) (Clemens, 2003).

More recently, the growing expansion of the branded meat product category provides a scope for brand assurances that incorporate traceability either explicitly in a quality claim (as in the example above) or implicitly as a means for food retailers or manufacturers to reduce the transaction costs of verifying quality assurances within their supply chains. Ultimately firms' incentives to adopt private sector meat traceability systems will depend on the strength of market-based incentives. The "story meats" type of traceability system exhibits a high degree of depth and precision, but at a considerable cost, and a firm's decision to implement traceability assurances of this detail will depend on the extent to which consumer demand and willingness to pay for traceability information justifies these costs. Consumer attitudes toward traceability have been the subject of a growing body of literature which is discussed in the next section.

4. Consumer Demand for Meat Traceability

Meat traceability systems deliver a number of potential benefits to consumers. As a component of ex ante quality verification, traceability systems can reduce information costs for consumers and enhance consumer confidence in quality assurances. While traceability, in and of itself, is not synonymous with food safety, it is a component of good food safety risk management systems, and as such, traceability systems may also play a role in reassuring consumers.

A host of studies have emerged investigating consumers' attitudes toward meat traceability (see for example, Dickinson and Bailey, 2005; Hobbs et al., 2005; Cicia and Colantuoni, 2010; Lee et al., 2011; van Rijswijk and Frewer, 2012; Lim et al., 2013). Spanning a number of countries, these studies use a variety of methods to assess attitudes toward meat traceability assurances and to elicit willingness-to-pay estimates, including stated preference survey methods, focus groups, and experimental auctions.

While the studies are quite diverse, some general themes emerge from the literature. As might be expected, consumers in different countries have differing degrees of familiarity with the concept of traceability; however, in general the concept tends to be associated most closely with food safety and risk reduction, particularly in the United States, Canada, and Asia. Although food safety and risk factors are also important to consumers in Europe, European studies find that traceability is also strongly associated with assurances of the origin and authenticity of food products. It is difficult to compare willingness-to-pay estimates across different contexts and countries; nevertheless, estimates of willingness to pay for traceability-related assurances tend to be higher in Europe and in Asia than in North America. Traceability as a vehicle to deliver credible quality assurances about production methods, origin, and safety tends to elicit a positive-stated willingness to pay across a number of countries. Even in the absence of a direct consumer willingness to pay for meat traceability itself, however, the demand for credence quality attributes provides an incentive for firms to invest in traceability as a means of delivering product differentiation (Boecker et al., 2013).

5. Conclusions

This chapter has drawn a number of distinctions that are relevant for considering the future trends and directions of meat traceability. Fig. 17.1 summarizes the key concepts examined in the chapter. First, it is important to recognize that traceability has many facets and can be viewed from a number of perspectives. It may be limited to a simple need to trace products backward in the supply chain or encompass a broader notion of tracking forward, along with tracking information on additional product attributes. Traceability differs in terms of its breadth (how many attributes are tracked), its depth (how far through the supply chain products can be traced), and its precision (what degree of accuracy is required). Finally, traceability systems perform a number of functions which may not be mutually exclusive, from ex post cost reduction, to facilitating the allocation of liability, to ex ante quality verification. It is difficult to talk about meat traceability without also addressing livestock

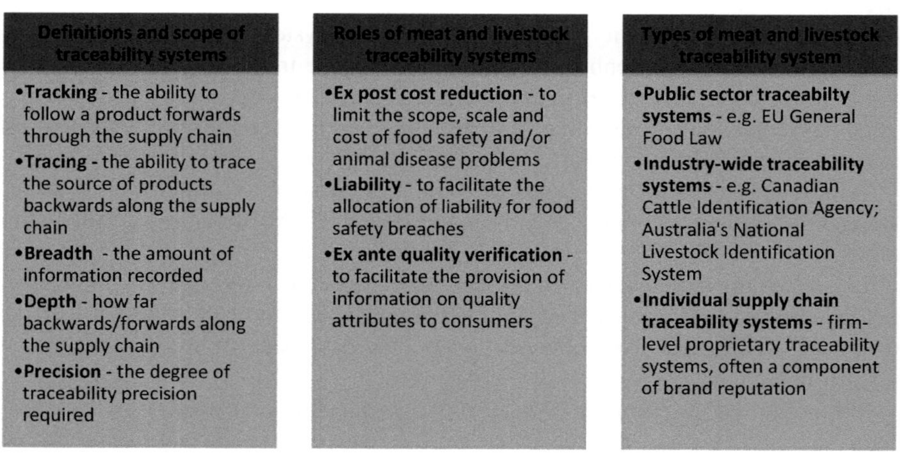

Figure 17.1 Summary of key concepts.

traceability, and as the examples discussed in this chapter illustrate, livestock traceability systems in many cases form the foundation upon which meat traceability builds.

A number of countries currently have, or are contemplating, mandatory traceability requirements for livestock and/or for food products. Mandatory public sector traceability regulations set the minimum requirements for traceability to address broad societal goals with respect to risk management. By and large, the focus of these public sector systems is (and should be) limited to an ex post cost reduction or reaction function to facilitate management of livestock diseases and expedite food recalls. The future growth in meat traceability systems will come in the private sector, in the delivery of ex ante quality verifications. As identification, measuring, monitoring, and information management technologies advance, so does the potential for meat traceability systems to encompass greater degrees of breadth, depth, and precision. Traceability in this context is a component of product differentiation and competitive positioning within specific supply chains. Ultimately, consumers will determine the success of new meat traceability and quality verification initiatives. How consumers will respond to new traceability technologies and traceability assurances in the meat sector provides considerable scope for further research.

References

Boecker, A., Hobbs, J.E., Kerr, W.A., Yeung, M.T., 2013. Traceability: regulatory requirements and consumer acceptance. Canadian Food Insights 1 (1), 35–37.

CAC/GL 60-2006, 2006. Principles for Traceability/Product Tracing as a Tool within a Food Inspection and Certification System. Codex Alimentarius Commission. Available from: http://www.codexalimentarius.org/ (28.08.15.).

CFIA, 2012. ARCHIVED – a Legislative Framework for Traceability – Proposed Elements. Canadian Food Inspection Agency. Available from: http://www.inspection.gc.ca/animals/terrestrial-animals/traceability/proposed-elements/eng/1325170775384/1325170880037 (accessed 28.08.15.).

Cicia, G., Colantuoni, F., 2010. Willingness to pay for traceable meat attributes: a meta-analysis. International Journal of Food Systems Dynamics 3, 252–263.

Clemens, R., 2003. Meat Traceability and Consumer Assurance in Japan. MATRIC briefing paper 03-MBP 5, September 2003 Midwest Agribusiness Trade and Research and Information Centre (MATRIC), Iowa State University. 26pp. Available from: http://www.card.iastate.edu/ (07.09.15.).

Dickinson, D.L., Bailey, D., 2005. Experimental evidence on willingness to pay for red meat traceability in the United States, Canada, the United Kingdom, and Japan. Journal of Agricultural and Applied Economics 37 (3), 537–548.

EC Regulation 1760/2000, 2000. Regulation (EC) No. 1760/2000 of the European Parliament and of the Council of 17 July 2000 Establishing a System for the Identification and Registration of Bovine Animals and Regarding the Labelling of Beef and Beef Products and Repealing Council Regulation (EC) No 820/97. The European Parliament and the Council of the European Union. Official Journal of the European Union L 204, 11/08/2000 P. 0001 – 0010. Available from: http://eur-lex.europa.eu/legal-content/EN/TXT/?uri=celex:32000R1760 (07.09.15.).

EC Regulation 178/02 2002, 2002. Regulation (EC) No. 178/2002 of the European Parliament and of the Council of 28 January 2002 Laying Down the General Principles and Requirements of Food Law, Establishing the European Food Safety Authority and Laying Down Procedures in Matters of Food Safety. The European Parliament and the Council of the European Union (24pp.). Available from: http://eur-lex.europa.eu/legal-content/EN/ALL/?uri=celex:32002R0178 (28.08.15.).

EC Regulation 1337/2013, 2013. Commission Implementing Regulation (EU) No 1337/2013 of 13 December 2013 Laying Down Rules for the Application of Regulation (EU) No. 1169/2011 of the European Parliament and of the Council as Regards the Indication of the Country of Origin or Place of Provenance for Fresh, Chilled and Frozen Meat of Swine, Sheep, Goats and Poultry. Official Journal of the European Union. The European Parliament and the Council of the European Union. Available from: http://eur-lex.europa.eu/legal-content/EN/TXT/?qid=1441655028928&uri=CELEX:32013R1337 (07.09.15.).

Fearne, A., 1998. The evolution of partnerships in the meat supply chain: insights from the British beef industry. Supply Chain Management: An International Journal 3 (4), 214–231.

Golan, E., Krissoff, B., Kuchler, F., Nelson, K., Price, G., Calvin, L., 2003. Traceability in the US food supply: dead end or superhighway? Choices 18 (2), 17–20. Available from: http://www.choicesmagazine.org/choices-magazine (31.08.15.).

Hobbs, J.E., 2004. Information asymmetry and the role of traceability systems. Agribusiness 20 (4), 397–415.

Hobbs, J.E., 2006. Traceability in the agri-food sector: issues, insights and implications. CAB Reviews: Perspective in Agriculture, Veterinary Science, Nutrition and Natural Resources 1 (029). 7pp. http://www.cabi.org/cabreviews.

Hobbs, J.E., Bailey, D., Dickinson, D.L., Haghiri, M., 2005. Traceability in the Canadian red meat sector: do consumers care? Canadian Journal of Agricultural Economics 53 (1), 47–65.

Hobbs, J.E., Fearne, A., Spriggs, J., 2002. Incentive structures for food safety and quality assurance: an international comparison. Food Control 13 (2), 77–81.

Hobbs, J.E., Kerr, W.A., 1992. The cost of monitoring food safety and vertical coordination in agribusiness: what can be learned from the British Food Safety Act 1990? Agribusiness 8 (6), 575–584.

Independent Review of XL Foods Inc, May 2013. Beef Recall 2012, 2013. Canadian Food Inspection Agency. Available from: http://foodsafety.gc.ca/english/xl_reprt-rapprte.asp (31.08.15.).

ISO, 2007. ISO 22005:2007 Traceability in the Feed and Food Chain – General Principles and Basic Requirements for System Design and Implementation. International Organization for Standardization. Available from: https://www.iso.org/obp/ui/#iso:std:iso:22005:ed-1:v1:en (28.08.15.).

Lee, J.Y., Han, D.B., Nayga, R.M., Lim, S.S., 2011. Valuing traceability of imported beef in Korea: an experimental auction approach. Australian Journal of Agricultural and Resource Economics 55, 360–373.

Liddell, S., Bailey, D., 2001. Market opportunities and threats to the U.S. pork industry posed by traceability systems. International Food and Agribusiness Management Review 4, 287–302.

Lim, K.H., Hu, W., Maynard, L.J., Goddard, E., 2013. 'U.S. Consumers' preference and willingness to pay for country-of-origin-labeled beef steak and food safety enhancements. Canadian Journal of Agricultural Economics 61 (1), 93–118.

Meuwissen, M.P.M., Velthuis, A.G.J., Hogeveen, H., Huirne, B.M., 2003. Traceability and certification in meat supply chains. Journal of Agribusiness 21 (2), 167–181.

OECD, 2003. Costs and Benefits of Food Safety Legislation. Working Party on Agricultural Policies and Markets, Organisation for Agricultural Co-operation and Development, Paris, p. 48. Available from: http://www.oecd.org/officialdocuments/ (28.08.15.).

Opara, L.U., 2003. Traceability in agriculture and food supply chain: a review of basic concepts, technological implications, and future prospects. Food, Agriculture & Environment 1 (1), 101–106.

van Rijswijk, W., Frewer, L.J., 2012. Consumer needs and requirements for food and ingredient traceability information. International Journal of Consumer Studies 36, 282–290.

Scharff, R.L., 2012. Economic burden from health losses due to foodborne illness in the United States. Journal of Food Protection 75 (1), 121–131.

Schroeder, T., Tonsor, G., 2012. International cattle ID and traceability: competitive implications for the US. Food Policy 37, 31–40.

Smith, G.C., Tatum, J.H., Belk, K.E., Scanga, J.A., Grandin, T., Sofos, J.N., 2005. Traceability from a US perspective. Meat Science 71, 174–195.

Souza-Monteiro, D.M., Caswell, J.A., 2004. The Economics of Implementing Traceability in Beef Supply Chains: Trends in Major Producing and Trading Countries, Working Paper No. 2004-6. Department of Resource Economics, University of Massachusetts-Amherst. Available from: http://people.umass.edu/resec/workingpapers/documents/resecworkingpaper2004-6.pdf (07.09.15.).

The Guardian, 2013. Horsemeat Scandal: The ABP and Comigel Connections. The Guardian. February 13. Available from: http://www.theguardian.com/uk/graphic/2013/feb/15/horsemeat-scandal-food-safety1 (31.08.15.).

Tonsor, G.T., Schroeder, T.C., 2006. Livestock identification: lessons for the U.S. beef industry from the Australian industry. Journal of International Food & Agribusiness Marketing 18 (3/4), 103–118.

USDA, 2013. Final Rule: Traceability for Livestock Moved Interstate: Summary of General Requirements by Species, Animal and Plant Health Inspection Service. United States Department of Agriculture. March 11, 2013. Available from: https://www.aphis.usda.gov/traceability/downloads/ADT_summary_species.pdf (07.09.15.).

A Closed-Loop Traceability System to Improve Logistics Decisions in Food Supply Chains: A Case Study on Dairy Products

18

R. Accorsi, E. Ferrari, M. Gamberi, R. Manzini, A. Regattieri
Alma Mater Studiorum - University of Bologna, Bologna, Italy

1. Introduction

The food industry has experienced profound transformation. Companies are increasingly facing new challenges and issues: on one hand, the access to new and attractive markets, and on the other, the increasing number of products, the competition in a global market, and the consumers' requirements in terms of quality and environmental sustainability of the products (Li et al., 2014). Furthermore, the recent scandals and frauds affecting food safety requirements elicit the attention of consumers and compel politicians to assess the responsiveness of food supply chains and their tracking and monitoring systems.

Because of the leading role of the food industry in the European manufacturing sector, the food companies are more and more resilient and strive to adapt quickly and responsively to the incoming issues and challenges. The food and drink industry turnover is 1048 billion euro in 2014, with an increasing value of both agro-food and beverage export and import respectively of 86.2 billion euro and 63.2 billion euro (World Bank, 2014). The demand of food specialties by foreign markets grows with the importance of logistics processes and operations in food supply chains.

To meet the growing trends of food demand (Ahumada and Villalobos, 2009) food products intensively travel across continents and oceans, thereby shifting the focus from the management of local agriculture models to the optimization of the global food supply chain. These trends are increasing the relevance of considering the related functions of the food supply chain: not only cultivation and food processing but also the packaging, storage, distribution, and control and monitoring of the product quality across its entire life cycle. These processes result in more complex supply networks, which emphasize the distance between the grower and the consumer and affect consumer awareness of the supply chain stages from farm-to-fork. Beyond the appealing shelves of the grocery store, food is grown, harvested, handled, stored, and shipped, and each of step may affect the product's quality and safety.

In order to assess how food supply chains affect the safety and the quality of food products, the prediction of food degradation (Labuza, 1982; Man and Jones, 1994), combined with the tracking and the environmental conditions experienced by products from farm-to-fork are fundamental (Manzini and Accorsi, 2013). Typically, next to

Advances in Food Traceability Techniques and Technologies. http://dx.doi.org/10.1016/B978-0-08-100310-7.00018-1

biological variations, food quality is determined by time and environmental conditions, which may be influenced by the type of packaging, way of loading, and the availability of temperature-controlled packages, vehicles, and warehouses and generally by the complete factors configuring the supply chain.

To properly control the safety and quality of food products across processing, packaging, storage and distribution, the development of accurate traceability and monitoring systems is recommended, particularly when the supply chain in out of the control of the producer. Regattieri et al. (2007) defined traceability as "the history of a product in terms of the direct properties of that product and/or properties that are associated with that product once these products have been subject to particular value-adding processes using associated production means and in associated environmental conditions." An effective food traceability system is an important tool not only to manage food quality and safety risks, but also to promote the development of effective supply chains.

Two main categories of traceability technologies and devices can be distinguished: identification tags (ie, barcode, label, RFID tag) which address a product or a general item with a specific code for identification purposes and data loggers (sometimes called "black boxes"), whose aim is to trace and record the environmental conditions and profiles experienced by a product (Piazzi et al., 2011). In the food industry, radio-frequency identification (RFID) systems can also embed sensors and thereby working as an identification black box for traceability and logistics, as well as for anticounterfeit purposes (Barge et al., 2014).

The development of a traceability system based on an RFID black box for a food supply chain is an effective way to realize the so-called *real-time* control of the state of the food products (de las Morenas et al., 2014), which goes in agreement with safety and quality standards and regulations and customer satisfaction. When gathering the conservation conditions of a product in *real time*, a set of operative logistic decisions can be handled in *run time* to correct, improve, or modify the observed process. This approach would suggest to monitor those parameters (eg, time, temperature, humidity) having an impact on the quality of products. Unfortunately, the costs of tags and the investment required to build up real-time traceability infrastructures over the entire supply chain are not revenue-generating for companies, because of the low market price of food products, the extension of global supply chains, and the cost of primary packaging is often not reusable (Accorsi et al., 2014b).

The so-called *ex-post* monitoring systems are less expensive and are built up through proper black box or data logger devices, tracking the environmental profiles of storage and distribution activities. The *ex-post* monitoring system protocol is based on inserting data loggers within the product primary packaging, gathering the data loggers at the end of distribution, and collecting the distribution profile data. Finally, an *ex-post* detailed analysis of the collected data identifies the most critical steps of food distribution and aids operative decisions (eg, packaging improvements) to improve the control of the food safety and quality.

The main difference between a *real-time* and an *ex-post* monitoring systems, both illustrated in Fig. 18.1, is based on the rapid response of the former system enabling operative *real-time* decisions in response to events or stresses in the supply chain operations. This undoubtedly results in making the supply chain more resilient and responsive. Conversely, the *ex-post* monitoring system tracks the supply chain steps and gives feedback to the involved actors about what happened along the product storage, transport, and delivery after the product reached the customer. Therefore any possible corrective actions and improvement of the supply chain operations are handled ex-post driven by the findings from the tracking system. Highlighting what are indeed the issues affecting the products, assessing them in an objective and quantitative

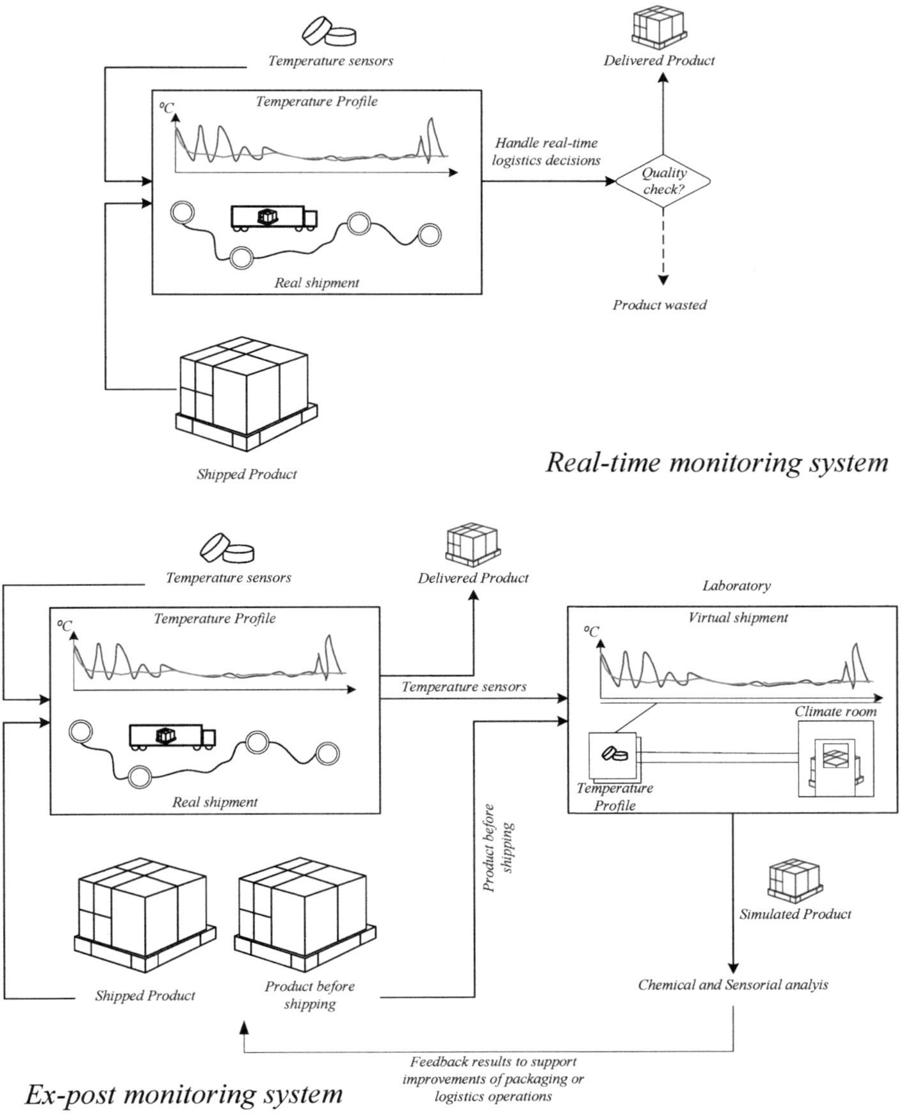

Figure 18.1 Comparison between *real-time* versus *ex-post* monitoring system.

manner helps in redesigning the packaging, the truck loads, the storage conditions in a warehouse, the shipping container (Accorsi et al., 2014a), up to choosing more careful carriers and logistics providers and renegotiating with them the transport standards and rules. This process can be particularly onerous for a single company. The Food Supply Chain project aims to bring together knowledge, experience, and tools of investigation to improve the distribution logistics of food products, thereby protecting a very important sector of the regional and national economy.

The *ex-post* monitoring system can be even more exhaustive whether combined with a closed-loop protocol, which includes the reconstruction and simulation in laboratory of the monitored environmental conditions (eg, temperature, humidity, vibration) experienced by the products, and the organoleptic and chemical analysis of the simulated products to assess the effects on product quality due to distribution. The closed-loop protocol was first introduced by Bartholdi and Mac Cawley (2011), as a main result of the *Wine Supply Chain Council* (http://wscc.scl.gatech.edu/), to analyze the international shipments of bottled wine and was extended by Manzini and Accorsi (2013) to other supply chains and products, as extra-virgin olive oil (EVOO) (Valli et al., 2013; Manzini et al., 2014a), fine chocolate (Manzini et al., 2014b), kiwi fruits (Manzini et al., 2014c), and others within the Food Supply Chain Project.

This international project put together a number of international research institutions, such as Georgia Institute of Technology (USA), San Francisco State University (USA), the Commonwealth Scientific and Industrial Research Organization (CSIRO) (Australia), the Pontificia Universidad Catolica de Cile (Chile), the Mendoza Universidad (Argentina), the Council for Scientific and Industrial Research (CSIR) (South Africa), and the University of Bologna (Italy), whose aim is to analyze how the variation of certain factors (eg, temperature, humidity, vibrations) can affect the products during storage and distribution phases. Through the direct involvement of companies of the food industry, this project aims to highlight the critical issues affecting the current global food supply chain and to assess possible improvements to meet food safety and quality standards.

As a methodological finding of this project, the monitoring closed-loop protocol (Manzini and Accorsi, 2013), illustrated in Fig. 18.2, consists on coupling the *ex-post* monitoring system to track the profile experienced by the products, with the reproduction in the laboratory of such a profile and the assessment of the resulting effects on the product quality and safety. The reproduction of the thermal and humidity profiles experienced by the products has carried out through climate-controlled chambers, illustrated in Fig. 18.3, properly designed and controlled by a Proportional-Integral (PI) control (Accorsi et al., 2014c).

The involved companies thereby have the opportunity to collect accurate data on the conditions experienced by their products when leave the production facility and are shipped to the consumer site and also to assess the effect of the supply chain on their products in terms of quality, taste, flavor, appearance, nutritional value, and safety.

Given the flexibility of the applied technologies, including the traceability data loggers and the climate controlled chambers, the close-loop protocol can be widely applied to different products and supply chains, for example, meat supply chains or dairy products.

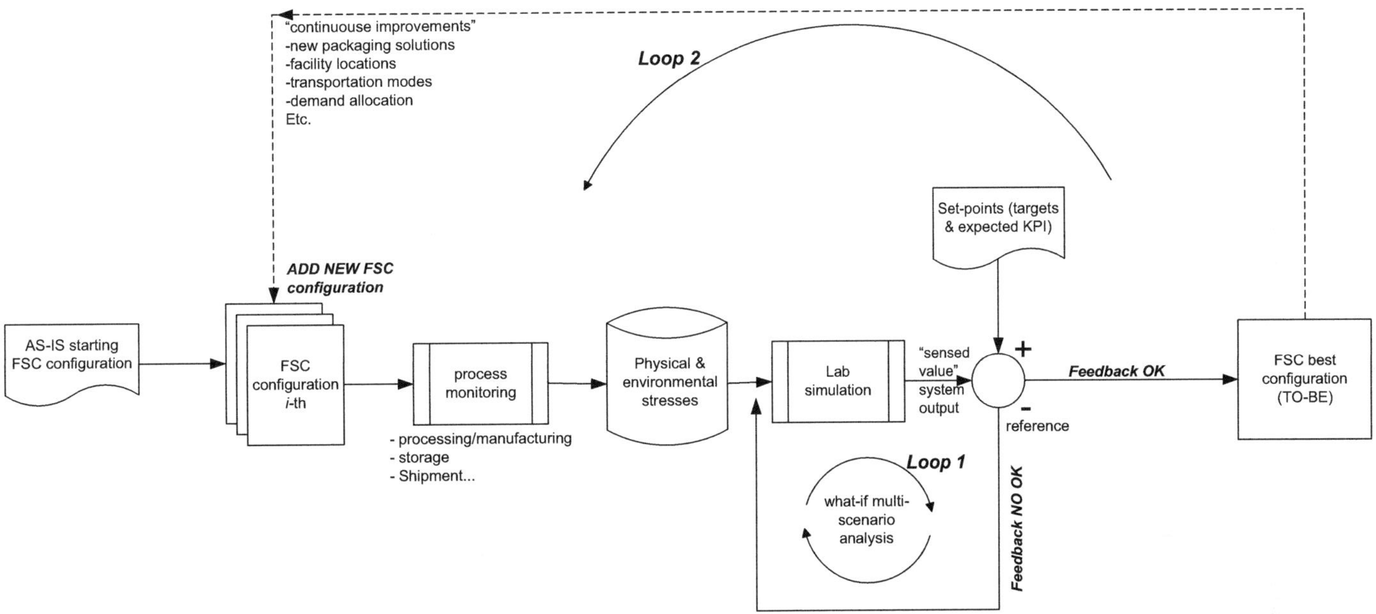

Figure 18.2 Closed-loop monitoring system.
From Manzini, R., Accorsi, R., 2013. The new conceptual framework for food supply chain assessment. Journal of Food Engineering 115 (2), 251–263.

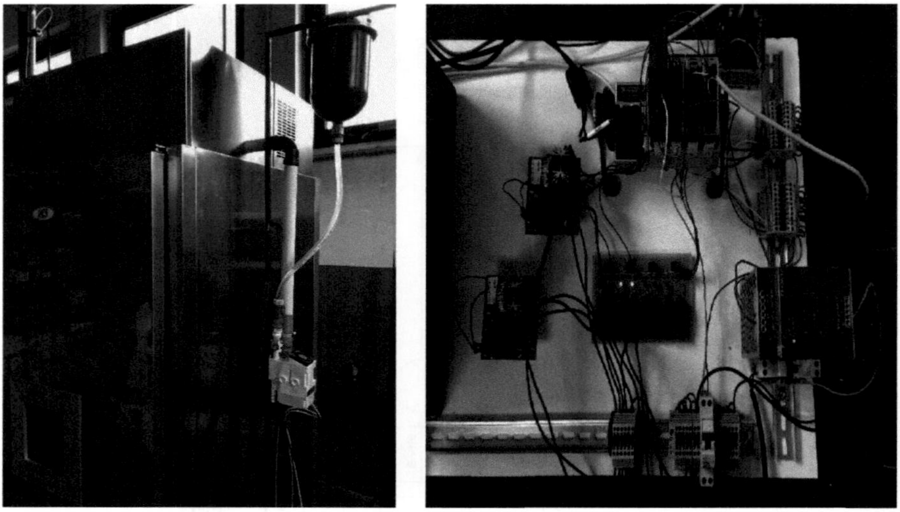

Figure 18.3 Climate-controlled chamber and related control panel.

This chapter presents the proposed closed-loop monitoring system applied to a case study of the dairy supply chain. The aim of this chapter is to assess the environmental and physical conditions (ie, temperature) experienced by dairy products during the transport and storage activities, identify criticalities, and measure the effectiveness of logistics and material handling operations in addressing the quality and the shelf life of the product. The ex-post monitoring system is applied and discussed, while the reproduction of the distribution profile in the laboratory is postponed to further developments of this research.

The remainder of the chapter is arranged as follows. Section 2 presents the adopted closed-loop monitoring methodology applied to the supply chain of the observed dairy products. Section 3 illustrates the case study of the observed dairy supply chain and resulting distribution profiles. Finally, Section 4 presents the conclusions and some notes and suggestions for further research.

2. Closed-Loop Traceability System: Methodology

While Fig. 18.2 illustrates the entire closed-loop monitoring system, the proposed chapter deals with the monitoring campaign of the environmental conditions affecting the distribution activities of the observed dairy products. The dairy sector accounts for 14% of the whole European food industry, with more than 750,000 dairy farms across Europe and about 12,000 production sites. Furthermore, it contributes significantly with more than 350 POD cheeses and dairy products to the European nutritional and cultural heritage (EDA, 2014; F&D Europe, 2014).

The dairy production is extremely diffused in Italy, and particularly in the Emilia-Romagna region, with renowned POD specialties as Parmigiano-Reggiano cheese. While the production phase within the producer facility is strictly controlled to comply with safety standards, this chapter highlights what could happen during product distribution. Given the fact that the Italian dairy products represent a significant opportunity for exporting, the attention to the maintenance of the optimal conservation conditions during the transport phase is crucial. For the Parmigiano-Reggiano cheese the quality of the raw materials (ie, milk) is certified by the consortium which puts together the cheese factories and regulates the brand as well as the rules against counterfeiting.

2.1 The Characteristics of Cheese Supply Chains

This chapter collects the monitoring campaign of the national distribution of dairy products obtained by Parmigiano-Reggiano cheese (ie, butter, snack, cuts) through one the most critical steps of the supply chains: the fractionated transport from the vendor facility to the consumer points. The horizon of time of such shipments is usually about three to five days. When imposed by the customer, the shipments are realized via reefer container maintaining the shipment temperature between 4–8°C. Because of the less constraining requirements in fresh product distribution than for frozen, the carriers behave arbitrarily given the contract and the client. When products are shipped to the large-scale retailers, the carriers belong to the retailer fleet, and temperature control is guaranteed until the retailer warehouse or the sales point is achieved.

In the case that clients are grocery shops, restaurants, or cafes, the order is generally small in quantity and not insufficient to fill the truck. Therefore the carrier shares the cargo with other food products, not necessarily dairy. Such fractioned cargo carries out a number of stops, as a delivery tour, where the truck doors are opened and the internal storage condition altered. During transportation, temperature should vary within the range of 4–8°C (DM 01/04/88 n. 178, Section C Part II DPR 327/80). While temperature up to 14°C are allowed for short periods during the fractioned transport (DM 01/04/88 n. 178, Section C Part II DPR 327/80), the taste, the flavor and the nutritional properties of the cheese products can be significantly compromised when the conservation temperature passes the threshold of 8°C. In particular, the most sensitive cheese varieties to the temperature are the soft cheeses (which are not treated in this study), butter, and cut or grated cheese.

2.2 The Production Company

The enterprise observed in this study is a leading brand in the national and international distribution of Parmigiano-Reggiano cheese and related by-products (eg, cheese cuts, grated cheese, snacks). Belonging to the Parmigiano-Reggiano consortium, it distributes the products with three renowned brands and is the 3P producers for a popular Italian large scale retailer cooperative. The company owns nine cheese factories and two bug processing and warehousing facilities in Modena and Reggio-Emilia.

2.3 The Tracking Methodology

When the products are packed, the warehouse operators at the production facility put the temperature data logger, illustrated in Fig. 18.5, within the secondary packaging, fill the tracking label of Fig. 18.4 and load the pallet into the cargo. Once received at destination, the product is unloaded, the label is filled by the client and the data logger

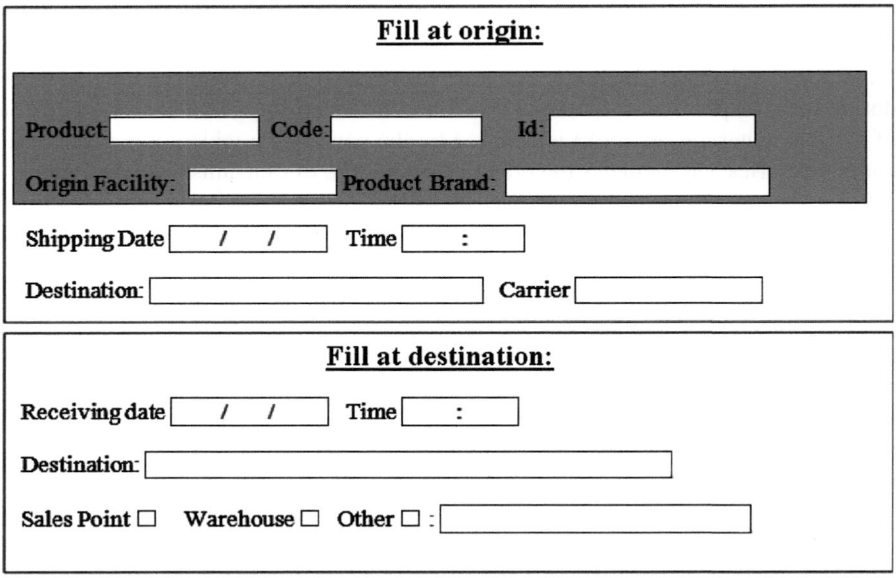

Figure 18.4 The ex-post tracking label.

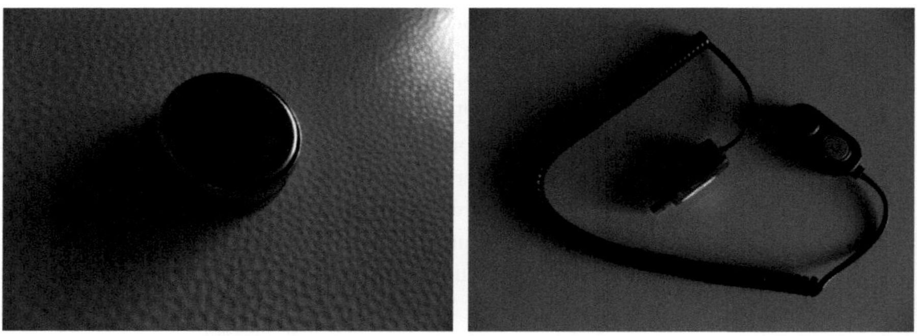

Figure 18.5 The tracking temperature data logger and reader.

retrieved and shipped back for data analysis and manipulation. Fig. 18.6 illustrates the four steps of the handling protocol which consists on sealing the secondary packaging with the sensor embedded, loading the carton in the pallet, and collecting the temperature sensor at the destination point.

3. Data Collection and Analysis

Resulting from the transportation monitoring campaign, a set of different temperature profiles is illustrated. Different key performance indices (KPI) are introduced and evaluated to identify the aforementioned issues in the fractioned national distribution

Figure 18.6 Enclosing and retrieving the sensor to/from the product packaging.

Table 18.1 Tracked Shipments

Client		Expected Departure	Expected Arrival	Destination	Carrier
Shipment 1	Consorzio Europrogea	18/5	22/5	Rutigliano (Ba)	Carrier 1
Shipment 2	Komedis	18/5	23/5	Catania	Carrier 2
Shipment 3	Autogrill mag. Anagni	18/5	22/5	Anagni (Fr)	Carrier 3
Shipment 4	Comby service	22/5	24/5	Roseto d.A. (Te)	Carrier 4

of dairy products and to aid improvements of the supply chain operations and packaging. Table 18.1 and Table 18.2 report the characteristics of the four milk-run shipping tours tracked with the introduced ex-post monitoring system.

For each shipment, the temperature logger has been collected and data manipulated and analyzed. The following graphs and related statistics summarize the temperature profiles experienced by the dairy products during the fractioned transport from the producer facility to the client site.

The graphs from Figs. 18.7–18.10 respond to the following legend: the green belt represents the optimal temperature condition for dairy conservation (ie, between 4–8°C);

Table 18.2 **Ordered Dairy Products**

Clients/Order Products	Lot	Expiring Date
Client 1		
Parmigiano-Reggiano grated Unigrana g.100	07L12225	04/08/2012
Client 2		
Parmigiano-Reggiano stick 30m 100g×12	27L12231	15/10/2012
Parmigiano-Reggiano Cuor di mix grated g.70doy×20	11L12223	04/08/2012
Parmigiano-Reggiano slices 150g×18	L141112	14/11/2012
Parmigiano-Reggiano grated 60g×20	11L12235	18/08/2012
Client 3		
Cheese Alpina Cuor di Fette g.140	25L12229	21/07/2012
Scamorza Affumicata Cuor di Fette g.140	25L12229	21/07/2012
Parmigiano-Reggiano snack 20g×5×10	24L12235	15/10/2012
Client 4		
Parmareggio butter POL.200×16	L12241	25/08/2012
Parmigiano-Reggiano snack 20g×5×10	24L12230	15/10/2012
Parmigiano-Reggiano snack 20g×5×10	24L12231	15/10/2012

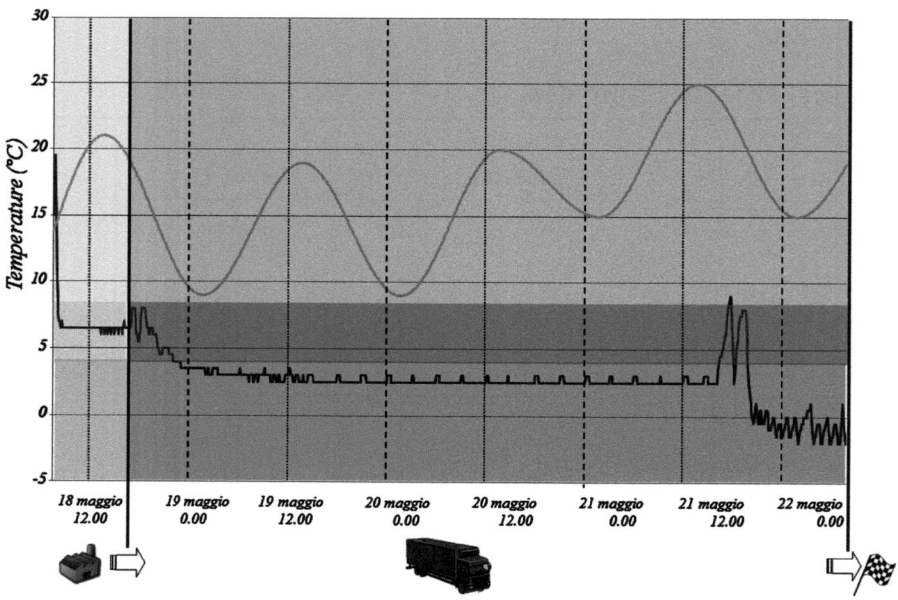

Figure 18.7 Temperature profile of shipment 1.

the temperatures below this belt are colored in blue, while the temperatures above are colored in red. The blue line represents the profile experienced by the dairy product along the transport, while the orange line indicates the external environmental temperature measured outside the truck (Table 18.3).

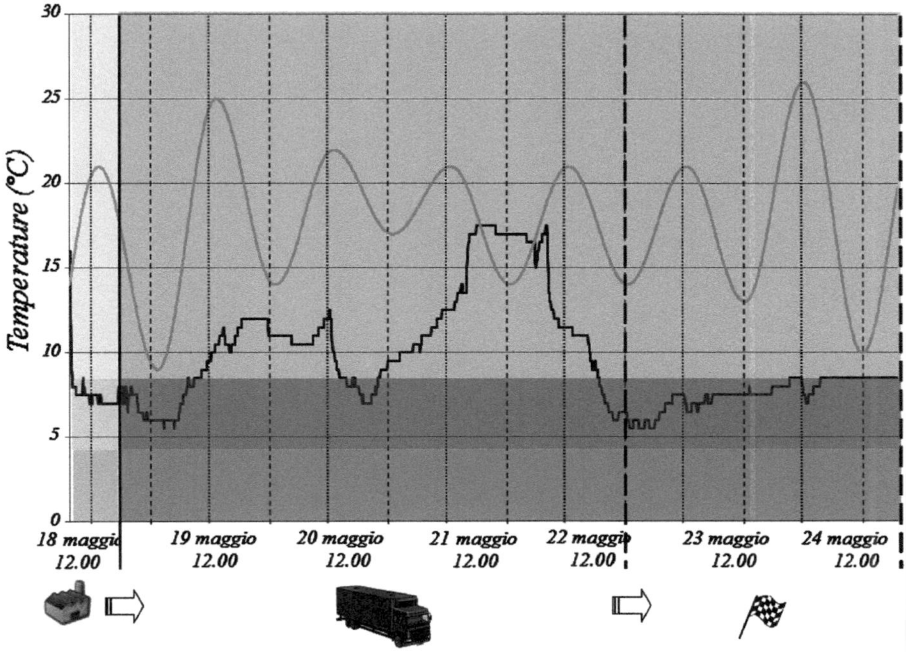

Figure 18.8 Temperature profile of shipment 2.

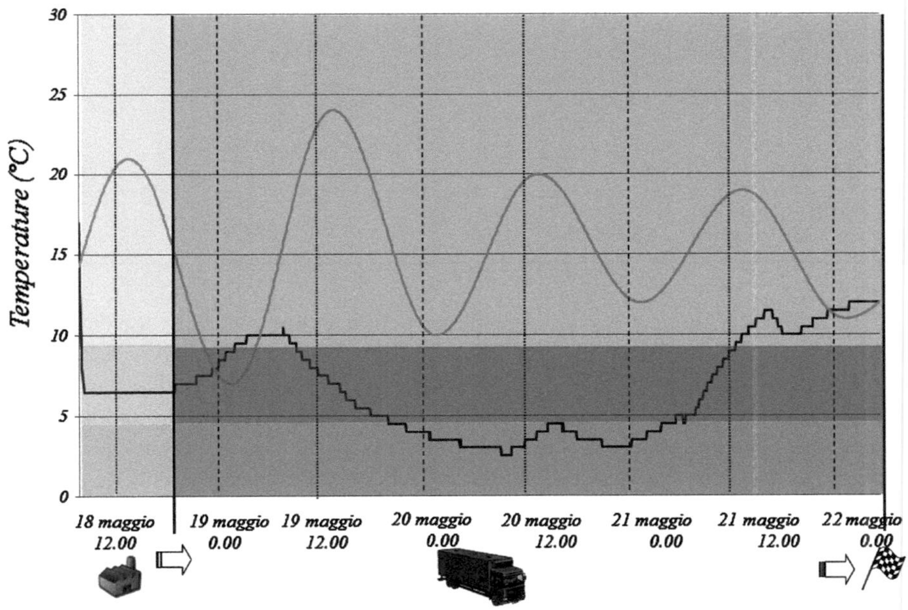

Figure 18.9 Temperature profile of shipment 3.

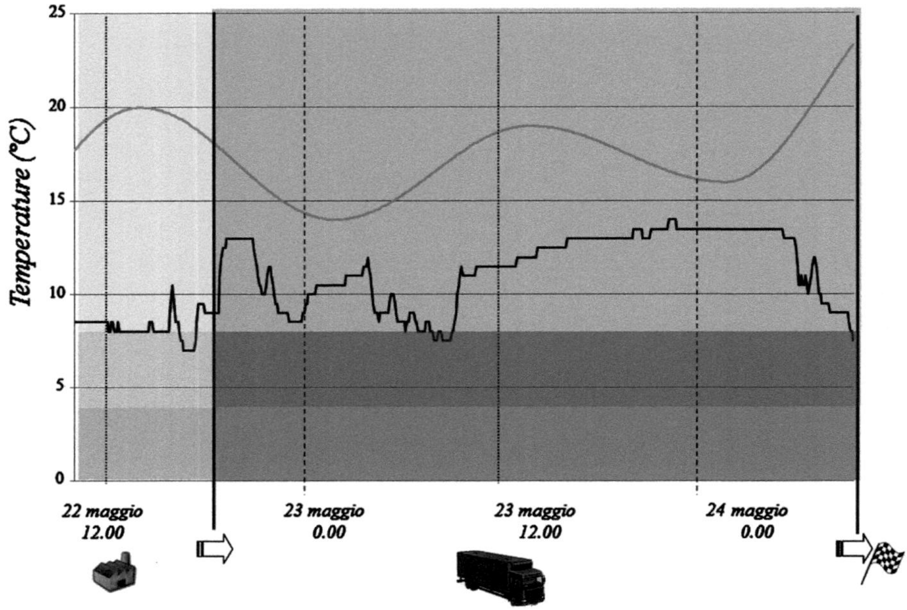

Figure 18.10 Temperature profile of shipment 4.

Table 18.3 Shipment 1 Summary

Shipment Summary	
Expected shipping date	18/05/2012
Packaging enclosure	18/05/2012 08:00
Shipping date	18/05/2012 17:25
Arrival date	22/05/2012 08:00
Transport duration	85 h 35 min
Sampling time	5 min
Samples	1040
Max temperature (°C)	9
Min temperature (°C)	−2
Average temperature (°C)	2.5975
Samples above 8°C	5 out of 1040

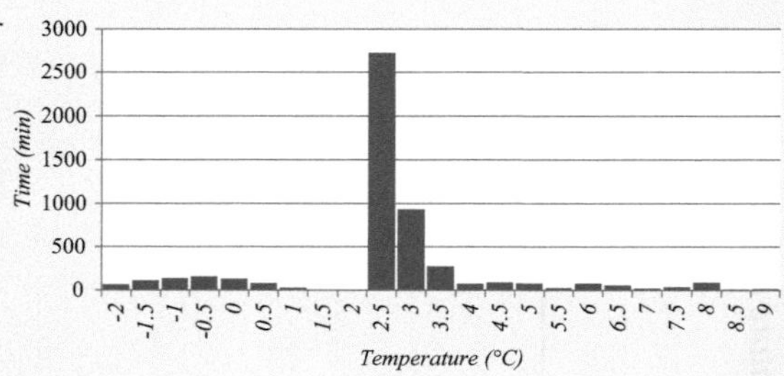

The profile of Fig. 18.7 reveals how the temperature experienced by the daily products is constantly maintained below the critical threshold of 8°C. The average temperature is indeed around 2.5°C, even below the recommended range. For a part of the first samples, measured when the sensor was not yet coupled with the product, only two samples overcome the threshold, in compliance with the recommended conservation standards. In the final part of distribution, when the products are unloaded and put away at the client dock, the temperature falls further, tracking values even below 0°C with a minimum of −2°C (Table 18.4).

The profile of Fig. 18.8 highlights an increasing trend of the temperature that rises up to 17.5°C until the products are delivered to the client warehouse on May 22nd and are thereby conserved within the optimal thermal belt. A number of samples are measured above the external environmental temperature, which is unusual and is probably

Table 18.4 Shipment 2 Summary

Shipment Summary	
Expected shipping date	18/05/2012
Packaging enclosure	18/05/2012 08:00
Shipping date	18/05/2012 17:51
Arrival date	23/05/2012 08:00
Travel distance (km)	1187
Transport duration	170 h 40 min
Sampling time	5 min
Samples	1912
Max temperature (°C)	17.5
Min temperature (°C)	5.5
Average temperature (°C)	9.57
Samples above 8°C	1145 out of 1912 (59.88%)

Table 18.5 Shipment 3 Summary

Shipment Summary	
Expected shipping date	18/05/2012
Packaging enclosure	18/05/2012 08:15
Shipping date	18/05/2012 19:08
Arrival date	22/05/2012 06:00
Transport duration	82 h 52 min
Sampling time	5 min
Samples	995
Max temperature (°C)	12
Min temperature (°C)	2.5
Average temperature (°C)	6.61
Samples above 8°C	350 out of 995 (35%)

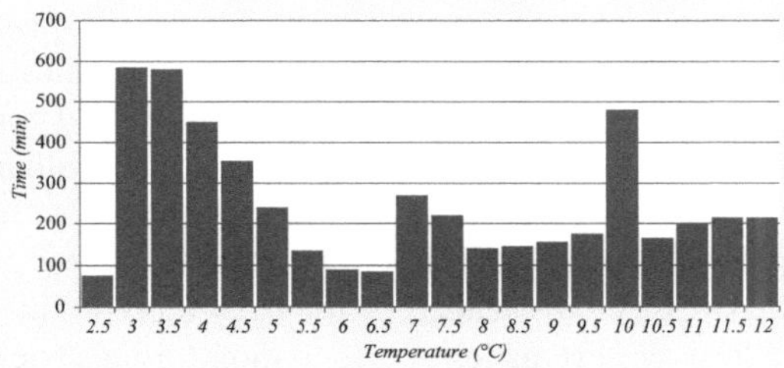

Table 18.6 Shipment 4 Summary

Shipment Summary	
Expected shipping date	22/05/2012
Packaging enclosure	22/05/2012 10:00
Shipping date	22/05/2012 18:24
Arrival date	24/05/2012 09:45
Transport duration	39 h 21 min
Sampling time	5 min
Samples	474
Max temperature (°C)	14
Min temperature (°C)	7.5
Average temperature (°C)	11.45
Samples above 8°C	446 out of 474 (94%)

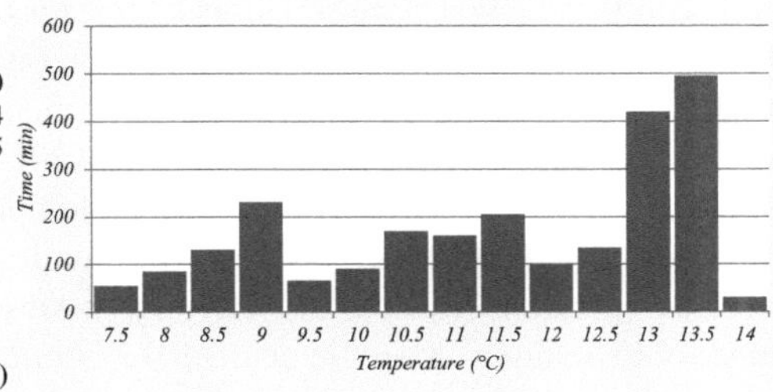

due the temporary storage of the products within an intermediate warehouse that is not climate-controlled. On average, the products are conserved at the temperature of 9.67°C, but suffer numerous critical thermal stresses with about 60% of tracked time above the optimal temperature belt.

The graph of Fig. 18.9 shows that the temperature is maintained within the optimal belt for a large part of the transportation phase. Along the transport time the temperature rises above the optimal threshold for two batches, for a total time of about 30 h, representing approximately 35% of the total travel time (Tables 18.5 and 18.6).

More than in the other shipments, the tracked profile illustrated in Fig. 18.10 shows that the dairy products can significantly suffer thermal stresses during the transport phase. Indeed, while the average temperature is about 11°C, the dairy products stay above their optimal conservation threshold for a period corresponding to 94% of the total travel time.

Although the regulation that defines the standards for dairy conservation allows a maximum temperature of 14°C during the fractioned transport (DM 01/04/88 n. 178, Section C Part II DPR 327/80), it also indicates that such temperature should be maintained for a short time.

In order to study the impact of the tracked shipments on the quality and safety of dairy products, the closed-loop monitoring system should be implemented and the tracked profiles should be reproduced in a laboratory to assess whether and how they affect the products, in terms of both chemical and organoleptic properties.

4. Conclusions and Further Research

This manuscript gives an overview of the differences between a *real-time* and an *ex-post* monitoring system and applies an ex-post monitoring protocol to a case study from the dairy industry. While real-time monitoring infrastructures are more effective and make the supply chain more resilient and controlled (de las Morenas et al., 2014; Barge et al., 2014), they require large investments and can be difficult to implement on a large or global scale. Conversely, the *ex-post* monitoring system is less expensive and can be easily applied to monitor the effect of transport and logistics activities on the stresses experienced by the agro-food products.

The *ex-post* and closed-loop monitoring system can be applied finally to determine and identify the most sensitive and critical phase of the supply chain supporting oriented investment for the development of a *real-time* monitoring architecture to control a critical phase of the product distribution process.

The aim of further developments is to assess the effect of the environmental stresses due to the logistics and the transport phase on the quality and the safety of the dairy products. The tracked profiles will be reproduced in a laboratory within properly controlled climate chambers (Accorsi et al., 2014c) able to reproduce the storage and shipping profiles in terms of experienced temperature and humidity. The aim of these simulations will be to assess how the quality and safety of dairy products can be affected by transportation and delivery processes and identify operative solutions in terms of packaging (Manzini et al., 2013), containment, and delivery routing, ensuring compliance with quality and safety standards.

References

Accorsi, R., Manzini, R., Ferrari, E., 2014a. A comparison of shipping containers from technical, economic and environmental perspectives. Transportation Research, Part D, Transport and Environment 26, 52–59.

Accorsi, R., Cascini, A., Cholette, S., Manzini, R., Mora, C., 2014b. Economic and environmental assessment of reusable plastic containers: a food catering supply chain case study. International Journal of Production Economics 152, 88–101.

Accorsi, R., Bortolini, M., Fabri, F., Gamberi, M., Manzini, R., Pareschi, A., 2014c. Closed-loop strategies to locally simulate food shipment conditions. In: Proceedings of the Third International Workshop on Food Supply Chain (IWFSC 2014). San Francisco State University.

Ahumada, O., Villalobos, R., 2009. Application of planning models in the agro-food supply chain: a review. European Journal of Operational Research 195, 1–20.

Barge, P., Gay, P., Merlino, V., Tortia, C., 2014. Item-level radio-frequency identification for the traceability of food products: application on a dairy product. Journal of Food Engineering 125 (2014), 119–130.

Bartholdi, J.J., Mac Cawley, A., 2011. Tracking the temperatures of international wine shipments: processes, information handling, and temperature simulation. In: Proceedings of the First International Workshop on Food Supply Chain (IWFSC 2011), 28–40. University of Bologna, Bologna. ISBN: 9788890650000.

European Dairy Association, EDA, 2014. Annual Report 2014. Avenue d'Auderghem, 22–28, 1040 Brussels.

Food and Drink Europe (F&D), 2014. Data & Trends of the European Food and Drink Industry 2013–2014. Avenue des Nerviens, 9–31, 1040, Brussels.

Li, D., Wang, X., Chan, H.K., Manzini, R., 2014. Sustainable food supply chain management. International Journal of Production Economics 152, 1–8.

Labuza, T.P., 1982. Shelf-Life Dating of Foods. Food & Nutrition Press, Westport, CT, USA.

de las Morenas, J., García, A., Blanco, J., 2014. Prototype traceability system for the dairy industry. Computers and Electronics in Agriculture 101 (2014), 34–41.

Man, C.M.D., Jones, A.A., 1994. Shelf Life Evaluation of Foods. Blackie Academic & Professionals, Glasgow, UK.

Manzini, R., Accorsi, R., 2013. The new conceptual framework for food supply chain assessment. Journal of Food Engineering 115 (2), 251–263.

Manzini, R., Accorsi, R., Ayyad, Z., Bendini, A., Bortolini, M., Gamberi, M., Valli, E., Gallina Toschi, T., 2014a. Sustainability and quality in the food supply chain. A case study of shipment of edible oils. British Food Journal 16 (12), 2069–2090.

Manzini, R., Accorsi, R., Bortolini, M., Ferrari, E., Gamberi, M., Trombini, M., 2014b. Quality assessment of fine Italian chocolate subject to time-varying stress of temperature during the logistic distribution. In: Proceedings of the Third International Workshop on Food Supply Chain (IWFSC 2014). San Francisco State University.

Manzini, R., Accorsi, R., Bortolini, M., Tampieri, F., Garbellini, F., Evangelisti, F., Gamberi, M., Soli, S., August, 2014c. The cool chain under x-ray. Fresh Point Magazine 3 (8).

Manzini, R., Accorsi, R., Ferrari, E., Mora, C., Regattieri, A., Santarelli, G., Versari, L., 2013. Accelerated life test analysis for packaging solutions. A case study of edible oils distribution. In: Proceedings of the Second International Workshop on Food Supply Chain (IWFSC 2013). Pontificia Universidad Catolica de Cile.

Piazzi, P., Adami, S., Bortolini, M., Gamberi, M., Accorsi, R., Manzini, R., 2011. Design, development and test of a vibration monitoring embedded system. In: Proceedings of the First International Workshop on Food Supply Chain (IWFSC 2011), 1–12. University of Bologna, Bologna. ISBN: 9788890650000.

Regattieri, A., Gamberi, M., Manzini, R., 2007. Traceability of food products: general framework and experimental evidence. Journal of Food Engineering 81, 347–356.

Valli, E., Manzini, R., Accorsi, R., Bortolini, M., Gamberi, M., Bendini, A., Lercker, G., Gallina Toschi, T., 2013. Quality at destination: simulating shipment of three bottled edible oils from Italy to Taiwan. La Rivista Italiana delle Sostanze Grasse 90 (3), 163–169.

World Bank, 2014. World Bank Data. <http://beta.data.worldbank.org>.

Effective Use of Food Traceability in Animal Feed

19

C. Morcia, G. Tumino, R. Ghizzoni, V. Terzi
Council for Agricultural Research and Economics, Genomics Research Center, CREA-GPG, Fiorenzuola d'Arda (PC), Italy

1. Authentication and Traceability as a Key Factor for Feed Safety and Quality

According to the International Organization for Standardization (ISO), traceability is the "ability to trace the history, application or location of that which is under consideration". This definition is necessarily very wide and results in multiple approaches depending on the sectors considered. Previous chapters of this book underlie the relevance and increasing interest in traceability applied to several food production chains. However, the interest in traceability is not limited to the food production chain. On the contrary, the stakeholders involved in feed production and utilization need strong and innovative traceability approaches. In function of the different products, in fact, there may be the need to identify the plant, animal, and microorganism species present in a feed. In assessing the authenticity of plant and animal species in a feed product, a link between the raw materials used as ingredients and the safety and quality of the final products is recognized. Even more relevant is the possibility to track the presence of microorganisms from raw materials to final products: for example, the traceability of mycotoxigenic fungi that can contaminate cereals in fields or during storage is of outstanding importance from the safety point of view of a feed. An important topic for organic feed is the possibility to verify the absence of genetically modified plants. Sometimes, even the geographic origin can be relevant for some DOP (Denominazione di Origine Controllata) products, strongly linked to the territory of production. Great interest is therefore now focused on traceability: farmers are in fact interested in geographic and genetic traceability of their productions because it can add value to costly agricultural practices that result in high yields of good quality produce. Many industrial producers have also voluntarily introduced traceability in quality control and for enhancing safety in the food and feed chains. Consumers, after some food-related health crises in recent years, are clearly interested in systems that can ensure the authenticity and safety of products of both animal and plant origin.

Briefly, for the feed industries, traceability can mean several different actions, eg, the possibility to track a feed during production phases and to follow the distribution routes until the final destinations. This action can be done, for example, using innovative electronic coding systems. However, a key point for the safety and the quality of a feed is represented by the type and characteristic of ingredients. For this purpose,

Advances in Food Traceability Techniques and Technologies. http://dx.doi.org/10.1016/B978-0-08-100310-7.00019-3

one of the most innovative and popular technologies is DNA-based traceability, now widely accepted as playing a key role. DNA-based techniques provide useful tools in identifying ingredients because DNA is a persistent molecule, is present in all tissue types, and can retain sequence-specific information obtainable by a simple amplification reaction. Polymerase chain reaction (PCR) is therefore currently massively used not only in food authentication, but even in feed controls. Currently, several classes of DNA markers based on PCR reaction, together with the PCR direct sequence analysis (FINS) are frequently used to identify plants and animals in feed and food. Moreover, the use of Real-Time PCR (RT-PCR) can provide quantification of different types of DNA (Teletchea, 2009).

In this paper, some examples of molecular traceability approaches applied to feed production chains are presented.

2. Determination of Animal Species in Feed

The regulations of the different countries and of the market require feed to be authentic in terms of species content. The presence of animal species in some feed classes plays a central role in their quality and safety. Methods, procedures, and reference materials developed for the food market can be applied even for the verification of the species type in feed. ELISA, qualitative PCR, quantitative real-time PCR, and sequencing of mitochondrial DNA are the methods currently used.

Several phylogenetic studies have indicated the mitochondrial DNA sequences, more specifically, the cytochrome genes (Lockley and Bardsley, 2000), as useful loci for genetic traceability. Mitochondrial DNA has several advantages over nuclear DNA: most eukaryotic cells contains mitochondria, mtDNA has a relatively fast mutation rate, the copy number of a mitochondrial genome exceeds that of a nuclear genome several fold, and ambiguities deriving from heterozygosity are avoided because mtDNA tends to be maternally inherited, and therefore individuals normally have only one allele (Balitzki-Korte et al., 2005). Moreover, mitochondrial loci are characterized by a large variation between species and small variation within a species. Techniques such as Restriction Fragment Length Polymorphism (PCR–RFLP techniques), sequencing specific gene assay, multiplex PCR assay, SNaPshot minisequencing analysis, and RT-PCR (Pegels et al., 2011), proposed for food control can be useful even for the feed market (Martin et al., 2007).

The massive use of these molecular analytical approaches come in fact from some negative past experiences of the feed industry. It is well-known that the outbreak of bovine spongiform encephalopathy (BSE) urged several countries to ban intraspecific recycling, ie, the ban of feeding animals with proteins (PAPs, Processed Animal Proteins) derived from the same species. In this frame, the animal by-products have been classified into three categories characterized by different safety levels (European Commission, 1994, http://ec.europa.eu/food/food/biosafety/tse_bse/docs/d96-239.pdf) in relation to transmissible spongiform encephalopathy. Only materials from animals allowed for human consumption (category three) could be used to feed farm animals: for example, only fish proteins can be used in pig and poultry feed. A softening of the feed ban for nonruminants has been recently requested. For example, EFPRA

(European Fat Processors and Renderers Association) proposed the reentry of nonruminant proteins in some feed sectors and the change of threshold limits of tolerance from 0 to 2%. In response to these positions, nonruminant PAPs have been recently reauthorized for aquafeed. Independently from further decisions, validated techniques are requested that enable the detection of animal species throughout the feed chain (Gizzi et al., 2004; EFSA Panel on Biological Hazards, 2011). The review of Fumiere et al. (2009) gives a state of the art of the different methods for detection, species identification, and quantification of processed proteins in foodstuff. The official analytical method is based on the classic microscopic examination for the detection of bones. Even Near Infrared Microscopy (NIRM) methods have been proposed as screening method: the advantage over microscopy is that it does not need experienced staff, but the disadvantage is that the required equipment is expensive. NIRS (Near Infrared Spectroscopy) is a widely used technology that can identify animal ingredients, with the major drawback of the limit of detection (LOD) that is higher than 1%. Several ELISA tests, organized as dipstick or in classical 96-wells, have been developed by different manufacturers. These analytical approaches are mainly used in screening steps. Powerful PCR and real-time PCR assays have been developed for animal species traceability. This panel of techniques has been recently evaluated in the frame of the European SAFEED-PAP project (http://safeedpap.feedsafety.org/) that has given a strong contribution to improve and harmonize adequate control methods for ensuring the correct implementation of the feed ban, taking into account that the tolerance levels must be related to reliable methods of quantification. In this respect, a combination of different and complementary analytical procedures can have interesting potentialities. An example of such a strategy has been proposed by Lecrenier et al. (2014) to overcome difficulties for correct animal species evaluation in some problematic samples, like compound feed for fish. Light microscopy can in fact distinguish terrestrial bones from fish bones, whereas PCR can identify the presence of animal DNA in feed. However, if in a sample, light microscopy detects particles of terrestrial animals and PCR detects ruminant presence, it is not possible to understand if the sample contains porcine PAPs and dairy products (allowed in aquaculture) or bovine PAPs (forbidden) and porcine blood (allowed). Lecrenier et al. (2014) proposed a combination of microscopy and DNA analysis to solve such problematic cases. They developed a new fluorescence in situ hybridization protocol, characterized by the following steps:

- development of fluorescently labeled probes specific for bovine and for ruminant species;
- preparation of nondemineralized bone particles; and
- hybridization of the probes to the bone particles and detection of hybridization results with fluorescence microscopy.

3. Botanical Origin and Geographical Origin of Plant Raw Materials

The composition of plant species can be very important not only for food quality, but even for feed quality because the feed composition can impact animal health and growth, as well as on nutritional and organoleptic properties of derived food animal products. EU Regulation EC767/2009 has defined "rules on the placing on the market and use of feed for both food-producing and non-food producing animals within the Community, including requirements for labelling, packaging and presentation". The new regulations require therefore that the feed composition is declared in the label and that the quantity of each ingredient can be made available on request.

The classical barcoding approach, efficient in the case of animal species traceability, has proven more difficult in plants than in animals. Fazekas et al. (2009) have reviewed several barcoding works in animals and plants and have hypothesized that animal species have larger intra- and interspecific genetic distances than plants, in which hybridization appears to have contributed significantly to variation in genetic discontinuity. Therefore for barcoding plants a combined use of different genomic regions has been suggested, such as DNA regions from the chloroplastic genome that is uniparentally inherited, nonrecombining, and structurally stable together with multiple nuclear genes, such as rbcL, encoding for ribulose 1,5 bisphosphate carboxylase-oxygenase, and noncoding regions such as the trnL intron and the trnL-F intergenic spacer. Referring specifically to plant ingredients, the FEED-CODE Project (http://www.feedcode-project.eu/) is working "to provide stakeholders along the food chain with an innovative animal feed certification platform and procedure to guarantee the quality of meat and dairy products through an automatic and simple identification method". FEED-CODE starts from the fact that presently there is no practical time and cost-efficient method to allow frequent and effective controls to mixtures such as feed. The FEED-CODE Project is developing an innovative and automated DNA barcode system based on plant tubulin-based polymorphisms (TBP). In more detail, the sources of DNA polymorphisms are the Intron 1 and 2 of beta-tubulin genes, and the protocol for their characterization is protected by European Patent No. 1144691, owned by the National Council for Research in Italy. The beta-tubulin gene family has in fact been demonstrated to be very useful not only for assaying genetic diversity in plants, but even for plant species and variety assessment (Breviario et al., 2007). Ponzoni et al. (2014) developed an oligonucleotide microarray for the molecular recognition of multiple plant components in commercial feeds based on the highly polymorphic first intron sequences from members of the plant β-tubulin gene family as a target for plant DNA identification. In some particular food production chains, such analytical tools can have a key role in the traceability of feed ingredients (Casazza et al., 2011). An example is the Parmigiano-Reggiano cheese, a high-value Italian cheese whose production regulation establishes a particular feeding program for cows that produce milk that is destined to the production of Parmigiano-Reggiano cheese (http://www.parmigianoreggiano.com/consortium/disciplinare_produzione_vigente_sino_agosto_2011/feeding_regulation_dairy_cows_1.aspx). The feed rationing of dairy cows is in fact based on the use of local forage (only some vegetal species are permitted), appropriately integrated with pellets capable of balancing the supply of the various diet nutrients. Milking cows can never be fed with some vegetal species that may confer anomalous aromas and flavors to milk, and the use of any type of silage is forbidden. This is just an example of the relevance for some high-quality products of the feed ingredients. Moreover, giving the globalization of the source of feed ingredients, the traceability of the botanical and geographical origin of raw materials is becoming very important for the whole feed sector. The European project "Quality and Safety of Feeds and Food for Europe, QSAFFE" (www.qsaffe.eu) contributed to the development of authentication approaches for product traceability, using different techniques, like stable isotope analysis (IRMS and SNIF-NMR), multielement analysis or spectroscopic techniques such as NIR, MIR, and Raman spectroscopy (Garrigues and de la Guardia, 2013; Gonzalvez et al., 2009). The isotope ratio mass spectrometry has been demonstrated to be useful for the determination of the geographical origin of feed ingredients (Nietner et al., 2014), whereas FT-IR spectroscopy can be applied to authenticate both the botanical and geographical origin (Nietner et al., 2013). Tena et al. (2015) proposed a nondestructive analytical method based on near infrared microscopy for the traceability of feed ingredients in a short time. The proposed method can be especially applied to classification against the botanical origin.

For feed quality, the presence of particular ingredients can be important, not only at the species level, but even at the variety level. In fact, the use of cereal genotypes that are characterized by peculiar nutritional content can be relevant for high-quality feed. For example, oat-specific breeding programs have been activated for the release of innovative cultivars characterized by high metabolic energy obtained by increasing oil content and decreasing the lignin husk. Great interest for ruminants is now in naked oats, but despite the naked seeds, additional traits are required, such as high yield, high protein and oil content, and low trichome density. Such genotypes can be superior both from a nutritional point of view and from agronomic one, increasing therefore their price on the raw materials market. This means that an analytical assay can be required to test the authenticity even at the cultivar level. Molecular fingerprinting can be applied to the identification of plant genotypes in raw materials and, in some cases, even in processed products (Terzi et al., 2008). Several classes of molecular markers can be used, with Simple Sequence Repeats (SSR) and Single Nucleotide Polymorphism (SNP) being the most popular for these applications.

4. Traceability of Genetically Modified Plants and Ingredients

The traceability of GM plants and ingredients in feed is even more relevant than in food chains. Maize and soybeans, the most important feed crops, are even the most important GM plants in terms of cultivation area worldwide (http://www.isaaa.org). In the last 20 years, in fact, there was a constant increase in GM maize and soybean adoption. In 2014, GM soybeans were cultivated on more than 90 million hectares, accounting for 82% of the worldwide soybean production, whereas GM maize covered more than 50 million hectares, corresponding to one-third of the worldwide maize production. The great majority of cultivated GM plants are improved for herbicide tolerance, insect resistance, or their combinations. However, biotechnology has been successfully applied in the development of crops improved for animal feed that are now under field trial or are already in the early commercialization stages. Examples are GM maize with increased lysine (LY038) content, rice expressing transgene OASAID with elevated

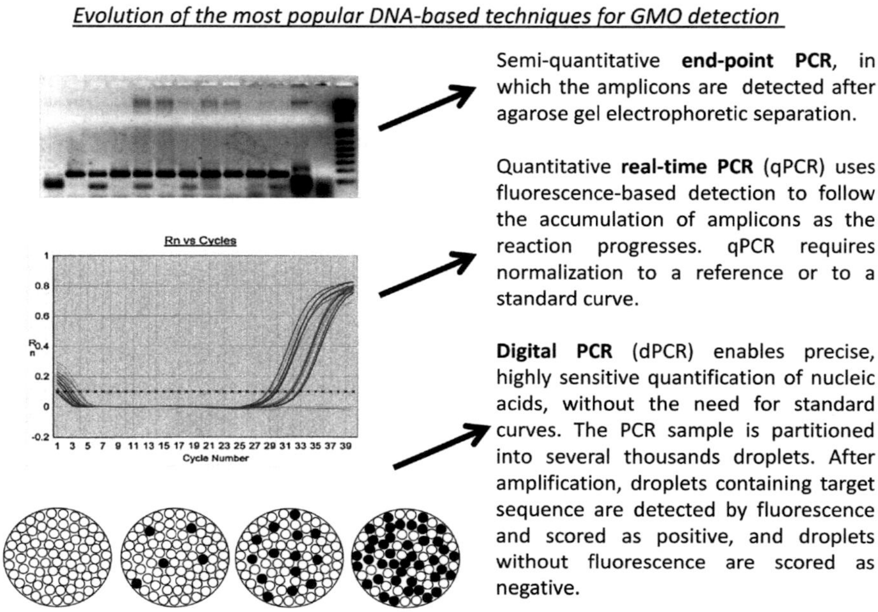

Figure 19.1 Examples of DNA-based approaches for GMO detection.

free tryptophan (Trp) in the seeds for poultry feeding, soybean event M703 characterized by more digestible amino acids content, and narrow-leafed lupin with a two-fold increase in methionine content. Other GM plants have been enriched with biologically active compounds: examples are barley expressing a β-glucanase that acts on barley glucans, rice with human lactoferrin (LF) and lysozyme (LZ) genes to increase antibacterial and immune-stimulating properties, and feed crops with improved phosphorus availability thanks to the introduction of the phytase encoding gene.

The use of genetically modified plants and derived products in feed is subjected to regulatory oversight in a number of countries. Analytical methods for the GM traceability are therefore required (Rodriguez-Lazaro et al., 2007; Ruttink et al., 2010). Among the different approaches, the methods of choice for GM detection are based on transgenic DNA traceability, as reported in Fig. 19.1 (Engel et al., 2006). Several questions are still open with respect to GMO controls, ranging from the reduction of the analytical costs to the harmonization of analysis, the interpretation of results, and the detection of stacked events, in which several transgenes are present, or nonauthorized ones. Interesting perspectives are now opened by the adoption of droplet digital PCR (ddPCR). This new approach works by partitioning a DNA sample into many single and parallel PCR reactions; some of these reactions will be positive for the presence of the target sequence, while others will be negative. To detect sequence-specific targets, dye-labeled probes are used. The fraction of negative reactions is used to generate an absolute count of the number of target molecules in the sample, without the need for standards or endogenous controls. In the work of Morisset et al. (2013)

the applicability of ddPCR has been evaluated for the routine analysis of feed samples. In this study, both MON810-certified reference materials and other samples containing MON810 maize event were used. DNA extracted was subjected to both qPCR and droplet digital PCR analysis, and the results were used to evaluate dynamic range, repeatability, limits of detection and quantification, specificity, and applicability. A wide dynamic range of quantification, close to five orders of magnitude, has been found, and the limits of quantification, trueness, and repeatability are comparable or superior to those of the validated qPCR assays. These results support therefore the idea to introduce ddPCR technology for the routine quantification of GMO in feed samples.

5. Mycotoxins and Biological Contaminants in Animal Feed

Mycotoxins are toxic compounds produced by the secondary metabolism of several filamentous fungi that can contaminate crops in the field and during storage. If contaminate plants are used directly for forage, mycotoxicosis may occur in the fed animals, their severity depending on several factors, ranging from the type of mycotoxin to the animal species and age. Moreover, some toxins can be transferred into milk and eggs, affecting consumers' health indirectly. Even if plants and grains contaminated with mycotoxins are used as ingredients for feed, several problems can rise. The mycotoxin concentration can in fact be hardly reduced during the feed production because of the stability of these toxic compounds to high temperatures and several technological treatments. More than 300 different mycotoxins have been isolated on plants and forage used routinely for feed; however, only a few of them are regulated by national legislations for animal feed. For example, in the EU the ruled mycotoxins are aflatoxin B1, deoxynivalenol, fumonisins, zearalenone, and ochratoxin A. However, many more toxic compounds can be found routinely in feed ingredients, as demonstrated by several monitoring activities. In the work of Zachariasova et al. (2014) the occurrence of 56 mycotoxins produced by Fusarium, Alternaria, Penicillium, Aspergillus, and Claviceps was evaluated in 18 classes of nonfermented or fermented foodstuffs, foodstuff supplements, and complex compound feed. The highest mycotoxin concentrations were found in dried distillers' grains with solubles (DDGS). The authors found that even the most significant exposure of animals occurred after the consumption of maize silage and complex compound feeds.

Several strategies can be activated to reduce and control the mycotoxin problem in feed production chains, such as the breeding for resistant varieties and management measures in fields, and during further production steps, the development of standardized analytical methods to determine mycotoxin content or fungal levels, the decontamination, and postharvest strategies to reduce contamination (Terzi et al., 2014). A key point for the mycotoxin problem control is the traceability of both mycotoxins and mycotoxigenic fungi from field to feed, taking into consideration even the new and emerging mycotoxigenic strains (Fig. 19.2). The classical diagnosis based on plant symptoms and on fungal morphophysiological trait evaluation has several limitations. Molecular traceability is now used for

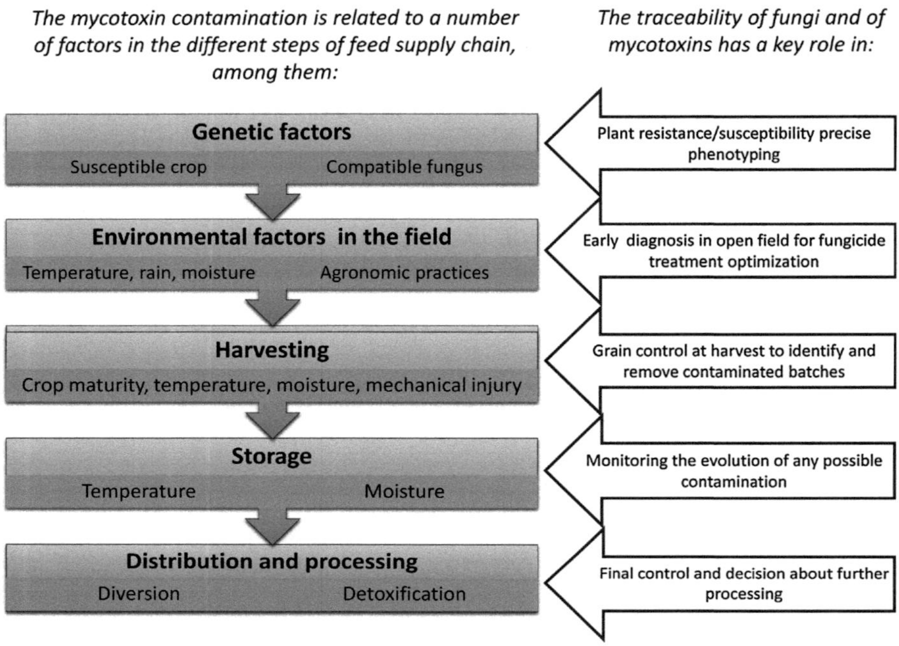

Figure 19.2 Examples of fungi and mycotoxin traceability applications to control key steps of feed supply chain.

fungal species identification and quantification, from field to the feed. Several classes of molecular markers have been proposed for fungal diagnostics. qPCR assays have been developed to target fungal genes that can be both involved in mycotoxin biosynthetic pathways or not. A new impetus for molecular diagnostics has been derived from the fungal genome sequencing projects. Molecular traceability can be very useful in epidemiological studies and can be very effective, in particular for forage and feed, when applied to a very early diagnosis in the field. Postharvest traceability is the other pillar to reduce the incidence of mycotoxins. Advanced analytical methods can be used, ranging from high-performance liquid chromatography and gas-chromatography to Near Infrared Reflectance, mass-spectrometry, electronic nose and tongues, biosensors, and receptors. Among these technologies, DNA-based assays are available, even for postharvest microorganisms, and this traceability technology can play an important role in an integrated approach to effectively reduce the mycotoxin incidence in feed chains.

6. Conclusions and Perspectives

The last food crises, that have underlined the importance of feed control, have further demonstrated the relevance of traceability for ex ante and ex post controls. In particular, genetic traceability can play a useful role in the more general frame of advanced traceability. Moreover, thanks to the fast advancement of genomics, genetic traceability can be considered a fast-growing application very useful not only in food, but even in feed control. Animal, plant, and microorganism traceability can gain from the genome sequencing and ri-sequencing projects. Next generation sequencing technologies can in fact produce, in a short time and with reasonable costs, a large amount of genetic data in the form of short sequences. Moreover, starting from the information deriving from resequencing, a "core genome" can be identified for each plant, animal, and microorganism species, and on this "core genome," robust assays can be designed for diagnostic purposes and in general for the identification of animals, plants, and microorganisms at the species level (Morgante et al., 2007; Marroni et al., 2011). Moreover, data on "dispensable genomes" (related to single genotype) can be used for the traceability at a level of specific variety or accessions.

Acknowledgments

This review was done within the activities planned in the "Sofia" project.

References

Balinzki-Korte, B., Anslinger, K., Bartsch, C., Rolf, B., 2005. Species identification by means of pyrosequencing the mitochondrial 12S rRNA gene. International Journal of Legal Medicine 119, 291–294.

Breviario, D., Vance Baird, W.M., Sangoi, S., Hilu, K., Blumetti, P., Gianì, S., 2007. High polymorphism and resolution in targeted fingerprinting with combined beta-tubulin introns. Molecular Breeding 3, 249–259.

Casazza, A.P., Gavazzi, F., Mastromauro, F., Gianì, S., Breviario, D., 2011. Certifying the feed to guarantee the quality of traditional food. An easy way to trace plant species in complex mixtures. Food Chemistry 124, 685–691.

European Commission, 1994. Commission Decision 1994/381/EC of 27 June 1994 concerning certain protection measures with regard to bovine spongiform encephalopathy and the feeding of mammalian derived protein. Official Journal of European Communities L172, 23–32.

Engel, K.-H., Moreano, F., Ehlert, A., Busch, U., 2006. Quantification of DNA from genetically modified organisms in composite and processed foods. Trends in Food Science and Technology 17, 490–497.

Fazekas, A.J., Kesanakurti, P.R., Burgess, K.S., Percy, D.M., Graham, S.W., Barrett, S.C.H., Newmaster, S.G., Haj Ibabaei, M., Husband, B.C., 2009. Are plant species inherently harder to discriminate than animal species using DNA barcoding markers? Molecular Ecology Research 9, 130–139.

Fumiere, O., Veys, P., Boix, A., von Holst, C., Baeten, V., Berben, V., 2009. Methods of detection, species identification and quantification of processed animal proteins in feedingstuffs. Base. 13. http:popups.ulg.ac.be/1780-4507/index.php?id=3525.

Garrigues, A., de la Guardia, M., 2013. Food protected designation of origin—methodologies and applications. Analytical Chemistry 60, 101–122.

Gizzi, G., von Holst, C., Baeten, V., Berben, G., Van Raamsdonk, L., 2004. Determination of processed animal proteins, including meat and bone meal, in animal feed. Journal of AOAC International 87 (6), 1334–1341.

Gonzalvez, A., Armenta, S., de la Guardia, M., 2009. Trace-element composition and stable-isotope ratio for discrimination of foods with protected designation of origin. Trends in Analytical Chemistry 28, 1295–1311.

Lecrenier, M.C., Ledoux, Q., Berben, G., Fumière, O., Saegerman, C., Baeten, V., Veys, P., 2014. Determination of the ruminant origin of bone particles using fluorescence in situ hybridization (FISH). Scientific Reports 4, 573. http://dx.doi.org/10.1038/srep05730.

Lockley, A.K., Bardsley, R.G., 2000. DNA-based methods for food authentication. Trends in Food Science & Technology 11, 67–77.

Marroni, F., Pinosio, S., Di Centa, E., Jurman, I., Boerjan, W., Felice, N., Cattonaro, F., Morgante, M., 2011. Large-scale detection of rare variants via pooled multiplexed next-generation sequencing: towards next-generation. The Plant Journal 67, 736–745.

Martin, I., Garcia, T., Fajardo, V., Lopez-Calleja, I., Hernandez, P.E., Gonzalez, I., Martin, R., 2007. Species-specific PCR for the identification of ruminant species in feedstuffs. Meat Science 75, 120–127.

Morgante, M., De Paoli, E., Radovic, S., 2007. Transposable elements and the plant pan genomes. Current Opinion in Plant Biology 10, 149–155.

Morisset, D., Štebih, D., Milavec, M., Gruden, K., Žel, J., 2013. Quantitative analysis of food and feed samples with droplet digital PCR. PLoS One 8 (5), e62583.

Nietner, T., Haughey, S.A., Ogle, N., Fauhl-Hassek, C., Elliott, C.T., 2014. Determination of geographical origin of distillers dried grains and solubles using isotope ratio mass spectrometry. Food Research International 60, 146–153.

Nietner, T., Pfister, M., Glomb, M., Fauhl-Hassek, C., 2013. Authentication of the botanical and geographical origin of DDGS by FT-IR spectroscopy. Journal of Agricultural and Food Chemistry 61 (30), 7225–7233.

Pegels, N., et al., 2011. Applicability assessment of a real-time PCR assay for the specific detection of bovine, ovine and caprine material in feedstuffs. Food Control 22 (8), 1189–1196.

Ponzoni, E., Morello, L., Gianì, S., Breviario, D., 2014. Traceback identification of plant components in commercial compound feed through an oligonucleotide microarray based on tubulin intron polymorphism. Food Chemistry 1 (162), 72–80.

Rodríguez-Lázaro, D., Lombard, B., Smith, H., Rzezutka, A., D'agostino, M., Helmuth, R., Schroeter, A., Malorny, B., Miko, A., Beatriz Guerra, B., Davison, J., Kobilinsky, A., Hernández, M., Bertheau, Y., Cook, N., 2007. Trends in analytical methodology in food safety and quality: monitoring microorganisms and genetically modified organisms. Trends in Food Science and Technology 18, 306–319.

Ruttink, T., Demeyer, R., Van Gulck, E., Van Droogenbroeck, B., Querci, M., Taverniers, I., De Loose, M., 2010. Molecular toolbox for the identification of unknown genetically modified organisms. Analytical and Bioanalytical Chemistry 396, 2073–2089.

Teletchea, F., 2009. Molecular identification methods of fish species: reassessment and possible applications. Reviews in Fish Biology and Fisheries 19, 265–293.

Tena, N., Boix, A., von Holst, C., 2015. Identification of botanical and geographical origin of distillers dried grains with solubles by near infrared microscopy. Food Control 54, 103–110.

Terzi, V., Morcia, C., Faccioli, P., 2008. Molecular traceability in the post-genomic era: an application of DNA technology to food science. In: Greco, L.V., Bruno, M.N. (Eds.), Food Science and Technology: New Research. Nova Science Publishers, Inc, New York USA, pp. 211–241.

Terzi, V., Tumino, G., Stanca, A.M., Morcia, C., 2014. Reducing the incidence of cereal head infection and mycotoxins in small grain cereal species. Journal of Cereal Science 59, 284–293.

Zachariasova, M., Dzuman, Z., Veprikova, Z., Hajkova, K., Jiru, M., Vaclavikova, M., Zachariasova, A., Pospichalova, M., Florian, M., Hajslova, J., 2014. Occurrence of multiple mycotoxins in European feeding stuffs, assessment of dietary intake by farm animals. Animal Feed Science and Technology 193, 124–140.

Index

'*Note:* Page numbers followed by "f" indicate figures and "t" indicate tables.'

H

I

J

K

L

R

S

T

Edwards Brothers Malloy
Ann Arbor MI. USA
May 16, 2017